ALTERNATIVE FOOD GEOGRAPHIES
REPRESENTATION AND PRACTICE

ALTERNATIVE FOOD GEOGRAPHIES

REPRESENTATION AND PRACTICE

EDITED BY

DAMIAN MAYE
Coventry University, UK

LEWIS HOLLOWAY
University of Hull, UK

MOYA KNEAFSEY
Coventry University, UK

United Kingdom – North America – Japan
India – Malaysia – China

Emerald Group Publishing Limited
Howard House, Wagon Lane, Bingley BD16 1WA, UK

Copyright © 2007 Emerald Group Publishing Limited

Reprints and permission service
Contact: booksandseries@emeraldinsight.com

No part of this book may be reproduced, stored in a retrieval system, transmitted in any form or by any means electronic, mechanical, photocopying, recording or otherwise without either the prior written permission of the publisher or a licence permitting restricted copying issued in the UK by The Copyright Licensing Agency and in the USA by The Copyright Clearance Center. No responsibility is accepted for the accuracy of information contained in the text, illustrations or advertisements. The opinions expressed in these chapters are not necessarily those of the Editor or the publisher.

British Library Cataloguing in Publication Data
A catalogue record for this book is available from the British Library

ISBN: 978-0-08-045018-6

Awarded in recognition of Emerald's production department's adherence to quality systems and processes when preparing scholarly journals for print

INVESTOR IN PEOPLE

Contents

List of Figures	ix
List of Tables	xi
Contributors	xiii
Acknowledgements	xvii

1 Introducing Alternative Food Geographies　　　　　　　　　　　　1
   Damian Maye, Moya Kneafsey and Lewis Holloway

PART I Alternative Food Geographies: Concepts and Debates

2 Localism, Livelihoods and the 'Post-Organic': Changing Perspectives on
   Alternative Food Networks in the United States　　　　　　　　　23
   David Goodman and Michael Goodman

3 Connecting Social Justice to Sustainability: Discourse and Practice in
   Sustainable Agriculture in Pennsylvania　　　　　　　　　　　　39
   Amy Trauger

4 From 'Alternative' to 'Sustainable' Food　　　　　　　　　　　　55
   Larch Maxey

5 Beyond the 'Alternative'–'Conventional' Divide? Thinking Differently
   About Food Production–Consumption Relationships　　　　　　　77
   *Lewis Holloway, Moya Kneafsey, Rosie Cox, Laura Venn,
   Elizabeth Dowler and Helena Tuomainen*

6 Globally Useful Conceptions of Alternative Food Networks in the
   Developing South: the Case of Johannesburg's Urban Food Supply System　95
   Caryn Abrahams

7 Justifying the 'Alternative': Renegotiating Conventions in the Yerba
   Mate Network, Brazil   115
   Christopher Rosin

8 Is Meat the New Militancy? Locating Vegetarianism within
   the Alternative Food Economy   133
   Carol Morris and James Kirwan

PART II Public Policy and Alternative Food Projects

9 Regionalisation, Local Foods and Supply Chain Governance:
   A Case Study from Northumberland, England   149
   Damian Maye and Brian Ilbery

10 Governing the Speciality Food Sector: Integrating Supply Chains,
   Sectors and Scales in West Wales   169
   Catherine Walkley

11 Public Sector Food Procurement in the United Kingdom: Examining
   the Creation of an 'Alternative' and Localised Network in Cornwall 185
   James Kirwan and Carolyn Foster

12 'Bending Science to Match Their Convictions': Hygienist Conceptions of
   Food Safety as a Challenge to Alternative Food Enterprises in Ireland 203
   Colin Sage

13 Market-Oriented Initiatives for Agri-Environmental Governance:
   Environmental Management Systems in Australia   223
   Vaughan Higgins, Jacqui Dibden and Chris Cocklin

PART III Practising Alternative Food Geographies

14 From the Ground Up: California Organics and the Making of
   'Yuppie Chow'   241
   Julie Guthman

15 Buying into 'Buy Local': Engagements of United States Local
   Food Initiatives   255
   Patricia Allen and Clare Hinrichs

16 Manufacturing Fear: the Role of Food Processors and Retailers
   in Constructing Alternative Food Geographies in Toronto, Canada   273
   Alison Blay-Palmer and Betsy Donald

17 Networking Practices Among 'Alternative' Food Producers in England's West Midlands Region *David Watts, Brian Ilbery and Gareth Jones*	289
18 The Appropriation of 'Alternative' Discourses by 'Mainstream' Food Retailers *Peter Jackson, Polly Russell and Neil Ward*	309
19 Sidestepping the Mainstream: Fairtrade *Rooibos* Tea Production in Wupperthal, South Africa *Tony Binns, David Bek, Etienne Nel and Brett Ellison*	331
Index	351

List of Figures

Figure 3.1	Migrant field crew: migrant Mexican labourers picking organic tomatoes	46
Figure 3.2	WAgN in action: WAgN Steering Committee at a strategic planning retreat	48
Figure 3.3	Tractor equipment workshop: WAgN members learning about equipment maintenance	50
Figure 4.1	Sixfold model of sustainability	62
Figure 4.2	The case study locations	63
Figure 6.1	The Johannesburg region in the Gauteng Province	99
Figure 6.2	The road-side sale of live farm chickens	100
Figure 6.3	Cultural chicken snacks	100
Figure 6.4	Traditional nuts sourced directly from a farm	101
Figure 6.5	Traditional African vegetables	101
Figure 7.1	Examples of packaging used by Santa Catarina yerba mate processors	117
Figure 7.2	Map of the study area: Santa Catarina, Brazil	123
Figure 9.1	The case study region: Northumberland, England	155
Figure 9.2	Institutional structures for local food economies in Northumberland	159
Figure 10.1	The West Wales study region	171
Figure 11.1	Support for the Cornwall Food Programme	194
Figure 13.1	A Gippsland EMS farm promoting its environmental credentials	226
Figure 13.2	Example of a Gippsland EMS producer profile	228
Figure 15.1	US 'Buy Local' promotional food label	257

x List of Figures

Figure 17.1 The West Midlands region: selected administrative zones, places and transport links 290

Figure 17.2 Business Network Diagram for interviewee 'a' 299

Figure 19.1 South Africa's West Coast Region 332

Figure 19.2 *Rooibos* tea growing in the wild 337

Figure 19.3 *Rooibos* tea harvested and drying in a tea court 340

List of Tables

Table 5.1	Analytical fields for describing food projects	81
Table 5.2	Earthshare community-supported agriculture, near Forres, Scotland	84
Table 5.3	Salop Drive market garden, West Midlands, United Kingdom	86
Table 5.4	Adopt a sheep scheme, Anversa, Abruzzo National Park, Italy	88
Table 7.1	Regimes of justification	119
Table 9.1	Institutional profiles for surveyed institutions	157
Table 12.1	Contrasting attitudes expressed in court represented as a hygienist approach to food safety and a qualitative approach to product quality	217
Table 15.1	Buy Local Food campaigns in the US	259
Table 15.2	Themes of programme objectives and promotions of FoodRoutes Buy Local Food campaigns	261
Table 15.3	Claims about buying local food in programmatic literature of 18 Buy Local Food campaigns in the US	262
Table 15.4	Themes with associated claims about the benefits of buying local foods	264
Table 16.1	Fear-based and positive descriptors from key informant advertising and websites	283
Table 17.1	Characteristics of interviewees' agricultural and non-agricultural enterprises	296

Contributors

Caryn Abrahams is a Tutor in the School of Geography, Archaeology and Environmental Studies, University of the Witwatersrand, Johannesburg, South Africa.

Patricia Allen is Associate Director of the Center for Agroecology and Sustainable Food Systems, University of California, Santa Cruz, USA.

David Bek is a freelance researcher currently working on projects for the Geography Departments at the Universities of Otago and Durham and the Judge Business School, University of Cambridge, UK.

Tony Binns is Professor of Geography in the Department of Geography, University of Otago, Dunedin, New Zealand.

Alison Blay-Palmer is Assistant Professor in the Department of Geography, Queen's University, Kingston, Ontario, Canada.

Chris Cocklin is Pro-Vice-Chancellor of the Faculty of Science, Engineering and Information Technology, James Cook University, Queensland, Australia.

Rosie Cox is Lecturer in London Studies at Birkbeck, University of London, UK.

Jacqui Dibden is a Research Fellow in the School of Geography and Environmental Science, Monash University, Victoria, Australia.

Betsy Donald is Assistant Professor in the Department of Geography, Queen's University, Kingston, Ontario, Canada.

Elizabeth Dowler is Reader in Food and Social Policy in the Department of Sociology, University of Warwick, Coventry, UK.

Brett Ellison is a postgraduate researcher in the Department of Geography, University of Otago, Dunedin, New Zealand.

Carolyn Foster was formerly a Research Fellow in the Countryside and Community Research Unit, University of Gloucestershire, Cheltenham, UK, and is now a self-employed consultant.

David Goodman is Professor of Environmental Studies in the Department of Environmental Studies, University of California, Santa Cruz, USA.

Michael Goodman is Lecturer in Geography at King's College London, UK.

Julie Guthman is Associate Professor in the Department of Community Studies, University of California, Santa Cruz, USA.

Vaughan Higgins is Senior Lecturer in Sociology at Monash University, Victoria, Australia.

Clare Hinrichs is Associate Professor of Rural Sociology in the Department of Agricultural Economics and Rural Sociology, Pennsylvania State University, USA.

Lewis Holloway is Lecturer in Geography at the University of Hull, UK.

Brian Ilbery is Professor of Rural Studies in the Countryside and Community Research Unit, University of Gloucestershire, Cheltenham, UK.

Peter Jackson is Professor of Human Geography and University Director of Research (Social Sciences) at the University of Sheffield, UK.

Gareth Jones is the Managing Agent for the National Farmers' Retail and Markets Association (FARMA), based in Southampton, UK.

James Kirwan is a Senior Research Fellow in the Countryside and Community Research Unit, University of Gloucestershire, Cheltenham, UK.

Moya Kneafsey is a Research Fellow in the Department of Geography, Environment and Disaster Management, Coventry University, UK.

Damian Maye is Senior Lecturer in the Department of Geography, Environment and Disaster Management, Coventry University, UK.

Larch Maxey is a Research Fellow in the Department of Geography, Swansea University, UK.

Carol Morris is Lecturer in the School of Geography, University of Nottingham, UK.

Etienne Nel is Professor of Human Geography at Rhodes University, Grahamstown, South Africa.

Christopher Rosin is a Research Fellow at the Centre for the Study of Agriculture, Food and the Environment, University of Otago, Dunedin, New Zealand.

Polly Russell is a Research Associate based at The British Library, London, UK.

Colin Sage is Senior Lecturer in Geography at University College Cork, Republic of Ireland.

Helena Tuomainen is Research Fellow in the Department of Sociology, University of Warwick, Coventry, UK.

Amy Trauger is a Postdoctoral Research Fellow in the Department of Agricultural Economics and Rural Sociology, Pennsylvania State University, USA.

Laura Venn is a cultural research analyst for the West Midlands Regional Observatory, Birmingham, UK.

Catherine Walkley is a Research Associate in the Institute of Geography and Earth Sciences, The University of Wales, Aberystwyth, UK.

Neil Ward is Professor of Rural and Regional Development and Director of the Centre for Rural Economy at the University of Newcastle, UK.

David Watts is a Research Fellow at the Institute for Transport and Rural Research, School of Geosciences, University of Aberdeen, UK.

Acknowledgements

Most of the chapters in this book are based on papers presented as part of three sessions on 'alternative food systems' at the annual Royal Geographical Society/Institute of British Geography (RGS/IBG) conference in London, on the 1st and 2nd of September, 2005. The editors would like to thank the Rural Geography Research Group for supporting the sessions and especially Professor Henry Buller for encouraging us to draw up the original call for papers. The chapters were all subjected to an anonymous refereeing process, and we would like to thank all the reviewers who kindly gave up their time to comment on them. Joanna Scott, Tony Roche, Mary Malin and Helen Collins at Elsevier have provided excellent technical advice and continued encouragement throughout the course of the project. We are grateful to Stuart Gill of the Cartography Unit at Coventry University for helping to draw some of the figures and appreciate the support received from colleagues in the Geography Departments at Coventry and Hull Universities. Special thanks go to our partners, families and friends who have supported us during the project. We end with a final note of thanks to the contributors for their support and interest in the book, which, due to their collective efforts, helps to reflect the diversity of debates, representations and practices surrounding attempts to reform food provision in different places and spaces around the world.

Chapter 1

Introducing Alternative Food Geographies

Damian Maye[1], **Moya Kneafsey**[1] **and Lewis Holloway**[2]
[1] Department of Geography, Environment and Disaster Management, Coventry University, UK
[2] Department of Geography, University of Hull, UK

Introduction

This book began life as a session on 'alternative' food systems, convened on behalf of the Rural Geography Research Group held at the 2005 annual conference of the Royal Geographical Society/Institute of British Geographers (RGS/IBG) in London. Most of the chapters are based on papers presented at the conference, with additional contributions from authors based in North America, Australia and New Zealand. Each chapter has been externally refereed. By way of introducing the collection, this chapter first outlines the key questions which drove the original call for papers from which the book emerged; secondly, it reviews in more detail the current body of work dedicated to research on 'alternative' food geographies; finally, it identifies some of the new research agendas which emerge from this edited collection, including a summary of each of the eighteen chapters.

Since the late 1990s, discussions about the evolving character and implications of 'alternative' food geographies have become prominent (Goodman 2003, 2004; Winter 2003a, 2004). The debates have generally been couched in terms of a perceived trend towards the emergence of food production–consumption relationships which offer an 'alternative' set of possibilities to those provided by the 'conventional' industrialised agro-food complex. Much of the literature has concerned possibilities for agricultural producers in marginalised, northern hemisphere contexts to carve out market niches for themselves by selling high-quality products embedded with information about product, process and place (Ilbery et al. 2005; Sonnino and Marsden 2006). These market niches are constructed through 'short' supply chains (Renting et al. 2003) and more direct, 're-connected' and/or 're-localised' relationships with consumers. The possibilities for consumers, whilst having received less attention, exist in terms of the creation of access to better quality, more healthful and more ethically – and aesthetically – satisfying foodstuffs within a context of more transparent – and hence 'trustable' – relationships with food producers. Examples of the possibilities which have been researched include farmers' markets, community-supported agriculture, community food projects, box schemes and food link schemes and fair trade (see, for e.g. Hinrichs 2000; Holloway and Kneafsey 2000; Hendrickson and Heffernan 2002; Renard 2003; Bryant and Goodman 2004; Kirwan 2004).

The mushrooming of research into these possibilities has raised as many questions as it has answered. This book sets out to address two broad sets of questions which have been stimulated by this work. The first set of questions concerns the utility of the word 'alternative'. For example, recent empirical research has questioned the extent to which small businesses engaged in activities grouped under the 'alternative' food networks (AFNs) umbrella can really be seen as 'alternative'. As noted by Ilbery and Maye (2005), in the UK context, such businesses have to 'dip in' and 'dip out' of 'conventional' supply chains due to the ways in which the dominant agro-food system is currently structured. Research by Venn et al. (2006) also suggests that the actors involved in projects which are described as 'alternative' very rarely describe themselves as 'alternative' and indeed, would baulk at the term. This evidence of disjuncture between academic and lay discourses prompted us to ask how useful the concept actually is. This question has only become more insistent as research under the banner of the 'alternative' proliferates to include healthy living, community development, sustainability, organic, ethical and green consumption and more. There is a sense that the term is in danger of being 'emptied out' of meaning. Thus, one of the key questions that this edited collection aims to address is: How can 'alternative' food systems be conceptualised? All of the chapters here are therefore concerned with understanding 'alternativeness' and its related geographies, but all adopt varying perspectives to understand the nature of such relations. As such, they clearly begin to show that the notion of the 'alternative' is contested and has to be understood in relation to an equally contestable notion of the 'conventional'. They also demonstrate that the utility and meaning of the term are context-dependent and that there is much to be gained from pursuing detailed, theoretically informed, empirical research into the complex geography of 'alternatives'.

Whilst research to date has revealed much about the ways in which 'alternative' geographies of food are being performed in certain contexts and by particular actors, it has also stimulated a range of research questions about the places, processes and people 'missing' from accounts so far. This edited collection thus begins to address a second set of questions concerning the discourses, representations and practices involved in the construction of 'alternative' food geographies in different geographical contexts. Insights are drawn from a whole range of actors involved in the construction of 'alternatives', from producers, to consumers, processors, retailers and the institutions engaged in the governance of such activities. Contributors draw on case study research covering a range of products (including fair-trade tea, unpasteurised cheese, Pembrokeshire potatoes, organic salad mix, beef and chicken) and places (including examples drawn from the UK, Ireland, Italy, the USA, Canada, Brazil, South Africa and Australia). Crucially, we are delighted that the collection includes accounts from developing economies in the southern hemisphere. We welcome these contributions and recognise that they begin to expand understandings of the complex geographies of AFNs in ways which lay down important avenues for continued future research.

The need to understand the values and potential of 'alternative' geographies of food has been prompted by a range of factors that extend well beyond the theoretical preoccupations of academics interested in revealing novel modes of food supply. Within a developed world context, food occupies a prominent place in the public conscience, especially in terms of anxieties about the health, environmental and ethical implications of the ways in which food is produced and processed. In terms of health, there is evidence that consumers harbour fears about the long-term impacts of ingesting the artificial additives and chemicals in their food (Kneafsey *et al.* 2004; Sassatelli 2004; see also the essays in Lien and Nerlich (2004) which examine consumer anxieties about food). In many wealthy countries concerns are also increasingly raised over people's eating habits as rates of obesity, heart disease and diabetes continue to rise. Rapidly developing economies such as those of Japan and China are also experiencing similar problems as consumers adopt 'western' eating habits. In developing world economies, problems of malnutrition and hunger persist, and critics continue to point to the unjust and unequal relations of trade which operate at the expense of the world's poor and to the benefit of global corporations. Not only does this raise ethical concerns in the minds of many consumers in developed market economies, but awareness of the environmental impacts of 'food miles' have also lodged in the public consciousness. These comments, although necessarily brief, are intended to show why 'alternative' food initiatives have become a focal point for political and social advocacy, promoted by activists and academics alike as possible strategies that can resist the dominance of global and industrial food systems (see Hines 2000). However, as suggested above, these 'alternatives' are complex, contested and ambivalent, encompassing different ideologies, institutional discourses and production and consumption practices. Based on empirically rich and critically alert contributions, *Alternative Food Geographies* aims to capture and explore some of this complexity whilst at the same time prompting further research and debate on the possibilities for constructing more progressive politics and practices of food. The next part of the chapter reviews, in more detail, how research in agro-food studies, and especially human geography, has developed to interpret AFNs and their related geographies.

Alternative Food Geographies: A Short Review

Whatmore and Thorne (1997) were amongst the first to write about 'alternative food geographies', both as a conceptual device and a tangible feature of the contemporary agro-food landscape, defined in this case through the international fair-trade movement. Since then, debates within agro-food studies have been thriving (Winter 2003a).[1] Following Watts *et al.* (2005), we have divided this review of AFNs into two parts: 'product and place' and 'process and place'. The main geographical focus is developed market economies, especially the UK, other parts of Western Europe and North America; although for a fuller review of US alternatives see Chapter 2. This bias reflects the prime focus for much AFNs research to date. Having organised the review in this way,

---

[1] See, for example, themed issues in *International Planning Studies* 4 (3), 1999; *Sociologia Ruralis* 40 (2), 2000; 40 (4), 2000; 41 (1), 2001; 42 (4), 2002; *Journal of Rural Studies* 19 (1), 2003; *Environment and Planning A* 35 (3) and the *British Food Journal* 105 (8), 2003.

we are aware that it may imply a sense of linearity to developments over time and that the different alternatives are coherent, rational and codified entities that belong to one of two camps. This is not the intention. Instead, the aim is to show how one can identify different AFNs and therefore question the conflating ways in which AFNs are sometimes defined and labelled, including within academic discourses.

Goodman (2003) suggests that geographical comparisons can be drawn between American and European 'models' of alternative food practice, with US alternatives viewed more as social and oppositional movements (e.g. Allen *et al.* 2003), in contrast to the European Union's (EUs) more endogenous claims on historical and cultural traditions of product and place (e.g. Ilbery and Kneafsey 1998). The material reviewed here reflects this broad trend. In drawing these distinctions between different alternatives and different alternative geographies, it is also possible to demonstrate philosophical differences that influence interpretations of AFNs. The product and place alternatives, distinctly European in origin and with a clear emphasis on quality production, thus offer politico-economic interrogations of territorially embedded, agrarian-centred strategies to adding value to food commodities and products, with an aim to identify new development pathways for the wider rural economy (Marsden *et al.* 2002; Goodman 2004). The process and place alternatives, meanwhile, have more emphasis on supply chain relations and offer more distinctive oppositional socio-economic interrogations to improve livelihoods and local well-being (Watts *et al.* 2005). While not specific to North America, it is notable that a number of papers from this context are more process-orientated, with a more radical political agenda related to production–consumption, emphasising, for example, questions about social justice or 'alternative' economic relations, rather than focusing solely on rural development (see, for e.g. Hendrickson and Heffernan 2002; Hassanein 2003). In the context of this book, the important point is that agro-food commentators are now asking the central question: What is alternative about the 'alternative' food economy? (Whatmore *et al.* 2003).

Alternative Economies and Agricultural Restructuring

To address this question it is useful to first outline the main ways in which geographers have so far sought to explain alternative economies and agricultural restructuring. A growing number of economic geographers have become interested in examining the significance of 'alternative' economies that 'perform' outside the normal market conventions (Leyshon *et al.* 2003; Hughes 2005; see also McCarthy (2006) on 'alternative rural economies'). These 'diverse economies', it is argued, often operate in the context of socially disadvantaged areas, marginal populations or simply include groups with alternative economic ideas. In a notable contribution, Lee (2000), for example, usefully shows how producers of ornamental hardy plants are able to remain in business by working closely with their consumers to create 'economic geographies of regard'. These producers should, in conventional economic terms, be bankrupt. However, by running their business through 'sub-capitalist economic practices', they are able to remain economically viable. Elsewhere, Lee *et al.* (2004) have mapped the emergence of local currency systems. While these alternative schemes appear small and insignificant as direct challenges to 'capitalocentrism', the fact that they exist at all reveals, they argue, the possibility of a 'pro-active response' to prevailing economic norms. Thus, their existence and positioning *within* the market economy, as alternatives, are seen as the significant factors, not their relative size.

In the introduction to *Alternative Economic Spaces*, Leyshon and Lee (2003: 3) contextualise these developments against a backdrop of global capitalism and neoliberalism. This is also useful for a survey of alternative food geographies. Critics of the neoliberal market-based model argue that structuring the global economy in this way is undemocratic, designed to favour rich economies at the expense of the poor, with mounting social inequalities at both national and global scales (Watts *et al.* 2005). The neoliberal model which constructs a uniform, global economic geography is, therefore, contrasted with alternative economies that are attentive to the specificities of space and place (Leyshon and Lee 2003). The key question here is how can these ideas about alternative economies more generally inform an understanding of alternative food geographies? Three things are important. First, polarisation between neoliberal and alternative economy ideologies has resonances with comparisons drawn between 'conventional' and 'alternative' systems of food supply (see Ilbery and Maye 2005). Second, and related to this, is recognition that these alternative initiatives in actuality operate *within* the 'Big C' of capitalism (see also Gibson-Graham 1996). So, while there may be some form of resistance to capitalist relations, this may well be fragmented and perhaps not even intended as a direct challenge to capitalism; it will more likely take shape as a 'little r' of resistance which enables people to wrest back some control over how their food is produced. Thirdly, the emphasis on social relations and the ability of alternatives to enable positive local economic impacts is significant and consistent with the benefits often associated with AFNs, especially in rural development contexts (Renting *et al.* 2003).

The World Trade Organisation (WTO) is the key global institution that sets the context for the governance of production and trade, with the overall intention of establishing a globalised, neo-liberal food economy (Potter and Burney 2002). This involves enabling 'emerging' market economies to export non-traditional agricultural exports more effectively, while reducing protectionism in developed market economies to establish an agricultural trading system that is fair and market-orientated. As noted above, many of the world's food sectors and supply chains are in fact controlled by a small number of multinational interests (Hendrickson and Heffernan 2002). Agricultural alternatives may then emerge because of the inability of the market economy to engineer market-orientated, fair-traded relationships between chain members and nation states. One significant consequence of food industrialisation in developed market economies has also been the steady decline in the revenue that farmers are able to derive from the sale of their produce (Ilbery *et al.* 2005). Not only, then, are prices falling as traditionally protected markets are opened up to global trade liberalisation, but the locus of added value has moved away from the farm and more proximal processing industries to the larger food processing and retail sectors. These are important points of reference that determine not only why AFNs come about but also the economic contexts in which they are defined. Against this context, the review next examines two types of alternatives: product and place and process and place.

Product and Place Alternatives

One of the dominant features of AFNs, particularly in Europe, has been the attempt to link 'product and place' in order to add value to agricultural outputs. This is often defined as a process of re-localisation, in which locally distinctive quality food products are transferred to regional and national markets (Ilbery and Kneafsey 1998). This is deemed to provide a valuable economic stimulant, a way of reducing the deleterious impact of national and EU subsidy reforms and increasing trade liberalisation. Evidence to support this comes from three inter-linked sources (Watts *et al.* 2005): (1) Food labelling and the 'quality turn', (2) the growth of regionalism in European economies and (3) changes to rural policy in the EU, including the emphasis on (spatially extended) short food supply chains (SFSCs). This part of the review thus reveals a very Euro-centric vision of AFNs which, in response to changing policy contexts and perceived global economic competition, aims to produce geographically specific food products that can be sold outside the region they were produced in as niche market commodities.

The first of the three sources is the various policy initiatives which have been developed to promote foods with territorial/regional associations. Most notable is the Protected Designation of Origin (PDO) and the Protected Geographical Indication (PGI) quality labels introduced by the EU in the early 1990s to 'protect' and 'promote' food and drink products with a recognisable geographical origin (Ilbery and Kneafsey 2000; Parrott *et al.* 2002). Collectively these changes have been interpreted as a symbol of a 'quality turn' (Goodman 2003), with particular emphasis on specialist food production, especially in 'marginal economies'.

Emphasis on specific territories through food labels links to the second source of evidence: regionalisation. In one of the first special journal issues on alternative food chains, Marsden *et al.* (1999) argue that agriculture needs to become a stronger part of the regional economy and the regionalisation agenda. These processes of governance carry notable implications for AFNs, especially in terms of how support is given and what is supported and deemed significant. In the UK, for instance, the power of regional institutions has recently increased. In particular, Regional Development Agencies have been awarded significant powers, including the lead role in rural economic development and support for the local specialist food sector (see Winter 2006 for a fuller review). To date, the relationship between AFNs and public policy has been relatively neglected as a research theme.

Thirdly, product and place alternatives are influenced by changes to the Common Agricultural Policy (CAP). The Agenda 2000 reforms in the EU, for example, introduced the Rural Development Regulation (1257/99) as a response to anticipated changes to the WTO rules. This broadened rural policy away from a narrow sectoral focus on agriculture and towards a more territorial agenda (Lowe *et al.* 2002). While economic objectives are still paramount, social, cultural and environmental dimensions of rural development are promoted in a move to encourage economic diversification, agri-environmental schemes and the local processing and marketing of agricultural products (Banks and Marsden 2000). Linked to this, several commentators have identified the emergence of a 'new rural development dynamic' (see Ploeg *et al.* 2000; Marsden *et al.* 2002). SFSCs are central to this dynamic and can be divided thus: first, the production and retail of food products within a county/region and secondly, the sale of 'locality foods' as value-added commodities for export outside the designated locale (Marsden *et al.* 2000). The former usually include direct marketing initiatives and the sale of products via local retailers. The latter are spatially extended and distinguished through label and accreditation schemes, often with a focus on 'traditional', 'speciality' and/or 'quality' food products. Crucially, these chains are constructed in opposition to the 'agro-industrial dynamic'. These modes of agrarian restructuring thus promote alternative ways of farming and in doing so have fallen under the 'AFNs' umbrella.

Questions have been asked about whether spatially extended SFSCs really are an alternative. The important point is that these 'AFNs' have been defined from a specific European context. The overall message is that 'fixing' products to places in this way engenders greater endogenous rural development. These accounts are very useful, but they also provide a very particular view of the alternative food economy, inspired as a response to political and economic changes affecting agricultural restructuring. This short review has so far then focused almost exclusively on farms/producers and the quality food economy. As noted below, these examples represent only a partial view of AFNs.

Process and Place Alternatives

Product and place alternatives are focused on 'locality' food products and niche market food production. For Watts *et al.* (2005) these examples represent 'weaker alternatives' because they focus on value-added products with a clear geographical provenance rather than focusing on the nature of the food supply chain. Process and place alternatives, in contrast then, are viewed by Watts *et al.* (2005) as 'stronger alternatives' because they emphasise social and ethical values associated with particular supply chains. Evidence for process and place alternatives comes from four sources: (1) the local food sector; (2) community food projects; (3) public procurement (4) and organic foods and fair-trade markets. These initiatives include, but also move beyond, the primary producer and also resonate more closely with North American examples.

The local food sector is the first source of evidence for process and place alternatives. Morris and Buller (2003) argue that, despite significant growth, the expansion of the local food sector has been neglected compared to its locality food counterpart. This is significant given the greater emphasis on 'alternative' types of supply chain. Studies of the local food sector typically include the archetypal examples noted earlier, such as farmers' markets, box schemes, community-supported agriculture, farm shops and on-farm butchers; with examples in the US, the UK and other parts of Europe (in addition to those noted earlier, see for e.g. Gilg and Battershill 1998; Cone and Myhre 2000; Ventura and Milone 2000; Holloway and Kneafsey 2000; Sage 2003; Winter 2003b). Significant in all these examples is the direct nature of the supply chain and the important social and economic benefits which accrue from these types of food transaction. They can be categorised as proximate/direct marketing channels according to Marsden *et al.*'s (2000) classification of SFSCs, but some of the examples extend beyond farm-based production. Many of these accounts also note that farmers and other businesses operating in the local food sector often operate at relatively small scales. Crucially, these chains also encompass a much wider range of products, unlike product and place alternatives that mostly concentrate on processed foods such as wines, cheeses and cooked meats and often emphasise quality-based production (Watts *et al.* 2005).

The second process and place alternative is local community food projects. These projects involve more than just the establishment of alternative pathways for agrarian restructuring. A number of community projects, many of them supported and established by local authorities, are actively trying to bring food into economically deprived areas (e.g. local community food co-ops and buying groups). These areas have also been described as 'food deserts' (Wrigley 2002) and are believed to be particularly prominent in urban areas, where access to cheap, healthy food is poor. Hendrickson and Heffernan (2002), for example, provide an analysis of the Kansas City Food Circle and its attempt to establish a locally controlled food system that supplies healthy food to the community. Local food projects that supply relatively cheap, fresh foods therefore aim to reduce social problems such as bad diet and poor health, as well as providing employment for people involved in supplying and running the schemes. Dowler and Caraher (2003) suggest that these projects retain 'philanthropic' overtones, echoing Lee et al.'s (2004) point that the 'pro-active voice' of alternative projects may be as important as actual latent effects.

The third source for process and place alternatives is public procurement. This refers to the supply of food to public institutions (e.g. schools, hospitals and prisons) and has recently attracted significant political attention in parts of Europe and North America. In Europe, for example, a major obstacle to the localisation of public foods has been the principle of 'non-discrimination', where EU law prohibits food procurers from specifying the term 'local' in their purchasing contracts. Cost has become the dominant criteria when awarding contracts to supply food to public institutions. This has resulted in the exclusion of small local producers and suppliers who have been unable to offer competitive prices in comparison to large suppliers. In an effort to re-localise supply, Morgan and Morley (2002) encourage public procurers to be creative in stipulating the need for more 'fresh', 'organic', 'seasonal' foods, small lot contracts and more precise delivery times. By stipulating such demands, public institutions can, they argue, increasingly take advantage of initiatives in public sector food policy to promote social and environmentally sustainable food systems through the purchase of local quality assured commodities. The attempt to establish 'alternative' public supply contracts has thus become part of agro-food research on AFNs, although so far only to a limited extent.

The fourth process and place alternative is organic foods and fair trade. They present a tricky proposition: they receive strong support for their ethical stances but have also been criticised for commercial success and expansion, and for their use of mainstream distribution chains and the growth in corporate-based ethical and organic food labels. These criticisms are well founded but may do an injustice to some of the founding principles and successes of organics and fair trade. Allen et al. (2003), for example, retrace the social and economic goals of organic farming pioneers in California and compared these with current leaders of alternative initiatives. They show how current leaders do not share the rhetoric of opposition; their preference is for ecologically sound food systems and entrepreneurial, market-based solutions. Fair trade is also built on a transformative imperative, a project originally designed to displace conventional channels of international trade. Like Allen et al. (2003), Renard (2003) unearths sharp contradictions between original transformative missions and current pragmatic goals for expansion and development.

In summary, this chapter started by demonstrating the ways in which AFNs are sometimes located within a linear narrative, interpreted as a response to generic concerns associated with mass consumerism, food chain verticalisation and agricultural productivism. This popular view also filters into academic interpretations which label 'AFNs' as a set of binary opposites against the 'conventional'. This is consistent with past distinctions drawn between, for example, 'productivism' and 'post-productivism'.

Like previous attempts, these distinctions, although initially useful, over time lose their sense of meaning and their ability to capture real life complexity and difference, with new research now challenging such binary logics. It is against this backdrop of, on the one hand, societal and political advocacy of AFNs and, on the other, significant academic debate about the nature of AFNs that this book is set. The next section outlines the structure of the book and how it contributes to these continuing debates.

The Structure of the Book: Advancing Debates

Alternative Food Geographies brings together 18 chapters which, in different ways, both challenge and advance existing debates, theoretical approaches, metaphors and methodologies used to explain AFNs. Overall, the book reveals a heterogeneous mix of perspectives which are intended to better understand the various ways AFNs are represented and practised. The book is organised into three parts: (1) Alternative food geographies: concepts and debates; (2) Public policy and alternative food projects and (3) Practising alternative food geographies.

Alternative Food Geographies: Concepts and Debates

The central focus of the first part of the book is on reviewing theoretical and conceptual issues and debates related to AFNs, especially in relation to power, representations and discourses of the 'alternative'. All seven chapters seek to challenge current conceptualisations of AFNs, by offering new theoretical lenses or re-examining current 'readings' from new perspectives or geographical contexts. In other words, how, where and why is the term 'alternative' deployed?

Chapter 2 by *David Goodman* and *Michael Goodman* traces the contemporary development of AFNs in the USA. They begin by charting the growth of organic foods and show how it has become dominated by 'techno-scientific discourses'. Sustainable Agriculture Movements (SAMs) responded to this trend by going 'beyond organics', adopting local food systems as a new space of resistance and mechanism to retain the social significance of the family farm. The chapter argues that proponents of local foods often, however, ignore the 'politics of the local' and place too much emphasis on transformative potential at the expense of material/discursive practices. The authors thus propose a livelihoods approach (drawn from political ecology and development studies literatures) as a conceptual tool to critically explore AFNs. Applying this to published case studies, they argue for an assessment of all actors engaged in local foods, including the relative experiences of other social classes (especially marginal and excluded groups), as a mechanism to demystify localness and 'open up' the real social geographies of AFNs.

The chapter by *Amy Trauger* also focuses on questions of food and social justice. More specifically it is concerned with marginalised groups in farming contexts, defined in terms of race and/or gender, and the ways in which AFNs offer possibilities for such groups to assert their identities and capabilities, and to move towards a more socially just agri-food system. Empirically the chapter is based on research with two farming organisations in Pennsylvania in the USA: the first is an organic co-operative and the second aims to give greater voice to women involved in farming. The research reveals the complex social relationships which exist in particular food production conditions, noting racial and gender politics as two areas where the negotiation of social justice is most problematic. In conclusion, the chapter concedes that in some ways discourses of sustainability and social justice can contradict each other in the actual practice of running a farm business. However, the incorporation of social justice into the discourse of sustainability is to be regarded as a necessary development in shifting towards modes of food production–consumption which are socially progressive as well as environmentally sound.

In his chapter *Larch Maxey* is equally critical of the 'alternative' food concept and argues instead that the concept of 'sustainable food' offers a better approach to food analysis and action. He suggests that current approaches tend to leave the binary distinctions between 'alternative' and 'conventional' systems intact and unchallenged. Moreover, in perpetuating analysis of 'alternative' food, researchers may inadvertently normalise and legitimise highly problematic practices which are labelled as 'conventional'. Whilst acknowledging that 'sustainable' is also one half of a dualism, Maxey argues that it has analytical potency when used in a critical and relational way. On the basis of fieldwork conducted with small-scale food projects in Wales and Canada, he proposes a sixfold model of sustainability in which he attempts to make the components of sustainability more explicit. He concludes that the notion of 'sustainable food' has progressive potential to transform producer–supplier–consumer engagements with food.

In a related vein, *Lewis Holloway, Moya Kneafsey, Rosie Cox, Laura Venn, Elizabeth Dowler* and *Helena Tuomainen* also attempt in Chapter 5 to move debate beyond the usual dualistic approach to thinking about the nature of AFNs. Instead, they argue for a relational approach which recognises the multidimensional, contested and dynamic nature of food production–consumption relationships. They develop a heuristic framework for analysing such relationships consisting of seven analytical fields. The framework is applied to three case studies: a community-supported agriculture project in Scotland, an urban market garden in England and an internet-based sheep adoption scheme in Italy. The analysis demonstrates that each has unique characteristics and that the arrangement of these across the seven fields enables each project to express different forms of resistance to dominant systems of food provision. They conclude that it is necessary to go beyond simply labelling practices as 'alternative' and to examine how the specific ordering and spatiality of particular projects can challenge centres of power in food supply.

Sustaining this critical tone, *Caryn Abrahams* makes a timely intervention from the southern hemisphere and argues that studying AFNs in the global south has much to offer evolving agro-food theorisations. Abrahams thus addresses the failure of current debates to transcend the celebration of alternative consumption and challenges northern and exclusionary conceptions of AFNs by arguing that theory needs to build on empirical evidence from arenas other than the dominant European and North American contexts. Based on ethnographic research in the Johannesburg area, she identifies four networks of food provisioning: (1) direct farm retail; (2) local-cultural food provisioning networks; (3) cultural and religious food networks and (4) supply chains for the urban poor. These examples are grassroots endeavours, which contest the dominance of large-scale retailers and have emerged primarily as a response to the need for culturally specific, affordable and accessible foodstuffs amongst urban communities. Abrahams uses the examples to also develop a case for their contribution to a globally useful theorisation of AFNs.

Chris Rosin's chapter presents a second developing world case study – the Brazilian yerba mate network. Yerba mate is a tea, recognised as an important cultural and nutritional staple in parts of Latin America. Rosin's account is grounded in Convention Theory and sees AFNs as 'routine features' of the global food system. Like other chapters in Part One, Rosin thus attempts to breakdown the alternative/conventional dichotomy. His central focus is the 'justifications' that frame the yerba mate network, and the case study shows how this particular alternative was established in response to a perceived threat from heightened competition from cheap imports from Argentina. The yerba mate network was initiated by tea processors who were the most powerful actors involved and also the most threatened. This process-orientated view of Convention Theory thus reveals a range of strategic actions. Crucially, the actions do not strictly challenge a conventional network; instead, they reveal a more contested process of negotiation *within* the yerba mate network, with tea processors leading, and at the same time, dependent on others for support.

The final chapter in Part One, by *Carol Morris* and *James Kirwan*, examines the relationships between vegetarianism as a set of dietary choices and AFNs. Arguing from the position that the study of diet has been relatively neglected in agro-food studies, they are concerned to identify the ways in which vegetarianism resonates with some aspects of AFNs while jarring with others, in pursuit of an examination of the extent to which vegetarianism can be aligned with the precepts of 'alternativeness' in food production and consumption. The review shows that in terms of the ethical concerns expressed through a vegetarian diet, and in terms of the oppositionality sometimes associated with vegetarianism, there are clear parallels between vegetarianism and AFNs. However, their work also indicates that there are some areas of dissonance and ambiguity regarding, for example, the sustainability and healthiness of vegetarian diets, and the extent to which vegetarians do in fact overtly express ethical concerns in their dietary choices. The chapter concludes with calls for more explicit attention to be paid to diet within agro-food studies in general, and in particular, for more detailed studies of the practices of vegetarianism.

Public Policy and Alternative Food Projects

The thematic focus of Part Two of the book is the relationship between public policy and alternative food projects. This resonates with wider debates in rural studies examining the impacts of institutionalisation in terms of shaping rural economies. The impact of these processes on AFNs has so far been subject to only limited research (Winter 2006). The five chapters presented here begin to fill this gap and examine, in particular, the ways that institutions enrol, represent, support and, in some cases, impede the development of AFNs.

Chapter 9 by *Damian Maye* and *Brian Ilbery* attempts to bring together previously separate literatures on government regionalisation in the UK and the emergence of local food networks. The chapter then explores the changing modes of governance affecting food system localisation, focusing on Northumberland, in the north-east region of England. From their surveys of national, regional and local institutions, the authors draw distinctions between national policymaking, the transformation of such policies into regional and local strategies and the establishment of particular projects at regional and/or local levels. Using these distinctions, the chapter then assesses a number of issues influencing the development of local foods, including decentralisation and 'institutional thickness', the economic potential offered by local food strategies and the tension between 'local' foods and 'locality' foods. They conclude that institutional support for local foods in the UK has so far tended to focus on supporting 'locality' foods, neglecting the possibilities held out by 'local' food networks to contribute towards more sustainable food systems.

Catherine Walkley's chapter takes the analysis employed by Maye and Ilbery a stage further, examining support for specialist foods in West Wales from the perspectives of institutions and different actors operating along the food chain. The case study uses Amin and Thrift's (1994) four components of 'institutional thickness' as an analytical tool. The results show that support is biased towards the producer end of the supply chain. There is also evidence of competition between institutions. The other significant finding is the way the specialist sector is promoted and the type of initiative in place to support small food businesses. While some schemes are intended to re-localise specialist food markets, other initiatives are designed to enable access to multiple retail markets. Whilst recognising the important role of the specialist food sector as a tool for rural development, Walkley argues that conflation between the terms 'specialist' and 'alternative' must be treated with caution.

Chapter 11 by *James Kirwan* and *Carolyn Foster* next examines a particular form of institutional support: public procurement. They provide a detailed case study, which shows how a public sector food supply chain in Cornwall, a peripheral and largely rural region in the south-west of England, was re-fashioned. Using Actor-Network Theory (ANT) to examine the Cornwall Food Programme, the authors analyse 'from within' the ways in which local level actors build alternative networks of public sector food provision that actively incorporate notions of sustainability. Making a strong case for the continued use of ANT as a mode of analysis, they highlight the complex and contingent nature of this successful localised food network and the need for a strong vision that resonates with the agendas of the key actors involved.

Whilst Chapter 11 shows the positive contribution public policy can make to AFNs, Chapter 12 by *Colin Sage* shows how, in some instances, institutions can impede their development. The chapter provides a detailed account of a court case which took place in County Cork in the Republic of Ireland, involving a raw milk cheese maker and the Irish Department of Agriculture and Food (DAF). The court case intended to establish whether a batch of cheese was to be considered fit for human consumption. After introducing some of the key issues associated with science and food safety regulation, the chapter summarises the main areas of scientific dispute between the two parties. The case reveals a clear willingness on the part of the DAF to assert the scientific superiority of food safety expertise, even when presented with contrary evidence. Science is used to provide endorsement for value-based judgements reached by regulatory authorities with political or other agendas. In the process, the chapter demonstrates the huge costs under which speciality producers are placed if they fall foul of stringent food safety regulations and identifies a contradiction between rural development policies which are positive about small artisan food businesses and the obstacles imposed by food safety regulations which are based on risk assessment and reductionist science.

Part Two ends with a case study from Australia by *Vaughan Higgins, Jacqui Dibden* and *Chris Cocklin*. It reports on attempts to establish market-oriented initiatives, such as Environmental Management Systems (EMS), that are designed to address environmental degradation and create marketing opportunities. The chapter is set within a context of neoliberalist promotion of entrepreneurialism and self-reliance, where Australian farmers have experienced increasing pressures to adopt improved environmental practices while striving to ensure their economic viability. Through analysis of one regional project, they examine the extent to which EMS enable farmers to better manage financial and environmental goals. Whilst accepting the increasing significance of EMS for sustainable food production, they conclude that EMS mostly interest those farmers already engaged in environmental initiatives. Moreover, they are not widely recognised by consumers and so are unlikely to contribute to a widespread 'greening' of Australian agriculture.

Practising Alternative Food Geographies

The third and final part of the book moves from a focus on wider institutional networks to examine more explicitly the nature and politics of AFNs 'on the ground'. Collectively, the six chapters consider perspectives and practices from different actors and spaces in the food chain, including producers, retailers, consumers and local communities.

Chapter 14 by *Julie Guthman* presents a case study of organic food production and consumption in California. Building on her earlier work, the analysis takes salad mix (or mesclun) as its focus and examines how the provision of this particular organic commodity has changed over time. The way organic food is made and eaten has changed dramatically, shifting from what Belasco termed 'counter cuisine' to what organic growers call 'yuppie chow'. More specifically, the analysis shows two key things. First, organic salad mix is a carrier of major changes in the organic system of provision, shifting from small-scale to 'scaled-up' industrial operations in response to market liberalisation. Second, organic salad mix reveals a troubling politics of class and gender. Eating organics has thus become a symbol of class distinction and has an underbelly that reveals social injustices in terms of employment on organic farms. Guthman urges new local food projects to learn from these historical geographies and to build new initiatives from the 'ground up'.

Patricia Allen and *Clare Hinrichs* retain this critical lens and examine strategies for food system localisation through a focus on 'Buy Local' schemes in the US. Their concern lies with the motivations behind adopting 'Buy Local' as a strategy for establishing sustainable food systems, and in their analysis they assess the discursive frameworks associated with 'Buy Local' schemes, attempting to define a set of common themes which emerge from texts (including promotional materials, websites, brochures and reports) connected with particular projects. The chapter provides a narrative of the development of 'Buy Local' in the US, and of the institutional framework, which supports the strategy. Following this, six themes (aesthetics, community, economics, environment, equity and health) are identified and discussed. The authors show how schemes such as 'Buy Local' are awash with tensions and complexities (not least surrounding definitions of 'local'), which tend to be underexamined by those participating in them. At the same time, the economic, environmental and social benefits claimed by many projects are unsubstantiated.

In Chapter 16, *Alison Blay-Palmer* and *Betsy Donald* start from the premise that attempts to extend the scope and range of AFNs, that is, to make them more mainstream, are inherently problematic. This is partly due to the confusion surrounding the term 'alternative', but it is also due to the negative associations which have become attached to the mainstream food sector. Specifically they argue that it is because of a climate of fear surrounding mainstream food production that niche alternatives, explicitly defined in opposition to the mainstream, have become popular. Their argument then moves to a discussion of how particular codifications of food quality and safety, developed as part of attempts by the 'alternative' food sector to allay consumers' fears, have, ironically, allowed the appropriation of 'alternative' strategies by mainstream food companies. The authors examine their thesis in urban Toronto, Canada, and focus particularly on their interview work with food processors, distributors and retailers. They conclude that a shift from an emphasis on 'fear' to an emphasis on 'hope' is required to allow AFNs to escape appropriation.

Chapter 17 by *David Watts, Brian Ilbery* and *Gareth Jones* shifts focus to the producer, examining the networking practices of 'alternative' food producers in the English West Midlands. The focus is on the supply side of AFNs and aims to achieve a better understanding of producers' network-building practices. Based on interviews with local and locality food producers, they note that the networking associated with so-called AFNs often draws together connections between 'alternative' and 'conventional', and 'local' and 'non-local' enterprises, and is important in terms of both the exchange of goods and services and the communication of perspectives and opinions. Four types of network relationship are identified, varying from the highly informal relationship involving the exchange of knowledge or views, to the formalised contractual relationship. The chapter concludes by considering the relationships between the varying ideas of 'alternativeness' which permeate their analysis and the capitalist economy, suggesting that the enterprises examined clearly exist within the capitalist economy and as such must be profitable to survive, but that they can also function in ways that are different to those of 'conventional' businesses. Relations of 'trust' and 'integrity' are identified as being of particular importance.

Drawing on life history interviews with food producers and retailers, Chapter 18 by *Peter Jackson, Polly Russell* and *Neil Ward* next examines the development and marketing of a particular brand of chicken (Oakham White) by a British high-street retailer (Marks & Spencer). The chapter argues that the development of the brand was a response to growing consumer anxieties about food safety and quality following recent public health scares which have beset the industry. In an attempt to restore consumer confidence and differentiate themselves from their competitors, 'mainstream' retailers like Marks & Spencer are appropriating the discourse of 'alternative' food producers (associated with organic, free-range and GM-free foods). The use of a life history methodology provides insights into the tensions and ambivalences at work in the production of the Oakham White brand. The authors argue that this was as much about 'manufacturing meaning' as it was about the agricultural, technical and economic development of a new product. Comparisons are drawn with free-range chicken producers, selling direct to consumers via farmers' markets and it is argued that a blurring of boundaries is taking place between what has previously been defined as 'good' and 'bad' food and between local, small-scale producers and national corporations.

Part Three ends with a final chapter by *Tony Binns, David Bek, Etienne Nel* and *Brett Ellison* which examines the ways in which a South African rural community is responding to external challenges by engaging in a form of production that may be labelled 'alternative'. A community-based co-operative in an isolated mission community is generating employment and income through the production of fair-trade *rooibos tee* (red-bush tea), which only grows successfully in the microclimate and physical environment of the West Coast mountains region. The co-operative has delivered various socio-economic dividends to the host community and has enabled producers to 'sidestep' mainstream *rooibos* supply chains in the region. The chapter explores the applied relevance of the concept of alternative food systems, especially in terms of community economic survival and development in the developing world, and in South Africa specifically. The chapter concludes with a discussion of the critical and practical implications of the research and sounds a warning about possible threats to the sustainability of such projects.

Concluding Remarks: Food for Thought

This book set out to answer two key questions: first, how can AFNs be conceptualised and second, what discourses, representations and practices are involved in the construction of 'alternative' food geographies in different geographical contexts. We end by returning to these two questions and in the process identify six themes which, for us, emerge from this edited collection. Given the range and heterogeneous nature of material explored in the chapters, the themes are not intended to be exhaustive, but do appear to offer, we would argue, some important ways forward and useful points for future debate. We begin with the second, more empirical question and end with some conceptual and methodological reflections.

In terms of the discourses, representations and practices involved with AFNs, the chapters identify a range of activities related to products, places and processes. The first theme we want to emphasise in relation to this is geographical. The short review of current work on AFNs revealed a concentration of efforts in Europe and North America. Reviews have since called for perspectives from 'outside' this geographical domain (Hughes 2005). This collection includes perspectives from the 'global north' and the 'global south'. In particular, the chapters by Abrahams, Binns *et al.* (both based in South Africa) and Rosin (Brazil) challenge the way 'alternativeness' is theorised when placed in majority world contexts. Abrahams, for instance, shows how alternatives translate very differently when examined in a deprived neighbourhood in Johannesburg. Here, traditional and long-established street stalls represent the 'alternative', selling foods (e.g. chicken feet) to poorer residents who cannot buy these products in supermarkets. More work is needed to explore the developmental potential of AFNs beyond the minority world.

A second theme to emerge under this question is the attempt by some of the authors to historicise alternative food geographies and to learn from the past. This point about learning from the past is echoed in our own review, as well as the review of US alternatives by David Goodman and Michael Goodman. Empirically, it is expressed most clearly by Julie Guthman, who provides a recent historical survey of organics in California to reveal its detrimental incorporation into mainstream supply markets, with production 'scaled-up' and industrialised in response to market liberalisation. This awareness of the historical trajectories of AFNs resonates more broadly with past agro-food accounts about the globalisation of food supply, especially 'food regimes' (Friedmann and McMichael 1989), and more directly with the work of food historian Joan Thirsk (1997) who argues that farmers have experimented with new ways of using their land since the fifteenth century, often during periods of agricultural depression.

This leads on to our third theme which concerns the range of different actors involved with AFNs, including producers (e.g. Watts *et al.*; see also Higgins *et al.*), intermediaries (e.g. Rosin; Blay-Palmer and Donald) and retailers (Jackson *et al.*). The chapter by Jackson *et al.* for instance, helps to show the role of branding which in this case is used by a high-street retailer to appropriate AFN discourses to 'manufacture meaning' and relieve consumer anxieties. Perhaps most significant of all, though, is the emergence of institutionalisation and the increasing role of public policy in AFNs. It is perhaps surprising that little work has so far explored this theme (for exceptions see Marsden and Sonnino 2006; Winter 2006). Above all, it is clear that institutions play a significant role in supporting AFNs, whether in terms of supporting public procurement initiatives or in terms of assisting the development of local/speciality food networks in the context of regionalisation. As the chapter by Colin Sage shows, these developments can also be hampered by regulatory frameworks and institutional dogmatism. More work on the role of public policy and AFNs is urgently needed.

The final question concerns how one then conceptualises AFNs? The first thing to note is that most chapters in the book challenge the dualism between 'alternative' and 'conventional' in favour of a more relational approach to AFNs. Some authors have responded to this by offering alternative frameworks (e.g. Maxey's relational framework for sustainable food; Rosin's process-orientated view of Convention Theory or Holloway *et al.*'s use of analytical fields). Others have offered more empirically sensitive assessments of specific AFNs. Notable examples include Allen and Hinrichs review of 'Buy Local' schemes, in keeping with DuPuis and Goodman's (2005) critical assessment of local foods more generally, and Watts *et al.*'s epistemologically grounded assessment of local and locality food producers network practices, using notions of 'trust' and 'integrity' to further interrogate previous conceptualisations of regard and social embeddedness (cf. Sage 2003).

In response to our question, the overall conclusion from reading the chapters is that a singular 'alternative food system' in fact does not exist. Such a conclusion may initially appear to counter the impulse which lies behind this book, but it does not deny the values associated with 'alternativeness', at least in an ideological sense. Instead, it suggests that categorising spaces of economic activity as part of either 'alternative' or 'conventional' systems of supply is too simplistic and arbitrary. Rather, we would argue that food provision in general is best understood in terms of the construction of complex, changing and multiple sets of relationships. These relationships, although often contested, are often mutually constitutive in the sense that the discourses and practices associated with one category (e.g. 'alternative') develop *in relation to* the discourses and practices associated with another (e.g. 'conventional'). Moreover, the discourses and practices adopted by actors operating within any particular field of activity may overlap, or interlink, with the discourses and practices adopted by those operating in a supposedly 'opposing' field of activity.

Against this important general point, three further themes emerge which we would briefly like to reflect upon. The first, and perhaps most significant in relation to the above, is the way AFNs sustain themselves in terms of their overall development pathway and relationship with the market. Most chapters seem to agree that AFNs operate within the 'Big C' of capitalism, but some disagree on the relationship with certain markets, especially the issue of 'scaling up' and the potential threats expansion and increased competition may present. For some, AFNs are best placed operating against conventional market conventions – a situation Sonnino and Marsden (2006) describe as a 'battleground'. Blay-Palmer and Donald, for instance, argue that making 'AFNs' mainstream in response to a climate of 'food fears' is problematic because it plays directly into the hands of larger corporate competitors. Likewise, Guthman is clear that Californian organics' incorporation into the market was essentially a 'bad thing', and Binns *et al.* warn about the threats posed to fair-trade red-bush tea due to corporate intrusion and possible future competition. In contrast, Higgins *et al.* note the potential to produce market-orientated 'public goods' as a mechanism to reconcile competing supply chain demands (cf. Buller and Morris 2004). This is a fascinating and clearly still unresolved debate.

The second theme to emerge in relation to this question of conceptualising AFNs is the emphasis on social justice and social geographies. Hitherto, emphasis has been placed on the possibilities for transformation – especially in terms of concerns about developing identifiable alternative systems that are distinct from the conventional. Some of the chapters in this book adopt a more pragmatic/realist tone, less concerned with claims about transformative potential and more with actual impacts. Two are notable here: first, Goodman and Goodman's call for a livelihoods approach and second, Trauger's analysis of social justice. It is remarkable that very little has so far been written about the wider social implications of AFNs, also including related issues concerning moral and ethical claims. We see this as an exciting future AFN research agenda.

Following on from this, the final theme we want to emphasise is the role of diet and eating. For some time, calls have been made for a move away from production-centred analyses, towards theorisations that encompass more explicitly the cultures and politics of consumption (Goodman and DuPuis 2002). Agro-food geographers have been urged to look 'beyond the farmgate', and a number of chapters in this book examine other parts of the food chain, including consumption. Out of this latter focus emerge important questions about diet and the potential role of AFNs as mechanisms to overcome food and health problems. In particular, the chapter by Morris and Kirwan examines the place of vegetarianism in AFN debates. Work on food and health still remains to be done, alongside important questions about participation in AFNs and social status.

Having addressed the two questions which frame this edited collection, we end this introductory chapter by briefly drawing reference to some important methodological questions raised by the chapters. It is clear that the food chain (or network) retains significant resonance as both a metaphorical and methodological device to identify the material and social contours of alternative food provision (see also Jackson *et al.* 2006 and Maye and Ilbery 2006). Chapters in this collection thus present business or institutional network diagrams drawn with producers or others connected to or part of AFNs as a way to uncover their related geographies (see for e.g. Ilbery and Maye; Watts *et al.*). The principle of 'following things' (Cook *et al.* 2006) also retains powerful value, with attempts to follow, for example, links in the chain to explore the nature of producer–consumer relations (see Holloway *et al.*) or to show how meaning is manufactured along specific retail supply chains (see Jackson *et al.*). Novel case study approaches also enliven not only what but also how we study AFNs and their related geographies (see in particular Sage's account of a court case in Cork). In this vein, Maxey also identifies the possibilities of participatory methods (see Pain 2004 for a general review) as a way to critically 'animate' AFNs. Thanks to these and other innovations, we anticipate, then, a scenario of increasing methodological richness and diversity in relation to AFNs. This parallels, and in some ways is bound up with, the deployment of different theoretical perspectives on AFNs. We hope that this volume, as well as reflecting the 'current state of the art' in AFN studies, goes some way towards stimulating further theoretical and methodological innovation in the field.

References

Allen, P., Fitzsimmons, M., Goodman, M. and Warner, K. (2003). Shifting plates in the agrifood landscape: the tectonics of alternative agrifood initiatives in California. *Journal of Rural Studies* 19, 61–75.
Amin, A. and Thrift, N. (Eds) (1994). Living in the global. *Globalization, Institutions, and Regional Development in Europe* (pp. 1–19). Oxford: Oxford University Press.
Banks, J. and Marsden, T. (2000). Integrating agri-environment policy, farming systems and rural development: Tir Cymen in Wales. *Sociologia Ruralis* 40, 466–80.
Bryant, R.L. and Goodman, M.K. (2004). Consuming narratives: the political ecology of 'alternative' consumption. *Transactions of the Institute of British Geographers* NS 29, 344–66.
Buller, H. and Morris, C. (2004). Growing goods: the market, the state and sustainable food production. *Environment and Planning A* 36, 1065–1084.
Cone, C. and Myhre, A. (2000). Community supported agriculture: a sustainable alternative to industrial agriculture? *Human Organisation* 59, 187–197.
Cook, I. et al. (2006). Geographies of food: following. *Progress in Human Geography* 30 (5), 655–666.
Dowler, E. and Caraher, M. (2003). Local food projects: the new philanthropy? *The Political Quarterly* 74, 57–65.
DuPuis, E.M. and Goodman, D. (2005). Shall we go 'home' to eat?: towards a reflexive politics of localism. *Journal of Rural Studies* 21, 359–371.
Friedmann, H. and McMichael, P. (1989). Agriculture and the state system: the rise and fall of national agricultures, 1870 to the present. *Sociologia Ruralis* 29, 93–117.
Gibson-Graham, J.K. (1996). *The End of Capitalism (As We Knew It): A Feminist Critique of Political Economy*. Oxford: Blackwell.
Gilg, A.W. and Battershill, M. (1998). Quality farm food in Europe: a possible alternative to the industrialised food market and to current agri-environmental policies: lessons from France. *Food Policy* 23, 25–40.
Goodman, D. (2003). Editorial: the quality 'turn' and alternative food practices: reflections and agenda. *Journal of Rural Studies* 19, 1–7.
Goodman, D. (2004). Rural Europe redux? Reflections on alternative agro-food networks and paradigm change. *Sociologia Ruralis* 44, 3–16.
Goodman, D. and DuPuis, E.M. (2002). Knowing food and growing food: beyond the production-consumption debate in the sociology of agriculture. *Sociologia Ruralis* 42, 5–22.
Hassanein, N. (2003). Practicing food democracy: a pragmatic politics of transformation. *Journal of Rural Studies* 19, 77–86.
Hendrickson, M. and Heffernan, W. (2002). Opening spaces through relocalisation: locating potential resistance in the weaknesses of the global food system. *Sociologia Ruralis* 42, 347–369.
Hines, C. (2000). *Localization: A Global Manifesto*. London: Earthscan.
Hinrichs, C.C. (2000). Embeddedness and local food systems: notes on two types of direct agricultural market. *Journal of Rural Studies* 16, 295–303.
Holloway, L. and Kneafsey, M. (2000). Reading the space of the farmers' market: a preliminary investigation from the UK. *Sociologia Ruralis* 40, 285–299.
Hughes, A. (2005). Geographies of exchange and circulation: alternative trading spaces. *Progress in Human Geography* 29 (4), 496–504.
Ilbery, B. and Kneafsey, M. (1998). Product and place: promoting quality products and services in the lagging regions of the European Union. *European Urban and Regional Studies* 5, 329–341.
Ilbery, B. and Kneafsey, M. (2000). Producer constructions of quality in regional speciality food production: a case study from south west England. *Journal of Rural Studies* 16, 217–230.
Ilbery, B. and Maye, D. (2005). Alternative (shorter) food supply chains and specialist livestock products in the Scottish-English border. *Environment and Planning A* 37, 823–844.

Ilbery, B., Morris, C., Buller, H., Maye, D. and Kneafsey, M. (2005). Product, process and place: an examination of food marketing and labelling schemes in Europe and North America. *European Urban and Regional Studies* 20, 331–344.

Jackson, P., Ward, N. and Russell, P. (2006). Mobilising the commodity chain concept in the politics of food and farming. *Journal of Rural Studies* 22, 129–141.

Kirwan, J. (2004). Alternative strategies in the UK agro-food system: interrogating the alterity of farmers' markets. *Sociologia Ruralis* 44, 395–415.

Kneafsey, M., Holloway, L., Venn, L., Cox, R., Dowler, E. and Tuomainen, H. (2004). *Consumers and Producers: Coping with Food Anxieties through 'Reconnection'?* Cultures of Consumption and ESRC-AHRB Research Programme Working Paper Series, Working Paper No. 19. Available online at: http://www.consume.bbk.ac.uk, accessed 22 December 2006.

Lee, R. (2000). Shelter from the storm? Geographies of regard in the worlds of horticultural production and consumption. *Geoforum* 31, 137–157.

Lee, R., Leyshon, A., Aldridge, T., Tooke, J., Williams, C. and Thrift, N. (2004). Making geographies and histories? Constructing local circuits of value. *Environment and Planning D: Society and Space* 22, 595–617.

Leyshon, A. and Lee, R. (2003). Introduction: Alternative economic geographies. In A. Leyshon, R. Lee and C. Williams (Eds), *Alternative Economic Spaces* (pp. 1–26). London: Sage.

Leyshon, A., Lee, R. and Williams, C.C. (Eds) (2003). *Alternative Economic Spaces*. London: Sage.

Lien, M.E. and Nerlich, B. (Eds) (2004). *The Politics of Food*. Oxford: Berg.

Lowe, P., Buller, H. and Ward, N. (2002). Setting the agenda? British and French approaches to the second pillar of the Common Agricultural Policy. *Journal of Rural Studies* 18, 1–17.

Marsden, T. and Sonnino, R. (2006). Rural development and agri-food governance in Europe: tracing the development of alternatives. In V. Higgins and G. Lawrence (Eds), *Agricultural Governance: Globalisation and the New Politics of Regulation* (pp. 50–68). London: Routledge.

Marsden, T., Murdoch, J and Morgan, K. (1999). Sustainable agriculture, food supply chains and regional development: editorial introduction. *International Planning Studies* 4, 295–301.

Marsden, T., Banks, J. and Bristow, G. (2000). Food supply chain approaches: exploring their role in rural development. *Sociologia Ruralis* 40, 424–438.

Marsden, T., Banks, J. and Bristow, G. (2002). The social management of rural nature: understanding agrarian-based rural development. *Environment and Planning A* 34, 809–825.

Maye, D. and Ilbery, B. (2006). Regional economies of local food production: tracing food chain links between 'specialist' producers and intermediaries in the Scottish-English borders. *European Urban and Regional Studies* 13, 337–354.

McCarthy, J. (2006). Rural geography: alternative rural economies – the search for alterity in forests, fisheries, food, and fair trade. *Progress in Human Geography* 30 (6), 803–811.

Morgan, K. and Morley, A. (2002). *Relocalising The Food Chain: The Role of Creative Public Procurement*. Cardiff: The Regeneration Institute, Cardiff University, in association with Powys Food Links, the Soil Association and Sustain.

Morris, C. and Buller, H. (2003). The local food sector: a preliminary assessment of its form and impact in Gloucestershire. *British Food Journal* 105, 559–566.

Pain, R. (2004). Social geography: participatory research. *Progress in Human Geography* 28 (5), 652–663.

Parrott, N., Wilson, N. and Murdoch, J. (2002). Spatializing quality: regional protection and the alternative geography of food. *European Urban and Regional Studies* 9, 241–261.

Ploeg, J.D. van der, Renting, H., Brunori, G., Knickel, K., Mannion, J., Marsden, T., de Roest, K., Sevilla-Guzman, E. and Ventura, F. (2000). Rural development: from practices and policies towards theory. *Sociologia Ruralis* 40, 391–408.

Potter, C. and Burney, J. (2002). Agricultural multifunctionality in the WTO – legitimate non-trade concern or disguised protectionism? *Journal of Rural Studies* 18, 35–47.

Renard, M.C. (2003). Fair trade: quality, market and conventions. *Journal of Rural Studies* 19, 87–96.

Renting, H., Marsden, T.K. and Banks, J. (2003). Understanding alternative food networks: exploring the role of short food supply chains in rural development. *Environment and Planning A* 35, 393–411.

Sage, C. (2003). Social embeddedness and relations of regard: alternative 'good food' networks in south-west Ireland. *Journal of Rural Studies* 19, 47–60.

Sassatelli, R. (2004). The political morality of food: discourses, contestation and alternative consumption. In M. Harvey, A. McMeekin and A. Warde (Eds), *Qualities of Food: Alternative Theoretical and Empirical Approaches* (pp. 176–191). Manchester: Manchester University Press.

Sonnino, R. and Marsden, T. (2006). Beyond the divide: rethinking relationships between alternative and conventional food networks in Europe. *Journal of Economic Geography* 2 (6), 181–199.

Thirsk, J. (1997). *Alternative Agriculture: A History*. Oxford: Oxford University Press.

Venn, L., Kneafsey, M., Holloway, L., Cox, R., Dowler, E. and Tuomainen, H. (2006). Researching European 'alternative' food networks: some methodological considerations. *Area* 38, 248–258.

Ventura, F. and Milone, P. (2000). Theory and practice of multi-product farms: farm butcheries in Umbria. *Sociologia Ruralis* 40, 452–465.

Watts, D.C.H., Ilbery, B. and Maye, D. (2005). Making re-connections in agro-food geography: alternative systems of food provision. *Progress in Human Geography* 29, 22–40.

Whatmore, S. and Thorne, L. (1997). Nourishing networks: alternative geographies of food. In D. Goodman and M.J. Watts (Eds), *Globalising Food: Agrarian Questions and Global Restructuring* (pp. 287–304). London: Routledge.

Whatmore, S., Stassart, P. and Renting, H. (2003). What's alternative about alternative food networks? *Environment and Planning A* 35, 389–391.

Winter, M. (2003a). Geographies of food: agro-food geographies – making reconnections. *Progress in Human Geography* 27 (4), 505–513.

Winter, M. (2003b). Embeddedness, the new food economy and defensive localism. *Journal of Rural Studies* 19, 23–32.

Winter, M. (2004). Geographies of food: agro-food geographies – farming, food and politics. *Progress in Human Geography* 28 (5), 664–670.

Winter, M. (2006). Rescaling rurality: multilevel governance of the agro-food sector. *Political Geography* 25 (7), 735–751.

Wrigley, N. (2002). 'Food deserts' in British cities: policy context and research priorities. *Urban Studies* 39, 2029–2040.

Part I

Alternative Food Geographies: Concepts and Debates

Chapter 2

Localism, Livelihoods and the 'Post-Organic': Changing Perspectives on Alternative Food Networks in the United States

David Goodman[1] and Michael Goodman[2]
[1]Department of Environmental Studies, University of California, Santa Cruz, USA
[2]Department of Geography, King's College London, UK

Introduction: AFNs Working in the Local

This chapter traces the contemporary development of alternative food networks (AFNs) in the US against the background of changes in the dominant framings and eco-social imaginaries articulated by social movements, academics and other civil society actors promoting sustainable agriculture and localised food systems. It explores the declining political tractability of the 'organic' and the growing acceptance of the 'local' as the discursive and material site of alternative agro-food projects. We argue that this discursive shift, although significant, emphasises rupture or transition rather than continuity and so fails to acknowledge a deeper tradition of family-farmer activism and a long-standing concern for viable rural livelihoods which lies at the heart of the early incarnations of this movement. Accordingly, we undertake an instrumentalist reading of farmer activism and suggest that elements of the livelihoods perspective found in political ecology and development studies and an emphasis on rural survival strategies might enrich the analysis of alternative agro-food networks and their changing configuration.

We begin by examining the theses of the conventionalisation and bifurcation of organic agriculture in the US, and more particularly in California, following the adoption of the federal organic standards. The technocentric regulatory structure used to circumscribe the practices of the USDA's 'organically grown' label has fragmented sustainable agriculture movements (SAMs) and dissipated hopes that organic agriculture would provide a militant base from which to mount a transformative challenge to the dominant industrial agro-food system. With these hopes thwarted, activist ambition has been invested increasingly in the development of localised food systems and spatially demarcated labels of origin. Smaller organic growers, marginalised by scale from interregional and export markets, have sought new sources of economic rent and livelihood by going 'beyond organic' – what we term the 'post-organic' – and finding refuge in local direct marketing and local food networks (LFNs). These themes and the associated ideologies of localism are explored below following the review of the recent literature on AFNs and several case studies.

A Brief History of Sustainable Agricultural Movements: Regulation and the Question of Transformative Potential

USDA Regulation and the Conventionalisation Thesis

Over the past three decades, *grosso modo* discourses of militancy in AFNs in the US have been superseded by more circumspect, incrementalist narratives of change, better adapted to the hegemonic notions of the market and consumer choice promulgated by the dominant neoliberal political economy. This accommodationist posture is closely related to, and informed by, the ascendancy of a codified, USDA-regulated organic agriculture and the concomitant decline in its potential significance as a catalyst of social justice in food production–consumption relations (Allen *et al.* 2003). Empirically, the rise of a mimetic organic agro-food system with attributes of its industrial counterpart has been explicated by the conventionalisation and bifurcation theses. These envision a binary trajectory of change in the organic sector, with large-scale, specialised producers integrated into multiply-scaled commodity networks and a rump of holistic, 'movement' farmers or 'artisanal' growers serving localised markets (Vos 2000; Guthman 2004a; Lockie and Halpin 2005).

In an early statement of this conventionalisation thesis, Buck *et al.* (1997: 4) suggest that "[i]n California, where organic standards already emphasise inputs over processes, conventional agribusiness firms are commandeering the 'organic' label and its associated price premiums by using only allowable inputs, but otherwise employing an industrial mode of agriculture which avoids the more costly sustainable agriculture practices". This regulatory capture has forced progressive, more politicised organic, imaginaries to the margins. These embraced not only agro-ecological sustainability, involving such practices as cover cropping, rotations and biological pest control, but also non-exploitative labour relations and affordable access for all to locally produced, nutritious food. Nevertheless, the leading SAMs lobbied vigorously for the technocentric 'rules of the game' now institutionalised in the USDA Organic Standards. In this respect, the highly controversial inclusion of genetically modified organisms, irradiation and sewage sludge under the USDA's National Organic Programme Proposed Rule in 1998 proved to be an astute discursive gambit. The 'Big Three' drew heated opposition from the leadership and rank and file members of SAMs and provided an impressive display of public participation (Vos 2000). Yet this gambit effectively confined the debate to the terrain of production and allowable inputs, drowning out more progressive voices. In pursuing a technocentric production politics, epitomised by the USDA standard and regulatory structure, over a progressive social politics, SAMs have displaced debates on the social agenda of sustainable agriculture, of how human needs are to be met and inequities resolved, to other, more localised arenas.

This brief detour into the recent federal regulation of organic agriculture is a vital element of the conventionalisation analysis. The 'allowable inputs' basis of USDA federal organic certification underlies "a bifurcation among organic growers, with many large operations becoming specialised in the production of a few high-growth, high-profit crops, while smaller farms continue to diversify their strategies employing artisanal methods to grow a variety of marketable crops..." (Buck et al. 1997: 8). These authors also find that conventional patterns of operation are even more pronounced in marketing and distribution, accentuating the dualistic features of production (Buck et al. 1997: 12). That is, small growers are relegated to the "economically marginal distribution channel" of direct marketing, exemplified by farmers' markets, due to their inability to "meet the volume, timing and quality requirements of intermediaries, especially as retailing and wholesaling become increasingly integrated" (Buck et al. 1997: 12-13). In their view, "local and direct marketing arrangements... are effectively default choices for growers with few resources" (Buck et al. 1997: 14). However, as Lockie and Halpin (2005) perceptively observe, the conventionalisation thesis can project a problematic binary division between 'industrial' and 'artisanal' producers, which may conceal unwarranted normative and ideological assumptions. One of these implicit assumptions is that 'artisanal' maps directly on to more progressive, movement-oriented ideologies and practices, and that 'industrial' equates with both economic behavioural and conservative ideological elements of the conventionalisation thesis. Although these conflations demand empirical verification, their normative implications are scrutinised below.

The Question of 'Transformative Potential'

As the preceding discussion suggests, the received wisdom finds a strong association between organic codification, the entry of new, resource-rich producers, and the decline of organic agriculture as the focus of a politically radical social movement. The symbiosis between codification and conventionalisation, it is argued, has corroded the potential of this movement to transform the industrial agro-food system. The agro-industrial dynamics given full rein by the federal organic regulatory apparatus, in turn, have accentuated the livelihood pressures experienced by smaller, more ecologically oriented producers, leading to new 'post-organic' movements polarised around local food systems.

This scenario of displaced militancy, now reconfigured and re-energised at the local level, begs several questions about the characterisation of SAMs as bearers of progressive politics. In academic analyses particularly, SAMs in the US typically have been assessed against the standard of their 'transformative potential' to create an alternative, ecologically sustainable and socially just food system. Arguably, however, the internal ambivalences, divisions and contradictions that have beset SAMs from their inception suggest that this was an excessively ambitious and unrealistic yardstick. Indeed, their short history reveals the primacy of a technocentric alternative managerialism to the point that the social agenda of sustainable agriculture has passed by default to other social movements, such as anti-hunger organisations following entitlements and human rights-based approaches and community food security movements, with their roots in agrarian populism, rural–urban networks and localism.

Support for organic agriculture in the USA only began to take shape as a recognisable social movement in the late 1960s, when it was known as the 'back-to-the-land' movement and comprised a disparate assortment of romantics, hippies and peaceniks. In a classic process of innovation embedded in learning-by-doing and informal mechanisms of knowledge transmission, this unlikely coalition gave rise to marketed organic produce, urban food cooperatives and natural food stores. This nascent sector also developed novel forms of governance, articulating regulatory standards of organic practice, which were administered by farm organisations, such as the California Alliance of Family Farmers (CAFF) and Oregon Tilth. Organic farmers and their organisations drew inspiration from the pioneering work of Lady Eve Balfour, author of *The Living Soil* (1943) and co-founder in 1946 of the UK Soil Association, Sir Alfred Balfour, the so-called 'inventor' of compost (Reed 2001: 134), Rudolf Steiner and biodynamic farming, and the indefatigable propagandising in the USA of J.I. Rodale and the Rodale Press (Vos 2000).

Organic food activists also drew succour from the radical politics of the 1960s and 1970s, formed by the confluences of the civil rights movement, anti-war activism and the embryonic environmental movement. In this milieu, sustainable agriculture advocates honed their social and ecological critique of industrial agriculture and the giants of American agribusiness. These formative elements and convergences were joined by agricultural scientists whose activism focused on achieving scientific legitimacy and federal funding for organic agriculture research in the land grant university (LGU) system. This 'technologically-led vision' (Buttel 1997: 355) has prevailed over radical social critique and a transformative politics, which at once reveals both the central tension within the US SAM and its historical trajectory. That said, the political history of the SAM remains to be written.

The primacy of production politics emerged in crystalline form during the institutional negotiation of federal codification and the conflict provoked in 1998 by the USDA's Proposed Rule. As Vos (2000) notes, the leading SAMs accepted the narrowly defined parameters of negotiation and welcomed federal regulation, which effectively translates 'organic' into a market brand and product-differentiation strategy. However, the centrality of production has much deeper roots in SAM discourse, and these can be traced back to the 1970s and 1980s.

In his analysis of 'Hightowerism', Buttel (2005) discusses the social movement activism unleashed by Jim Hightower's (1973) *Hard Tomatoes, Hard Times* and efforts to mobilise small or family farmers to challenge the inequitable consequences of publicly funded LGU agricultural research. This book mounted a swingeing *social* critique of the priorities embedded in LGU research and identified agribusiness and large-scale farmers as the principal beneficiaries of public funds. SAMs were caught up in this contestation of public agricultural research priorities, but the original focus on social justice gave way to competing claims of scientific authority and demands for greater recognition of agro-ecology and sustainable agriculture in the LGU system. This campaign gained some success in the form of the low-input sustainable agriculture (LISA) initiative included in the 1985 Farm Bill. This focus on the 'point of production' characterised subsequent federal sustainable agriculture programmes in the 1990s, even though the social role of farming received some recognition (Allen 2004: 82–84).

In Buttel's (1997: 355) typology of environmental ideologies and discourses, this form of agricultural research activism is categorised as 'alternative technologism', which is "an increasingly influential expression of environmentalism in agriculture". In this imaginary, conventional agricultural technoscientific rationalism is displaced by a science-based alternative rationalism in order to influence "public research institutions to emphasise 'sustainable,' 'low-input,' or 'alternative' agriculture" (Buttel 1997: 355). In short, this technocratic current of the SAM sought legitimacy and political purchase by advancing sustainable agriculture as a scientific 'knowledge claim'.

Pursuit of this instrumentalist rationalism, Buttel (1993: 32) argues, has meant that sustainable agriculture has been defined largely in LISA-like terms of reducing chemical use and represented as a technology which can readily be "grafted onto the knowledge base of otherwise conventional agronomy". The profile of the USDA's Organic Rule could hardly be clearer. In Allen's (2004) blunt assessment, the "[f]raming of sustainable agriculture in a natural science discourse" (104) and adoption of its epistemological positivism have resulted in the failure to interrogate the assumption "that achieving agricultural sustainability is possible without changing social relations" (99).

Dissatisfaction with both rationalist and 're-rationalist' scientific discourses has led other activist groups to promote local farmer knowledge as a cornerstone of sustainable agriculture. Buttel (1997: 356) characterises this approach as 'indigenism', which he associates with the "rapidly growing community-supported agriculture (CSA) movement and a related community food systems or foodshed movement". The heirs of "Hightower-type agricultural research activism" thus are identified with localism/agricultural sustainability movements, which "arose in a limited way in the late 1970s, largely in continuity with the organic farming movement" (Buttel 2005: 279). These pluralistic currents within contemporary SAMs are described in broad terms by Buttel (2005: 280), who observes that activists increasingly are attempting "to build an alternative locally-based food system involving more direct linkages . . . between farmers and consumers. Other important segments of the sustainable agriculture activist community stress the development of green/'value-added' labelling and marketing strategies, while others stress issues such as community food security". In particular, community food security movements, centred on the Community Food Security Coalition in Los Angeles, have expanded this remit of indigenism to explicitly urban food spaces, often through the inclusion of low-income and marginalised consumers and/or the forging of rural and urban food networks between small-scale farmers and urban consumers (Gottlieb and Fisher 1996a,b; Anderson and Cook 1999; Bellows and Hamm 2002).

This 30-year trajectory of SAMs from socially progressive politics to contemporary accommodation with neoliberal discourse is traced in empirical detail in a study of 37 alternative food initiatives (AFIs) in California (Allen et al. 2003). In a strong statement, they argue that "[w]here in the early years AFIs combined the search for alternatives with a direct critique of existing agricultural practices, . . . the loss of this structural critique and the rise of a political culture of entrepreneurialism appear to have left these organisations with only neo-populism to explain the politics of their engagement" (Allen et al. 2003: 65). This shift is attributed to the new socio-political realities accompanying the rise of neoliberalism in the 1980s, which prompted AFIs to replace social justice demands and frame their programmes in terms of "the rights of consumers to choose alternatives, rather than in their rights as citizens" (Allen et al. 2003: 68). Correspondingly, "[f]or many current California AFIs, changing the food system means increasing the diversity of alternative markets such that consumers have more choice, rather than making deep structural changes that could reconfigure who gets to make which kinds of food choices" (Allen et al. 2003: 72).

This revealing analysis drives home the point that, in these neoliberal times, California AFIs have become 'alternative' rather than 'oppositional' organisations. In the course of this transition, the meaning of social justice has been transmuted from concern to redress exploitative relations in the farm labour process to "questions of food access, urban community empowerment, and support for small farmers" (Allen *et al.* 2003: 73). In terms of David Harvey's (1996) categories, California AFIs have lost not only their 'global ambition', articulated in the 1960s and 1970s through links with the national civil rights and environmental movements, but also their militant politics of justice. This now leaves the 'particularisms', rhetorically expressed in the modest and decidedly focused claims of localism, entrepreneurialism, community food security and sustainable local agriculture.

Going Native Through Local Food Networks: Recent Literature and Case Studies

This historical overview reveals the transformations in the imaginaries of SAMs and AFIs arising from the confluence of resilient neopopulist discourse and the hegemonic neoliberal impulses of market forces and consumer choice. With federal codification, the organic imaginary has lost its political purchase as a progressive, structuralist eco-social project, and localism has assumed talismanic importance among food activists in the USA. Recent analyses frame these changes in terms of 'resistance' to the 'time-space distantiation' characteristic of industrial agro-food systems (Allen *et al.* 2003: 73) and the 'food miles' arising from the socio-spatial separation of production and consumption (Buttel 2005: 280). Neopopulist influences catalysed by the social resonance of the family farm and rural community have found renewed vigour in the discourse of localism, as have local and 'community' rural/urban linkages in the community food security movement (Johnston and Baker 2005).

Framed in opposition to their larger-scale counterparts, both conventional and 'minimalist' organic producers, smallholder or 'artisanal' growers are cast as the 'cultural brokers' linking local production and local consumption. In drawing this equation, it is implied that local smallholder agriculture, even though it responds to capitalist logic and market imperatives for its social reproduction, is nevertheless more benign in its eco-social relations and its practitioners "more committed to the preservation of community, tradition, environment, and other non-market values" (Lockie and Halpin 2005: 287). These normative assumptions, so firmly entrenched in agrarian localist discourse, raise contentious questions about the process of commodification, particularly in political economies where, as Guthman (2004b) reminds us, agro-industrial dynamics are pervasive. In short, agrarian localist discourses can be seen as a strategy to engage these 'boundary politics' between market values and non-market meanings by stressing such notions as food-in-social-context (Lind and Barham 2004) and 'commensal community' (Kloppenburg *et al.* 1996) and, more generally, food in 'place, power and provenance' (Morgan *et al.* 2006).

In contributing to these discourses, scholar-activists have portrayed LFNs as discursive and material sites of resistance to the time–space distantiation and corporate power manifest in the highly concentrated, globalising food system. Viewed in binary opposition to this hegemon, the local is represented as a potentially 'insulated space', where it is possible "to maintain or create alternatives that will eventually bring substantive change" (Kloppenburg *et al.* 1996: 37). The local provides opportunity to re-embed the food system in a revitalised 'moral economy' by restoring the ethical precepts of "mutuality, reciprocity, and equity that ought to characterise all elements of human interaction" (Kloppenburg *et al.* 1996: 36).

In envisioning such spaces, in uneasy coexistence with the industrial food system, localist narratives are liberally seeded with notions of 'secession', interstitial growth, disengagement and, what Kloppenburg *et al.* (1996) term, the 'principle of succession'. This "finds expression in a strategy of 'slowly moving over' from the food system to the foodshed" (Kloppenburg *et al.* 1996: 38; cf. also Hendrickson and Heffernan 2002).

In a recent paper, DuPuis and Goodman (2005: 359) observe that such "normative localism places a set of pure, conflict-free local values and local knowledges in resistance to anomic and contradictory capitalist forces". However, this purified, normative localism erases the politics of the local and consequently fails to acknowledge that contested issues of distributive justice, human rights and identity can arise in these idealised, insulated spaces. DuPuis and Goodman (2005) accordingly propose a 'reflexive localism' in which local initiatives are negotiated democratically in an open local politics. A 'reflexive localism' would draw attention to the social relations and the politics of power which are submerged in the vague notions of 'sustainable local development' and 'local control': by and for whom? Such vagueness means that they are open to appropriation and interpretation by political forces of very different hues. A reflexive approach also reinforces a key proposition of livelihoods research, which would recognise the 'differential (or adverse) incorporation' of some social classes into the market economy, civil society and the state (Murray 2001; Du Toit 2004). Social classes or strata are differentially positioned to benefit from 'sustainable local development', but these distributive effects are lost in the discursive trappings of normative localism. Similarly, consumers who are unable to frequent farmers' markets, box schemes (CSAs) or local food cooperatives do not experience the non-market benefits of place-making and 'commensal community'.

From a wider perspective, activist narratives of a normative localism embedded in a secessionist moral economy and interstitial spaces imply that scale construction at the local level is relatively unproblematic. This voluntaristic premise neglects potentially negative outcomes of the dynamic, contested interactions between local forms of socio-spatial organisation and translocal actors and institutions. That is, the local is framed as a social space where new economic forms and institutions incorporating ethical norms are allowed to grow and flourish. Again, the reification of the local obscures the contested socio-spatial processes involved in its construction and the practicalities of secession and local control. Not the least of these challenges to local economic forms is the danger of imitation and co-optation by mainstream corporate actors (Goodman 2004).

Questions of scale construction also emerge in a recent study of two predominantly agro-industrial counties in Central Washington state in the Pacific Northwest, where alternative producers confront severe obstacles to disengagement (Selfa and Qazi 2005). Local direct markets have been slow to develop in Chelan and Grant counties, where growers complain that consumers do not appreciate local alternative produce, with one producer observing that "consumers won't pay a premium for organic, except in the city" (Selfa and Qazi 2005: 456). These authors also note that "'local' food systems do not always or easily map on to a bounded or proximate geographical space"

(Selfa and Qazi 2005: 457). Thus in Central Washington, "[i]n order to survive, small organic farmers need to market to high-end consumers in the Seattle region", some 200 miles away (Selfa and Qazi 2005: 457). Scale and co-optation also concern Guptil and Wilkins (2002) in their analysis of local food retailing in upper New York state. On a cautionary note, they observe that "the promotion of locally-produced foods in mainstream outlets could weaken the capacity of local food flows to empower regular citizens to shape the local economy" (Guptil and Wilkins 2002: 49). They go on to "suggest that 'localness' is just as commodifiable as 'organic'" (Guptil and Wilkins 2002: 50).

In a similar vein, but surrounding themselves in a more emphatic cloak of 'possibilism', recent work by Johnston and Baker (2005) points to the ongoing struggles that community food security coalitions have in 'scaling-up' and 'scaling-out.' Looking at FoodShare Toronto's 'good food box' (GFB) through the lens of social reproduction, they remind us that the construction of the local is always and everywhere connected to and mediated by other scales, and, for them, this means predominantly the global scale of conventional food chains. These connections create substantial problems for FoodShare: cheap global food "makes it difficult for local organic farmers to price their food competitively in local markets, and it also makes it impossible for the GFB to deliver 100% organic local produce to low-income consumers… The organic GFB is relatively expensive for low-income people and is sold almost exclusively to middle-class patrons" (319). At the same time, they see the global-scale issue of "growing poverty, inequality, and food insecurity in Post-Fordist economies" as "frustrating local action" by dampening the ability of LFNs to get beyond the local and "achieve socio-ecological goals on a much broader scale" (319). For Johnston and Baker (2005), attainment of these goals requires LFNs to become 'extra-local' by scaling-up and scaling-out through progressive state political and economic support at local, regional and national scales, state accountability and LFNs as culturally inflected "long-term pedagogical projects for good food" (319–323). Remaining optimistic, if unreflexive, to the scaled and mediated processes by which the local is constructed in the first place, they conclude that "creative food projects developed at a local scale, like the [GFB] program, are a necessary, if not sufficient, component of greater food security" (322).

Livelihoods, Family-Farm Activism and Survival Strategies: New Realism, New Politics

As the preceding discussion suggests, the predominant focus of US scholarship on the 'movement' aspects and 'transformative potential' of LFNs has been overstated to the exclusion of more instrumentalist readings of their material and discursive practices. Continuing a long tradition of ideological agrarianism, the family farm of artisanal production is figured in neopopulist LFN rhetoric as the repository of moral values of community and the site of socio-cultural resistance to the anomic forces of distantiated agro-food systems. However, these contemporary renderings of moralistic agrarian ideology arguably are more accurately read as a discourse of livelihood and social reproduction. This pragmatism, already evident in several recent analyses, reflects disillusion with the limited achievements and narrow sectionalism of SAMs, now reinvented as LFNs. As potentially fruitful resources to extend this 'new realism', we would suggest livelihoods research in development studies and the political ecologies of development, more specifically, the concept of economic rent, most recently revived in value-chain analysis and the notion of 'survival strategies' found in European rural studies in the 1980s and early 1990s (cf. Marsden 1990). In this perspective and addressed more fully below, LFNs represent 'new spaces of possibility' and creatively amplify the repertoire of household reproduction strategies (Goodman 2004: 12) that give rise to what Gibson-Graham (2005a,b) refer to as 'diverse economies' operating in the interstitial spaces of the big 'C' sort of capitalism.

The sectionalist critique of SAMs/LFNs is hardly new but, until recently, such analyses typically went on to declaim their failure to extend their purview and constituency to encompass farm labour and food service workers and address broad social equity goals (Allen and Sachs 1991, 1993). Arguably, however, the farm centrism that is so resistant to more radical and inclusive directions of change reflects structural continuities deeply embedded in LFNs. Thus the contemporary tradition of family-farm activism, including Hightowerism and responses to the 1980s US farm crisis (Buttel 1989), was a major formative influence on SAMs and local food movements. These continuities have gained even greater prominence in the present conjuncture since ideological agrarianism and farm centrism resonate strongly with the entrepreneurial ethos of neoliberalism. The economistic tenor of this new realist critique is evident in assessments of community food security programmes (Allen 1999, 2004), box schemes or CSAs (DeLind 2003, 2006) and Allen *et al.*'s (2003) crushing observation that the politics of California AFIs are explicable only in terms of neopopulism.

Although cast in the mode of declamatory social justice critique, Allen's (2004) comprehensive analysis of sustainable agriculture and local food movements can readily be reformulated in a narrowly economistic farm livelihoods framework. That is, as presently configured, 'alternative agrifood practices' serve almost exclusively to enhance the social reproduction of family farms. For example, citing an evaluation of 13 community food security projects in California (Feenstra and Campbell 1998), Allen (2004: 111) observes that "the central focus is less on the food needs of low-income people than on the 'well-being' of a region.... Most (projects) are directed toward providing security of markets for local agricultural producers...". Or again, Allen (2004: 113) notes that "[i]n many alternative agrifood practices it appears that what is to be secured are markets for producers rather than food access *per se*".

While acknowledging the importance given to democratic processes by alternative agri-food movements and institutions, Allen (2004: 148) goes on to suggest that "in other ways, though, these privilege the interests of only one group of those who labour in the agrifood sector – farmers". This leads to a more general, dyspeptic indictment of the farm-centric productionism of alternative agri-food movements, which tend "to privilege farmers as agents of change, the rightful beneficiaries of that change, and the savants who know what is to be done and how to do it. Most alternative agrifood advocates see farmers as the central figures" (Allen 2004: 150). Such pre-eminence can equally be attributed to the coherence of agrarian neopopulist 'story lines' (Hajer 1995) articulated by alternative farming organisations and their skill in constructing discourse coalitions to support family-farm livelihoods in changing economic and political circumstances.

Thus, we would argue that current affirmations of the local are predicated on the existence of markets which are sufficiently robust to generate producer rents that can sustain local farm livelihoods. From the perspective of value-chain analysis (Kaplinsky 2000), Guthman's (2004b) account of California organic agriculture can be seen as tracing the trajectory of the organic as a barrier to entry and source of economic rent, their competitive erosion and the ensuing structural consequences for the sector. As Guthman (2004b) observes, direct marketing in local food systems developed in part as a response to these competitive dynamics. Current post-organic initiatives, we would suggest, represent the latest attempt to erect barriers to entry to maintain producer rents. Of course, moral and ideological motivations also underpin these initiatives, but their *sine qua non* is the capacity to sustain economically viable farm livelihoods.

The entry barriers created by local food networks and 'beyond organic' labelling schemes now being established also are a form of territoriality. That is, space is differentiated in ways which give differential opportunities to local economic agents and citizens by exploiting the attributes of place (Moran 1993). Local food labelling and beyond organic certification systems are a variant of the type of territorial regulation of space that Moran (1993), in his study of winemaking and regional appellations in France and California, refers to as sectorial territorialisation and which can establish monopoly rents. Such local food schemes also can be considered as strategies to confer a spatial or territorial form of intellectual property rights on local farmers and food processors, although as yet in the USA these lack the formal structure of national and supranational institutional support of the EU's system of protection for products of designated origin.

The emphasis in LFNs on labels of origin and spatial certification to create and protect rent-rich activities also mirrors the priority currently being given in other sectors to the more intangible elements in the value chain, such as brand names and forms of intellectual property (Kaplinsky 2000: 127). These local, post-organic labelling and certification schemes now emerging in the USA are exemplified by the activities of Protected Harvest, a third-party certifying NGO, the Food Alliance and its Food Routes' 'Buy Fresh, Buy Local' campaign in partnership with local organisations and the new organic/fair-trade label and certification system being developed by the Organic Consumers Association.

As Ponte and Gibbon (2005: 13) note, these initiatives represent and promote conventions of quality based on 'narratives of place' and expressions of product identity formed by repeated social interaction between actors in localised markets. In such schemes, quality management focuses on the intersection between place and livelihood to create mechanisms which offer some measure of protection and economic rent to local farming and food interests. In short, although these 'new economic spaces' arising at the intersection of place and livelihood may be glossed as local control, local development or even 'resistance' in LFN narratives, they embed very specific power dynamics and distributive patterns.

Opening a Wider Lens: Extending Economistic Farm-Centred Analyses

A livelihoods approach brings social relations, power asymmetries and their distributive consequences, and the role of discursive strategies in naturalising these relationships to the forefront of the analysis.[1] More specifically, as Murray (2001) and others (de Haan 2000; Bebbington 2000, 2003; Bebbington and Batterbury 2001) remind us, a livelihoods approach would analyse the actors engaged in LFNs, both producers and consumers, *in relation* to the experience of other social classes. Here a livelihoods perspective, giving normative priority to those who are economically and socially marginalised and excluded (de Haan and Zoomers 2003), would centre on the opportunities and constraints of those working in LFNs and AFNs. It would explore how these are shaped by and cut across embedded histories, processes and experiences of class, race and gender, and political economies and ecologies (cf. Slocum 2006).

As some recent contributions have sought to demonstrate, a focus on the social relations of production and consumption and their embedded inequalities would open up dimensions of LFNs that have long been reified and unexamined. A more *materialist perspective* thus might qualify and restate the 'resistance' discursively attributed to LFNs as a rearguard action to secure the social reproduction of small agrarian capitals against the tide of consolidation and concentration evident among large-scale agrarian and downstream commercial and retail capitals.

As Bebbington (2000: 515) has argued more generally, but applied here to LFNs, "more viable [food] livelihoods will not be romanced into existence, but must instead be built up from already existing, and however imperfect strategies. Understanding livelihood thus becomes critical for [agro-food] theory, in order to understand how places are produced and governed, and who participates in these processes. It is also critical for practice—to understand the ways in which people have created livelihood opportunities that foster accumulation as well as the obstacles to such accumulation". Drawing again on Bebbington (2000: 498), the livelihoods approach helps us to understand "how people make a living and mak[e] it meaningful". It draws attention to people's intertwined capacities as producers, consumers and citizens, and the complex relations among these with respect to their food livelihoods (cf. Goodman 2002; Goodman and DuPuis 2002). In turn, this highlights the processes by which these relations coalesce and act to 'exclude' and 'include' in LFNs and AFNs.

This analytical emphasis on connected and networked livelihoods in LFNs resonates with Massey's (1993, 1999) work on the 'power geometries' in operation in the construction(s) and connection(s) of place(s). Indeed, as Bebbington (2003: 302) observes, recalling much of the geographically inflected work in political ecology, "place and livelihood clearly intersect as, to a considerable extent, places are produced out of livelihoods of people, while at the same time structuring elements of those livelihoods. But clearly neither livelihood nor place is ring fenced. Thus any discussion of place and livelihood must also be infused with concerns for *scale* and *networks*" (original emphasis). Drawing further on Massey, it is imperative to understand the lines of power

---

[1] The livelihood concept emerged from innovative work on development, risk and vulnerability (e.g. Chambers and Conway 1992; Blaikie *et al*. 1994; Chambers 1995) – with a great deal from the Institute of Development Studies (IDS). Livelihood is conceptualised loosely as the "capabilities, assets (including the material and social resources) and activities required for a means of living" (Chambers and Conway 1992). The concept, its renewed expression as "sustainable livelihoods" (e.g. De Haan 2000; DfID 2001) and its component pentagon of "capitals" (human, social, natural, financial and physical) is clearly not without controversy or debate (see *Antipode* 2002) nor should it be in LFN research without a healthy dose of criticality.

constraining, intersecting with and emanating from places, livelihoods and localities. In this vein, de Haan and Zoomers (2003: 350) urge us to transcend the earlier narrow conceptual local focus of livelihoods research since "livelihoods both depend on and shape global forces... [and, livelihood research]... tries to unravel the fuzzy relationship between globalisation and local development from an actor point of view".

In studying LFNs, a fuller understanding of how these power geometries of people's livelihoods are constructed is critical. Recalling the discussion above, these are 'boundary-making' projects which beg multiple and complex empirical and conceptual research questions. For example, what indigenous and exogenous power lines coalesce to determine how and why the various boundaries are drawn around and within LFN projects and how are these lines of power related and connected? What are the processes by which these simultaneously inclusionary and exclusionary social, economic, political, spatial and discursive boundaries are drawn? What are the boundaries that have been drawn for LFN projects by the larger state of the US political and discursive economy, and how, in this era, can LFNs work to engage with wider publics outside of sympathetic middle-class consumers? Finally, in a suggestion that might sit uncomfortably with much of the critical agro-food movement, perhaps this is exactly the time to see how the current emphasis on entrepreneurialism, creativity and diversity works out in the 'mangle of practice'. This at least entertains the possibility that these narrow moves might become a fulcrum for wider 'boundary-expanding work' to promote a more sustained progressive engagement to hold the state accountable.

Feeding (in) the Niche?

In the preceding analysis, we have argued that LFNs represent specific institutional and discursive 'projects' articulated by particular groups of actors to protect and advance their interests. In these terms, LFNs can be seen as *alignments of power*, which invariably are in contention with other networked power alignments of varying spatial scope.

The socially differentiated structure of the LFN niche emerges strongly in a study of CSA members in the five-county Central Coast region of California undertaken by the Center for Agroecology and Sustainable Food Systems (CASFS), University of California, Santa Cruz, in 2001. Estimating that CSA members represent 0.2% of the region's population, the "survey results suggest that these members are similar to other CSA shareholders nationwide: they tend to be European-American (90%), highly educated, and middle-to-upper income" (Perez *et al.* 2003: 2). In addition to this highly specific demographic profile, this study also identifies consumers' time–space equations and habits of 'getting and spending' as significant constraints to CSA growth. These constraints involve questions of convenience since organic food and fresh local produce increasingly are available through 'alternative' retail outlets (farmers' markets, natural food stores) and conventional supermarkets, and require "more time preparing and eating what is seasonally available..." (Perez *et al.* 2003: 4).

These survey results resonate with Laura DeLind's (2003) experiences as a 'citizen-activist' on a CSA in Michigan. Her account serves to demystify the ideals of 'classical CSA rhetoric' embodied in notions of 'shared commitment', 'shared responsibility' and 'community'. Shorn of these flourishes, DeLind's analysis fits readily into a livelihoods perspective, whether emphasising that "farmers will often adopt CSA as one of several strategies to diversify their overall income" (197) or that "[s]tripped of its community mystique, CSA is a small business arrangement in which farmers and members negotiate their respective positions across a more personable market divide" (DeLind 2003: 203). This instrumental, individualistic assessment is tempered only when the CSA model is part of a 'sheltering environment' provided by non-profit institutions, such as churches, environmental trusts or anti-hunger programmes, where "it can operate less as a commercial, profit-making endeavour... and become... part of a wider set of concerns, frequently those of social justice and food security" (DeLind 2003: 204). A second, complementary point of entry to the analysis of LFNs as socially exclusive livelihood niches is to explore the linkages between the reflexive 'turn' in food consumption, the aestheticisation of the local and the food safety–health nexus. That is, where the corporeal 'metabolic reciprocities' connecting food production and consumption weigh more heavily in individual purchasing decisions than 'movement' activism, and priority is given to foods whose social and natural provenance or 'embedded relations' are known.

Given the predominance of the urban consumption of LFN and alternative foods, another fruitful entry point might involve taking the 'fragments of urban culinary culture' in the 'post-industrial/modern' metropolis (Bell 2002) and situating these in relation to LFNs and wider alternative food provisioning networks. This might enhance our analytical acuity in deciphering the constructions, needs and schisms in these movements and markets – indeed, it is in the best 'livelihood' interests of small-scale farmers to intimately understand urban food tastes and systems.

However, at present, the interests of producers and consumers aligned in LFNs coalesce in the nexus formed between farm livelihood niches and the highly valorised, embedded metabolic reciprocities of local foods. It is precisely here, in these neoliberal times, that we see the fruitfulness of a livelihoods approach to LFNs. Now more than ever, we need analytical perspectives that privilege issues of equality and access, the social relations of production and consumption and the asymmetries of power. As US research on AFNs amply demonstrates, agency on each side of this production–consumption equation currently is the prerogative primarily of actors with substantial economic and cognitive resources.

References

Allen, P. (1999). Reweaving the food security safety net: mediating entitlement and entrepreneurship. *Agriculture and Human Values* 16 (2), 117–129.
Allen, P. (2004). *Together at the Table: Sustainability and Sustenance in the American Agrifood System*. University Park: Pennsylvania State University.
Allen, P. and Sachs, C. (1991). The social side of sustainability: class, gender and race. *Science as Culture* 2 (3), 569–590.
Allen, P. and Sachs, C. (1993). Sustainable agriculture in the United States: engagements, silences and possibilities for transformation. In P. Allen (Ed.), *Food for the Future: Conditions and Contradictions of Sustainability* (pp. 136–167). New York: John Wiley and Sons.
Allen, P., FitzSimmons, M., Goodman, M. and Warner, K. (2003). Shifting plates in the agrifood landscape: the tectonics of alternative agrifood initiatives in California. *Journal of Rural Studies* 19, 61–75.
Anderson, M. and Cook, J. (1999). Community food security: practice in need of theory? *Agriculture and Human Values* 16, 141–150.

Antipode. (2002). Symposium on social capital, 34, 790–811.
Balfour, E. (1943). *The Living Soil*. London: Faber and Faber.
Bebbington, A. (2000). Reencountering development: livelihood transitions and place transformations in the Andes. *Annals of the Association of American Geographers* 90 (3), 495–520.
Bebbington, A. (2003). Global networks and local developments: agendas for development geography. *Tijdschrift voor Economische en Sociale Geografie* 94 (3), 297–309.
Bebbington, A. and Batterbury, S. (2001). Transnational livelihoods and landscapes: political ecologies of globalization. *Ecumene* 8 (4), 369–380.
Bell, D. (2002). Fragments for a new urban culinary geography. *Journal for the Study of Food and Society* 6 (1), 10–21.
Bellows, A. and Hamm, M. (2002). US-based community food security: influences, practice, debate. *Journal for the Study of Food and Society* 6 (1), 31–44.
Blaikie, P., Cannon, T., Davis, I. and Wisner, B. (1994). *At Risk: Natural Hazards, People's Vulnerability, and Disasters*. London: Routledge.
Buck, D., Getz, C. and Guthman, J. (1997). From farm to table: the organic vegetable commodity chain of Northern California. *Sociologia Ruralis* 37, 3–20.
Buttel, F. (1989). The US farm crisis and the restructuring of American agriculture: domestic and international issues. In D. Goodman and M. Redclift (Eds), *The International Farm Crisis* (pp. 46–83). London: Macmillan.
Buttel, F. (1993). The production of agricultural sustainability: observations from the sociology of science and technology. In P. Allen (Ed.), *Food for the Future: Conditions and Contradictions of Sustainability* (pp. 19–45). New York: John Wiley and Sons.
Buttel, F. (1997). Some observations on agro-food change and the future of agricultural sustainability movements. In D. Goodman and M. Watts (Eds.), *Globalising Food: Agrarian Questions and Global Restructuring* (pp. 344–365). London: Routledge.
Buttel, F. (2005). Ever since Hightower: the politics of agricultural research activism in the molecular age. *Agriculture and Human Values* 22 (3), 275–283.
Chambers, R. (1995). *Poverty and Livelihoods: Whose Reality Counts?* Brighton: IDS.
Chambers, R. and Conway, G. (1992). Sustainable rural livelihoods: practical concepts for the 21st Century. Institute of Development Studies, University of Sussex, Discussion Paper 296. Brighton: IDS.
de Haan, L. (2000). Globalization, localisation, and sustainable livelihood. *Sociologia Ruralis* 40 (3), 339–364.
de Haan, L. and Zoomers, A. (2003). Development geography at the crossroads of livelihood and globalisation. *Tijdschrift voor Economische en Sociale Geografie* 94 (3), 350–362.
DeLind, L. (2003). Considerably more than vegetables, a lot less community: the dilemmas of community supported agriculture. In J. Adams (Ed.), *Fighting for the Farm: Rural America Transformed* (pp. 192–206). Philadelphia: University of Pennsylvania Press.
DeLind, L. (2006). Of bodies, place and culture: re-situating local food. *Journal of Agricultural and Environmental Ethics* 19(2), 121–146.
DfID (Department for International Development) (2001). *Sustainable Livelihoods Guidance Sheets*. London: DfID. Available online at: http://www.livelihoods.org/(accessed: 28/7/06).
DuPuis, E.M. and Goodman, D. (2005). Should we go home to eat? Towards a reflexive politics of localism. *Journal of Rural Studies* 21 (3), 359–371.
Du Toit, A. (2004). 'Social exclusion' discourse and poverty: a South African case study. *Development and Change* 35 (5), 987–1010.
Feenstra, G. and Campbell, D. (1998). *Community Food Systems in California*. University of California Sustainable Agriculture Research and Extension Program Publication 21574. Oakland: Division of Agriculture and Natural Resources.
Gibson-Graham, J.K. (2005a). Surplus possibilities: postdevelopment and community economies. *Singapore Journal of Tropical Geography* 26 (1), 4–26.

Gibson-Graham, J.K. (2005b). Traversing the fantasy of sufficiency. *Singapore Journal of Tropical Geography* 26 (2), 119–126.
Goodman, D. (2002). Rethinking food production–consumption: integrative perspectives. *Sociologia Ruralis* 42 (4), 271–277.
Goodman, D. (2004). Rural Europe redux? Reflections on alternative agro-food networks and paradigm change. *Sociologia Ruralis* 44 (1), 3–16.
Goodman, D. and DuPuis, M. (2002). Knowing and growing food: beyond the production–consumption debate in the sociology of agriculture. *Sociologia Ruralis* 42 (1), 6–23.
Gottlieb, R. and Fisher, A. (1996a). Community food security and environmental justice: searching for common ground. *Agriculture and Human Values* 3 (3), 23–32.
Gottlieb, R. and Fisher, A. (1996b). "First feed the face": environmental justice and community food security. *Antipode* 28, 193–203.
Guptil, A. and Wilkins, J. (2002). Buying in the food system: trends in food retailing in the US and implications for local foods. *Agriculture and Human Values* 19, 32–51.
Guthman, J. (2004a). The trouble with "organic lite" in California: a rejoinder to the "conventionalisation" debate. *Sociologia Ruralis* 44 (3), 301–316.
Guthman, J. (2004b). *Agrarian Dreams? The Paradox of Organic Farming in California.* Berkeley: University of California Press.
Hajer, M. (1995). *The Politics of Environmental Discourse.* Oxford: Clarendon Press.
Harvey, D. (1996). *Justice, Nature, and the Geography of Difference.* London: Blackwell.
Hendrickson, M. and Heffernan, W. (2002). Opening spaces through relocalisation: locating potential resistance in the weaknesses of the global food system. *Sociologia Ruralis* 42 (4), 347–369.
Hightower, J. (1973). *Hard Tomatoes, Hard Times.* Cambridge, MA: Schenkman Publishing.
Johnston, J. and Baker, L. (2005). Eating outside the box: foodshare's good food box and the challenge of scale. *Agriculture and Human Values* 22, 313–325.
Kaplinsky, R. (2000). Globalization and unequalisation: what can be learned from value chain analysis? *Journal of Development Studies* 37 (2), 117–146.
Kloppenberg, J., Hendrickson, J. and Stevenson, G. (1996). Coming into the foodshed. *Agriculture and Human Values* 13, 33–42.
Lind, D. and Barham, E. (2004). The social life of the tortilla: food, cultural politics, and contested commodification. *Agriculture and Human Values* 21, 47–60.
Lockie, S. and Halpin, D. (2005). The conventionalisation thesis reconsidered: structural and ideological transformations of Australian organic agriculture. *Sociologia Ruralis* 45 (4), 284–307.
Marsden, T. (1990). Towards a political economy of pluriactivity. *Journal of Rural Studies* 6 (4), 375–382.
Massey, D. (1993). Power-geometry and a progressive sense of place. In J. Bird, B. Curtis, T. Putnam, G. Robertson and L. Tickner (Eds.), *Mapping the Futures: Local Cultures, Global Change* (pp. 59–69). London: Routledge.
Massey, D. (1999). Spaces of politics. In D. Massey, J. Allen and P. Sarre (Eds.), *Human Geography Today* (pp. 279–294). Cambridge: Polity Press.
Moran, W. (1993). The wine appellation as territory in France and California. *Annals of the Association of American Geographers* 83 (4), 694–717.
Morgan, K., Marsden, T. and Murdoch, J. (2006). *Worlds of Food: Place, Power, and Provenance in the Food Chain.* Oxford: Oxford University Press.
Murray, C. (2001). *Livelihoods Research: Some Conceptual and Methodological Issues.* Background Paper 5, Chronic Poverty Research Centre, Institute for Development Policy and Management, University of Manchester.
Perez, J., Allen, P. and Brown, M. (2003). *Community Supported Agriculture on the Central Coast: The CSA Member Experience.* Research Brief No. 1, CASFS, University of California, Santa Cruz. Available online at: http://repositories.cdlib.org/casfs/rb/brief_no1/(accessed: 28/7/06).

Ponte, S. and Gibbon, P. (2005). Quality standards, conventions and the governance of global value chains. *Economy and Society* 34 (1), 1–31.

Reed, M. (2001). Fight the future!: how the contemporary campaigns of the UK organic movement have arisen from their composting of the past. *Sociologia Ruralis* 41 (1), 131–145.

Selfa, T. and Qazi, J. (2005). Place, taste or face-to-face? Understanding producer–consumer networks in 'local' food systems in Washington State. *Agriculture and Human Values* 22 (4), 451–464.

Slocum, R. (2006). Anti-racist practice and the work of community food organisations. *Antipode* 38, 327–349.

Vos, T. (2000). Visions of the middle landscape: organic farming and the politics of nature. *Agriculture and Human Values* 17 (3), 245–256.

Chapter 3

Connecting Social Justice to Sustainability: Discourse and Practice in Sustainable Agriculture in Pennsylvania

Amy Trauger
Department of Agricultural Economics and Rural Sociology,
The Pennsylvania State University, USA

Introduction

Sustainability and sustainable agriculture are terms frequently used to position small-scale, regenerative or local agriculture as alternatives to so-called conventional agriculture. While often debated and contested, most discourses of sustainable agriculture (including those in mission statements of organisations, advertising campaigns, etc.) position these alternatives as social, economic and environmental justice (Pretty 1995; Barham 1997; Hassanein 1999; Ritchie 2003). These discourses, however, are most frequently put into practice to make farming more environmentally friendly or more profitable and often both. Both objectives can be pursued simultaneously without much conflict between them, and the logic driving the adoption of such alternatives is clear: soils that are not healthy will not produce premium crops, and farms that are not profitable will not stay in business.

Farmers and activists, thus, are quite comfortable rehearsing the logical benefits of profit and fertility regarding business transactions and farm operations, but are less well versed in discussing the benefits and relevance of social justice to the way sustainable agriculturalists produce and consume food. The justification for incorporating social justice into sustainability often relates to things not explicitly linked to sustainability, such as religious convictions, civil rights concerns or charitable tax deductions. Exactly, what constitutes social justice in the discourses of sustainable agriculture is not well defined either. Whether it regards labour, poverty, racism, hunger, all of these things or something else entirely, is not clear. Consequently, the logical connections between social justice and the sustainability of farms and rural communities are not well articulated, and thus, they are not often well practiced.

In this chapter, I seek to clarify the meaning of social justice to sustainable agriculture through an examination of social justice in the literature and to discuss the logics of social justice through the discourses and practices of two case examples. My goal is to articulate social justice as essential to the practice of sustainability and develop conceptual frameworks that connect social justice to sustainability in the same way that economic and environmental justice are connected to sustainable farming and business practices. The research is based on one year of ethnographic research in two network communities embedded in the sustainable agriculture community in Pennsylvania: the Tuscarora Organic Growers (TOG) and the Women's Agricultural Network (WAgN). These networks illustrate movements towards and away from social justice, and I use observations of their social justice discourses and practice to discuss how social justice and sustainability can and do relate to each other. This chapter is by no means an exhaustive account of social justice in sustainable agriculture, and is meant to continue, and perhaps clarify, the terms in the conversations begun by scholars and activists working towards a more socially just agriculture.

The Research Site

Pennsylvania, with over 2.8 million rural residents, has the largest rural population of any US state (NEMW 2004), and 25% (approximately 11 million ha.) of land in the state is devoted to agriculture (ERS 2004). Pennsylvania also has a thriving sustainable agriculture community with plenty of fertile land, relatively easy access to the urban markets of the Mid-Atlantic cities and a celebrated tradition of small-scale agricultural entrepreneurship. The Pennsylvania Association for Sustainable Agriculture (PASA), founded in 1991, is a regionally significant sustainable agriculture organisation with over 3000 members. PASA was formed explicitly as a 'sustainable' versus 'organic' farming organisation, because founding members felt organic farming was not sufficient to accomplish the broader goal of sustainability (Sachs, PASA Founding Member, Personal communication, March 2003).

The respondents involved with this study are generally: (1) members of PASA; (2) interested in or supportive of sustainable agriculture to varying degrees; and (3) directly involved with farming in some way. Criteria for inclusion in the study are membership or involvement in any of the two networks included in the study, or attendance at any of the events sponsored by either of the networks. The networks were chosen because of their articulation with sustainability in both theory and practice and their significance and visibility in the sustainable agriculture community in PA. Both networks are also historically and powerfully connected to this larger community, as both have founding members of PASA on their respective boards or steering committees.

The study area was chosen not only because of the large numbers of rural residents and the high level of interest in sustainable agriculture, but also because my residence in the area allowed for the establishment of long-term relationships and the use of ethnographic methods. Participant observation was used with all networks under investigation and was the primary vehicle for data collection. Interviews (structured and semi-structured) and surveys were also used to collect network histories, participant perceptions and demographic information. All fieldwork and data collection were conducted in the year between February 2003 and February 2004. I participated in the development of the WAgN as a founder, steering committee member and organiser, roles through which I was able to participate as both a member and a researcher.[1] I lived and worked on the farms of members of the TOG cooperative for 3 weeks during the growing season of 2003, which allowed me to interview farmers and observe the activities of the cooperative.

The research was conducted as part of a larger dissertation project, the focus of which was the discourse and practice of social, economic and environmental justice in sustainable agriculture. In this particular piece, I focus on what kinds of logical frameworks must be in place to incorporate social justice more fully into sustainable agriculture. In other words, why does it make good (economic or otherwise) sense to incorporate social justice into the sustainability paradigm? And how can social justice be incorporated as a key component to sustainability on the farm, especially when social justice can conflict with other imperatives? From this, what can actors in network communities practising and subscribing to sustainable agriculture tell us about the challenges and processes of actually putting this into practice? Before answering these questions, however, a review of social justice and sustainability in the literature is warranted.

Discourses of Justice: The Moral Logic of Sustainability

Sustainable agriculture is frequently defined as a social movement that draws together diverse groups (farmers and consumers) in pursuit of broad social, economic and environmental justice goals (Buttel 1993; Barham 1997; Hassanein 1999; Redclift 2000; Cocklin *et al.* 2002). There are many versions of these movements that articulate differently in different places, but broadly defined they include markets that shorten supply chains to the advantage of producers and/or production practices that aim to conserve resources, including and especially soil and water (Pretty 1995). While transforming the inequitable social, economic and environmental conditions produced by conventional agriculture are priorities for sustainable agriculture, environmental soundness is often privileged over other imperatives (Allen 1993).

The use of pesticides, chemical fertilisers and biotechnology and their associated environmental problems are cited by nearly all, and particularly early, activists for sustainable agriculture as reasons to change farming practices from chemical intensive to organic (see Carson 1962; Berry 1977; Jackson 1980; Pretty 1995). These new farming practices relied on local knowledge production for their development and sophistication, and as such, farmers form networks to facilitate information exchange (Hassanein 1997; Andrew 2003; Simpson *et al.* 2003). While better farming methods

---

[1] My role as a founder and facilitator of WAgN made my role as a researcher challenging, given the inherent conflict between directing the organisation, and 'objectively' observing the activities of the organisation. Two realities, however, helped maintain a creative, rather than a problematic, tension between these two roles. First, the research methods were similar in many ways to focus group interviews and participatory action research. I did not allow my role as a researcher to influence any decision about the organisation. Secondly, my role as the ultimate 'insider' made my research role almost invisible, and my presence at meetings, observing and asking questions rarely created the sometimes artificial, and potentially biased, contexts associated with more conventional qualitative research methods. In the interest of full disclosure, as well, I continue my research with WAgN as a paid employee.

might improve environmental conditions, farmers must still produce a profit to stay in business and to be economically sustainable. Direct marketing, community-supported agriculture (CSA),[2] organic certification and cooperatives are a few of the ways in which farmers can realise greater profits directly and help reduce the risks of farming (Hinrichs 2000; Morgan and Murdoch 2000).

Some scholars argue that because of this emphasis on the technical aspects of agriculture (production and marketing practices) the economic and environmental justice goals are being met, but the social justice goals are not (Allen 1993; Allen and Sachs 1993; DeLind 1994; Sachs 1996). Social justice in the context of agriculture typically emphasises the social provision of quality food and nutrition to all people but also concerns issues of labour, education and oppressive social/cultural relations (Allen et al. 1991; Allen and Sachs, 1993; Delind, 1994; Shiva 1999; Feenstra, 2002). Allen (1993:11) argues that sustainable agriculture requires the "elimination of patriarchy, racism, and class exploitation – all of which maintain systems of power that reinforce the contradictory social relations on which nonsustainable food and agriculture systems are based". This includes, but is not limited to, the marginalisation of women from knowledge exchange and decision-making roles (Sachs 1983; Whatmore 1991; Leckie 1996; Trauger 2004), the exploitation of farm workers (Allen et al. 2003) and the persistence of hunger in the midst of unparalleled levels of food production (Allen and Sachs 1993).

Movements to engage with social justice include incorporating labour regulations into organic standards or other similarly labelled codification of agricultural practices (Henderson et al. 2003; Shrader 2005). Connected to this are movements to merge 'fair trade' with organic standards to protect farm workers from exploitation in the same way that independent producers are granted some forms of market protection with 'fair trade' initiatives (Raynolds 2000). The price premium for organic products can produce social benefits for producers beyond the financial, such as a greater sense of security, increased access to education, increased access to health care and so on, especially for producers in the Global South (Bray et al. 2002). The organic price premium, while benefiting farmers, can create a 'two-class' food system, where only the wealthy can enjoy the benefits of fresh, healthy food, and there are increasingly calls to extend access to low-income communities through community food security and local food systems (Gottleib and Fisher 1996; Allen 1999). Rural development is also an outgrowth of the increased financial security of farms practising organic or sustainable agriculture (Rosset 2000; Hillocks 2002).

While all of these initiatives engage with ideas of social justice, few articulate sustainability *as contingent on* social justice in the same way that fair prices or soil fertility are essential to the long-term viability of agriculture and agricultural communities. Allen et al. (1991: 37) write, "sustainable agriculture is one that equitably balances concerns of environmental soundness, economic viability, and social justice among all sectors of society". If sustainable agriculture is to truly achieve the social, as well as, economic and environmental objectives to which it aspires, it will require a much wider understanding of justice, one that incorporates, but also moves off the farm and into the social fabric of many communities as well. How to do this, however, is not clear, and may not be

---

[2]Community supported agriculture (CSA) is a form of marketing and distribution of farm products that involves the customers buying a share in the farm in exchange for an amount of farm produce weekly or biweekly.

as straightforwardly connected to sustainability, as is profitability and soil fertility. The following is a discussion of social justice concepts in a general sense that can aid in connecting social justice to sustainability in essential ways.

Social Justice and Sustainability

Much of the literature on social justice discusses "distributive" justice, or that which is concerned with the equal distribution of "good and bads" in society (Miller 1999:1). This refers not only to social benefits such as access to education but also responsibilities, such as military service or care for the elderly. Another aspect of distributive justice is providing individuals with "rewards proportional to their contribution" (Tyler 2000: 118). As such, distributive justice is concerned with achieving equity and preventing exploitation. These two frameworks suggest two ways of thinking about social justice in sustainable agriculture. The first is to share the risks and benefits of agriculture more equally between producers and consumers, such as in the CSA model. The second is to provide workers with compensation proportional to their contributions, which could include profit-sharing and decision-making authority about labour conditions, as is the case with apprentice models.[3]

Individuals tend to use their own internal measures of what is just and fair when they identify something as unjust or unfair, even when it is not in their self-interest, or when they are not the targets of injustice (Tyler 2000). Tyler explains, however, that people tend to be more concerned with issues of fairness and justice when dealing with those inside their own social group. Thus, he argues, group boundaries hinder the expansion of social justice across social groups, and as such, expanding group boundaries is key to expanding the scope of social justice. The sustainable agriculture movement is inclusive of many kinds of difference, due in large part to its position in left of centre politics, but 'family farming' continues to be idealised as the vehicle towards sustainable food production (Berry 1977; Jackson 1980; Pretty 1995; Mariola 2005). The persistence of these cultural frames endangers the future of the movement as the typically white male-headed-family farm continues to disappear from the agricultural landscape (ERS 2001). A social movement cannot grow when the borders of its cultural and collective identity are impermeable. As such, expanding the scope of 'who belongs' to sustainable agriculture not only expands the sphere of social justice but also enrols new consumers and advocates on the network.

The literature on social justice in general tends to overlook cooperation as a concept or framework for facilitating social justice. The literature on social justice in the food systems, however, stresses cooperation and cooperative models, as opposed to more capitalist frameworks, as ways to achieve social justice (Pretty 1995; Allen and Kovach 2000; Murray and Raynolds 2000; Simpson and Rapone 2000). Fair trade coffee cooperatives, for example,

---

[3] Apprentice labour is a unique category of labourers in sustainable agriculture. They are typically around 20 middle-to-upper middle class suburbanites (of all races/ethnicities, but the majority are white) who are interested in farming and/or have a desire to experience farm life and practice an environmental ethic. They work for room and board and a monthly stipend that is typically well below minimum wage. They are sometimes given a stake in the profits in the farm and are often recruited to be managers of a crew of labourers or are responsible for a particular crop on the farm.

are ways in which small-scale peasant farmers obtain a fair price for their product and share the risks and benefits of farming in a socially just way (Simpson and Rapone 2000). Frances Moore Lappe (1990) writes that Darwin's observations on the benefits of cooperation and mutual aid to the survival of the fittest, are often overlooked in favour of discourses of competitive behaviours. While competition is purported to provide the highest quality product at the lowest possible price, the benefits that accrue from this strategy are tilted towards the consumer not the producer. Thus, cooperation amongst producers (in creative tension with competition from other cooperatives) can also provide some benefits to producers.

Three social justice interventions are clear for the sustainable agriculture social movement in this literature, and all have specific implications for the long-term growth and stability of the movement. First, distributive justice seeks to balance rewards with contributions. Currently, this philosophy is well articulated within the movement and manifested in the organic price premium, which usually benefits producers more than labourers. Labour exploitation makes organic production increasingly less expensive, and more widely practiced, which drives down the price premium. That this is happening is becoming well established, as Wal-Mart adds organic produce to its supermarkets, and organic production moves to the Global South (FAO 2001; Warner 2006). Distributing rewards proportional to contributions will make socially just agriculture more sustainable than simply 'organic' agriculture in the long term, as local food systems, CSAs, or community food security more equitably distribute the costs and benefits between producers and consumers in ways that certification schemes cannot do alone.

Secondly, the old models of rural, white-male-headed households are increasingly less viable forms of petty commodity production. Thus, new forms of agricultural production, such as cooperatives, urban farming and community food initiatives, are increasingly relevant and successful ways of producing food. 'Who belongs' and who practices agriculture is rapidly changing, and in the long term, diversity allows the movement to grow, expand and incorporate new constituencies, new consumers and new advocates.

Thirdly, farmers who cooperate with each other (and with natural systems) are increasingly more 'competitive' in the marketplace. This takes the form of marketing or producer cooperatives and/or the sharing of business, production and marketing practices in open educational settings. Sustainable agriculturalists increasingly share their 'secrets to success' with each other and increasingly reap the benefits of innovation. Cooperation in capitalist markets not only provides economies of scale to small-scale producers but also facilitates the spread of innovation, which makes sustainable agriculture increasingly capable of coping with rapidly changing consumer demands.

Social Justice in Network Communities in Sustainable Agriculture

Tuscarora Organic Growers

The growing season of 2003 was the worst season on record for vegetable growers in Pennsylvania, as there were record levels of rainfall throughout the state. As Ed[4] told me, however, the growers cooperative, TOG, to which he belongs was a source

---

[4] All first names only are pseudonyms. When first and last names are used, they are the real names of respondents, used with permission.

of security and a buffer for risks. "Smaller growers benefit from the other growers in the co-op in a year like this. The bigger growers can fill in some of the gaps and so the co-op can still make a profit, which benefits us all". The TOG form a marketing cooperative that delivers fresh, 'local' produce to a regional market in the Mid-Atlantic states. TOG was formed in 1988 by Jim and Moie Crawford of New Morning Farm and five other growers in south-central Pennsylvania. The central motivation for starting the cooperative was a need to expand and diversify the market for organic produce through wholesaling. The founding members felt that by acting cooperatively as a wholesaler they could capitalise on efficiencies of scale and shared resources. Currently, TOG has 13 member farms ranging in size from less than 1 ha. to more than 30 ha. All farms are certified organic and family owned and operated.

While the standards defining organic agriculture outline a set of farming practices that can help farmers obtain a price premium for their products, they provide no guidelines for rectifying economic inequality in the food system, and some would argue that they perpetuate inequality. It has been argued that the price premium on organic fruits and vegetables produces a two-class food system, where farmers and labourers produce food they cannot afford to purchase. To make food less expensive and still realise a profit, farmers try to find less expensive forms of labour. Due to the labour intensity of organic practices, large farms especially rely to a greater degree on 'cheaper' labour than smaller farms usually employ, and the cheapest labour to be had is migrant Mexican labour.[5] Apprentices are also a cheap source of labour, as they are paid by the month, not the hour, and often are responsible for decision-making about some aspect of the farm operation.

Three of the largest farms (> 6ha.) currently use migrant Mexican labour or have used migrant labour in the past. Three farms (two of them > 6ha.) also use apprentice labour. The majority of the migrant labourers work as 'field crews', and their primary work is picking produce in the field, but they also participate in transplanting, weeding, preparing fields for cultivation or other kinds of labour-intensive work. Apprentices also perform this work but are involved to a greater degree in decision-making and supervision of labour crews. All those I interviewed about the use of migrant labour, which included farmers, apprentice farmers, local wage labourers, truck drivers and customers, justified the use of migrant labour in organic agriculture with some derivative of "Americans just don't want to work this hard". Consistent with this message, all of the farms using migrant labour reserved the hardest work for them. This typically included hand harvesting, such as the tomato harvest featured in Figure 3.1. Farm apprentices also did this work, but on farms where apprentices performed the majority of the labour, the day was divided between picking in the cool of the day and packing produce in the packing shed during the heat of the day. Apprentice farmers were also able to negotiate more favourable working conditions for themselves, as they were often 'in charge', whereas migrant labour was always under the supervision of a crew leader, typically a white male.

---

[5]The majority of migrant labourers in the TOG network are recruited from the Mexican migrant labour community drawn to the area by work in the Chambersburg, PA fruit orchards. Chambersburg, in south-central Pennsylvania, is climatically well suited for fruit production and supplies the large mid-Atlantic consumer market with peaches, pears and apples. Labourers I spoke to started at $7.50/hour, which is above minimum wage, but still below the poverty level for a family of three.

Figure 3.1: Migrant field crew: migrant Mexican labourers picking organic tomatoes.

Because migrant labourers are perceived to be willing to work hard, are there to fill a demand for labour and have no recourse to change their situation on the farm, there is no need to change the working conditions that 'Americans' find intolerable. What the 'good work ethic' discourse glosses over is the fact that Mexican migrant labourers may not *want* to work this hard either. No one *wants* to work this hard, not even those who are willing to do it. The migrant labourers I spoke to told me that they felt they worked too hard because "we work so late every night, six days a week. We don't have time to have fun, go to the beach, relax" (Antonio). Apprentice farmers, who have a much greater influence on their working conditions, on the other hand, do not feel they work too hard. "We don't feel exploited because we are learning while we work. We also get all of our food and housing costs covered and we get a share of the profit if the farm does well" (Debbie). The irony of the "hard work" discourse is that the largest category of labour on farms in the network is family labour (36 people), followed by local wage labour (21). Migrant labourers (14) and apprentices (13) are actually the smallest categories of labour in the network. Apparently, Americans are willing to work this hard, even for little or no pay, as is the case for most family labour.

The scale of the operation or the diversity in crops often dictates the use and management of particular kinds of labour. However, cultural discourses often determine and/or legitimise certain labour practices as well. The stereotype of 'hard working' migrant labourers and 'lazy' Americans is an obvious example. A more subtle version of this was the shared sense of cultural identity between farm owner/operators and apprentices that was not evident between farm owners and migrant labourers. For example, apprentices frequently shared meals with owner/operators, while migrant labourers were invited to glean the fields for their meals. In addition, apprentices, ostensibly because of their aspirations to be farmers, were given decision-making authority for the farm, while the migrant labourers were not, in spite of their aspirations to have farms or continue farming in Mexico.

Overall, the economies of scale and high levels of market orientation (the largest farms also produced for large wholesale markets at a higher volume) drive some farms to employ a diversity of low-wage labour strategies. In some cases, labourers who are viewed as culturally distinct from the owner/operator of the farm are employed, and the labour practices reflected and reproduced this cultural distinction. As a result, very little control of the production, or over working conditions, is extended to these rural 'others'. In contrast, smaller-scale farms with more community-based markets (CSA) not only pursue other labour strategies that attempt to reduce the level of exploitation but also cultivate a sense of shared cultural identity between workers and employers. As such, the middle class, 'American' apprentice farmers (who are as likely to be 'migrant' as the Mexican labourers) are extended agency within the farm operation and are less likely to be constructed as 'others'.

The logical connections between sustainable agriculture and social justice are clear in this case. TOG increasingly use migrant labour as markets for organic produce become increasingly competitive, and the downward spiral of increased production and deflation of prices continue to force the use of cheaper labour. In this case, TOG moves away from concepts of distributive justice and, while still very successful, is practising the sort of competitive labour practices that may potentially undermine its own success, as well as the success of the movement in general when social justice for farmers is premised on organic price premiums. TOG, however, also practices other forms of labour management, such as apprenticeships, that ultimately produce more and better farmers. Apprentices come from all kinds of backgrounds, which expand the circle of 'who belongs' to women and racial minorities. As such, TOG strengthens and broadens the social community of sustainable agriculture. Also, TOG functions as a cooperative in a competitive and capitalist market, and the success of the cooperative is clearly linked to its ability to share resources, markets and expertise. Thus, the value of cooperation is a key component not only to the long-term sustainability of the organisation but also to the movement itself. I turn now to another organisation which struggles with these same issues but in an entirely different context.

Women's Agricultural Network

The Women's Agricultural Network (WAgN) is a cooperative extension-affiliated programme developed in Vermont in 1994 that has since diffused to Maine and Pennsylvania. All three organisations are dedicated to supporting women farm owners/operators with educational programmes. These include business planning workshops, online courses, discussion groups, technical assistance, newsletters and conferences. Women are generally seen as an underserved population in agriculture, as farming is still strongly associated with masculinity. WAgN is an organisation devoted to rectifying this marginalisation by providing the support and resources that traditional agricultural organisations do not provide to them. WAgN connects isolated farmers and functions as a support system and a source of information and shared resources. Pennsylvania WAgN has grown rapidly since it was founded in 2003, and as of this writing boasts 631 members.

By developing a dense network of opposition and opportunity, WAgN changes some of the patriarchal contexts of agriculture by enrolling individual actors, media and information technologies and commanding resources at various spatial scales. WAgN changed the context for educational programming for women in agriculture in Pennsylvania (as it has in Vermont and Maine), and this agency has been emergent from, and an effect of the network (for individuals and for the collective). No individual could have accomplished WAgN's objectives alone, but these accomplishments have been dependent on the emergence of leaders from within the collective. Leaders *assume* leadership: there

48 *Alternative Food Geographies: Concepts and Debates*

Figure 3.2: WAgN in action: WAgN Steering Committee at a strategic planning retreat.

are no elected positions or chairs, and all decision-making is made by consensus or by staff on advisement from members and is guided by periodic strategic planning by the steering committee (see the meeting, for example, featured in Figure 3.2). This loose framework opens multiple directions of progress and innovation, as those with expertise (i.e. farmers with knowledge to share) host field days, or faculty write grants to fund programming.

WAgN is self-consciously committed to providing social justice by resisting the patriarchal and misogynistic worldviews embedded in both sustainable and conventional agriculture.[6] However, identifying for whom WAgN works underscores the larger question of 'social justice for whom?'. The Pennsylvania WAgN chapter's mission statement identifies the broadest possible group: "women in agriculture", and departs from the Vermont and Maine chapters' missions explicitly stating "women farm operators". Pennsylvania WAgN's mission statement glosses over a relatively heated debate among the former steering committee members over who WAgN should serve, and what underlies this tension is a desire to target a community that does not conform to the patriarchal paradigm of conventional agriculture.

Liz, a farmer, has strong opinions on this issue and identifies the audience as "women who self-identify as directly involved in agricultural enterprises", and "women who want to become actively involved in farming of some kind". Angela, who is not a farmer, nevertheless also articulates a strong position on this issue: "The primary audience should

---

[6]WAgN is embedded within the sustainable agriculture community in Pennsylvania in various ways, but resists explicit identification with sustainable agriculture only. Currently, all of WAgN's funding comes from sustainable agriculture and 'small farm' programme areas of the USDA, all founding members are members (some founding members) of PASA, and WAgN's programmes generally all fall under sustainable agricultural practices, broadly defined, such as direct marketing, pastured livestock, CSA development, organic conversion, etc.

be farmers, because what is the point of ag educators or ag researchers without farmers? Farmers are agriculture and should be the primary focus for a women in agriculture group". Emmy, an apprentice farmer, who looks to WAgN as a source of mentors, echoes Angela. "I would like to see more full-time women farmers. It's disheartening to want to be a farmer and not see any examples of people doing it as a full time job". The positioning of women farmers as the primary constituency illustrates the construction of a woman farmer identity against the identity of 'farmwife', or a woman primarily in supportive roles who conforms to the patriarchal model of the family farm.

Another strongly and widely held opinion is that it should be "all inclusive" (Emmy), even among those who think it should be restricted, as Emmy's case illustrates. Laurie, along with several other members of the steering committee, had reservations about excluding any interested party:

> The population we should serve is females involved in agriculture. This will be farmers/producers (livestock, food, and fibre), farm managers, agri-business owners/employees, ag educators, and hobby farmers. I wouldn't want to exclude any female that has some tie to or involvement with agriculture.

In general the steering committee was split along the lines of those who wanted it to be clear that farmwives were not explicitly excluded, and those who wanted no part of the sexual and gender politics of traditional farm organisations that reify the subordinate roles of women on farms. All members, however, because of WAgN's conscious efforts against the explicit exclusion of women from education, knowledge and authority struggled with the idea of excluding anyone (even men) for reasons that may be entirely consistent with WAgN's project. Members have criticised programmes for tacitly excluding men, and WAgN was encouraged by partner organisations to advertise programmes with a 'men welcome' caveat. This is ironic given that the other activities of these organisations regularly support programmes that do not attempt to invite women at all. This tacit exclusion is precisely the kind of patriarchal paradigm that WAgN positions itself against.

WAgN's struggles over identity and inclusivity reflect a broader struggle in the women's movement over identity politics within women's groups. Identity politics tend to divide women along the lines of membership in racial, class or sexual identities. Within the community of women in agriculture, women identify themselves as 'farm women', 'farm partners', 'women farmers' and so on. As illustrated above, identity politics divides women who identify primarily as women farmers from women who identify primarily as wives. WAgN members are loath to reproduce the conditions of exclusion that marginalised women in the first place but fear welcoming those who conform to the gender roles that WAgN helps women resist. The politics of exclusion sits in uneasy tension with an ethic of inclusion, but WAgN officially attempts to subvert the injustices of exclusion by welcoming as wide an audience as possible.

Running counter to this conflict over identity is, rather paradoxically, an overarching ethic of cooperation. Women farmers, who often have much to learn about farming, are enthusiastic educators of each other. Far from being concerned about sharing their hard-earned secrets to success, women leap at the opportunity to network with other women and engage in peer learning events. Because they are often not taken seriously in largely male-dominated social contexts, they actively seek opportunities to build a

50 *Alternative Food Geographies: Concepts and Debates*

Figure 3.3: Tractor equipment workshop: WAgN members learning about equipment maintenance.

peer group with other women. Women farmers consistently prefer the kinds of learning environments that feature hands-on, intensive, multiple-direction learning, and WAgN is premised on a farmer-to-farmer education model where farmers teach most of the workshops, with assistance, when necessary from relevant 'experts'. The young women changing the oil of a tractor in an equipment workshop shown in Figure 3.3 illustrate the team learning WAgN helps foster.

This is well illustrated by feedback from women who attended the National Women in Sustainable Agriculture conference in Burlington VT in 2005. WAgN was able to fund the travel and attendance of 20 women farmers from Pennsylvania, and members (including myself) of PA-WAgN were part of the conference planning committee. One participant wrote in her post-conference evaluation: "The networking opportunities with receptions and dinners were exceptional. The best part of the conference was meeting other women". Another woman translated this interaction into innovations on the farm:

> There were many things about the conference that inspired me to want to [make changes to my operation]. These were: 1) how open, friendly and helpful most women were about sharing their experience and expertise; 2) stories of how women have struggled so hard to achieve what they have accomplished; and 3) how happy women in [agriculture] can be despite many challenges.

One participant made explicit connections between the social support activities of networks such as WagN and the long-term survival of agricultural businesses. "Watching all the women interact and discuss major issues, made me realise that without sustainable practices there will be no soil to grow food, [and] without a profitable business there will be no money to pay the bills and remain in agriculture". The general reaction to the (almost) women-only space of the gathering was one of wonder at the openness and willingness of women to share with each other, and the almost complete lack of competitiveness in both formal and informal interactions.

Again, movements towards and away from social justice are clear within WAgN, and the relevance of social justice to the future and the sustainability of the organisation and the social movement within which it is embedded are manifest. WAgN is explicit about sharing leadership, and no formal hierarchy exists within the organisation. Leadership is 'taken' by actors within the organisation, and as such, the distribution of rewards is typically proportional to the contribution. The rewards, however, are intended to be distributed throughout the organisation to the membership. Because of the fractures over 'who belongs' as a woman farmer, these rewards are not always evenly distributed to those who could benefit from them. Expanding the scope and scale of the community to a wider group is crucial to the long-term health and sustainability of the organisation, and WAgN risks its future by excluding potential constituencies. Despite fissures regarding its political identity, WAgN members enthusiastically share with each other all manner of information about their farm operations. This creates a situation where many, rather than few, can succeed, which is central to the long-term health and sustainability of agriculture, WAgN, and the sustainable agriculture social movement.

Conclusions

Social justice and sustainability are sometimes at cross purposes to one another. For example, using labour-intensive, environmentally friendly practices may require farmers to employ cheap labour to remain competitive in the marketplace and sustainable in an economic sense. Social justice, however, does not have to be mutually exclusive of sustainable practices, and the organisations discussed above illustrate well that 'everyone does better, when everyone does better'. As such, sustainability can and should incorporate social justice as a logical practice, rather than simply a moral obligation. Literature on social justice highlights equal distribution of responsibility and benefits, broadening the scale and diversity of the community and cooperating across differences and in competitive contexts. All three of these ideas have logical and tangible benefits to the sustainability of farms, organisations and ultimately the social movement that go beyond simply being the 'right thing to do'.

In both examples, the organisations are involved with practices that move simultaneously towards and away from a socially just sustainable agriculture but illustrate well the logical connections between social justice and long-term sustainability. In the TOG cooperative, farmers employing apprentices produce new generations of farmers and compensate them with skills and experience rather than capital. By working with farms who increasingly use migrant labour (the source of which may not be guaranteed in the future), TOG also moves away from sustainability and social justice by relying on systems of political, economic and cultural oppression that both produce migrant labour in the first place and perpetuate a downward spiral in prices that necessitates cheaper and cheaper labour. On the other hand, cooperation between farmers is clearly related to the success of the organisation, and the production of new, experienced young farmers through apprenticeships is clearly a boon to the long-term success of the social movement.

WAgN is also committed to producing a future of farmers, but of a different cultural stripe than most conventional farmers, and highlights the importance of expanding who belongs to the 'justice' community of sustainability. Expanding the social justice community expands the space in which to distribute the goods and bads of society. Expanding the community also helps grow the movement, expand the network and creates further and future nodes of change and agency. WAgN, however, struggles with internal divisions over identity and belonging, which if unresolved, can jeopardise the future of the organisation, and is not conducive to social justice or sustainability. On the other hand, cooperation and distribution of responsibilities and benefits throughout the organisation are hallmarks of WAgN's organisation and are central to its long-term success and relevance as an organisation. The beneficial logic of expanding the agricultural community to include 'others' cannot be writ larger on the landscape as women farmers with diverse operations grow in number at the same time that male-dominated conventional agriculture continues to be in crisis.

In both cases, the movements towards socially just sustainability involve changing conventional paradigms around the production of food, which include both labour and property relations. TOG use an apprenticeship model which 'produces' new and well-educated farmers, in addition to the crops planted, harvested and marketed with their labour, and the capital that accumulates to the farm operation from their labour. WAgN challenges the family farming model by introducing a new cultural frame for the category of 'farmer'. In spite of the struggles over who exactly qualifies as a farmer, WAgN's existence and popularity challenges the definition of the 'family farm', and the ownership structure and the divisions of labour that implies, and thus the way the farming household is organised. In both cases, TOG and WAgN offer alternatives to conventional views of the farm as a place of production, and thus impinge on what is meant by social justice on both the scale of the household and on the scale of the community of farmers.

Sustainability implies perpetuating something in the future, and thus a distribution of costs and benefits through time. Justice, however, implies sharing across human (and non-human) communities, and thus suggests a distribution across space. This tension lies at the heart of the difficulty with incorporating social justice into sustainability frameworks, because it requires that we expand the boundaries of our communities. The time dimensions of sustainability are presumably infinite, but where we draw the line in space around the social justice community is less clear. Is it the household, the neighbourhood, state, nation, the world? What is clear, however, is that as the scope and scale of the sustainable agriculture community grows, so do the responsibilities to and benefits of that community. Whether we can address the challenges this presents is difficult to say, and we will probably never know if we have accomplished our goals in the future. Given this, Barry (1999) suggests that the only way to go about this is to assess the present situation, make changes and find ways to extend these visions into the future. The examples outlined here illustrate that inclusivity, plurality, equality and cooperation are crucial and necessary aspects of a sustainable agriculture in the future.

References

Allen, P. (1993). Connecting the social and the ecological in sustainable agriculture. In P. Allen (Ed.), *Food for the Future: Conditions and Contradictions of Sustainability* (pp. 1–16). New York: John Wiley and Sons.

Allen, P. (1999). Reweaving the food security safety net: mediating entitlement and entrepreneurship. *Agriculture and Human Values* 16, 117–129.

Allen, P. and Sachs, C. (1993). Sustainable agriculture in the United States: engagements, silences, and possibilities for transformation. In P. Allen (Ed.), *Food for the Future: Conditions and Contradictions of Sustainability* (pp. 139–168). New York: John Wiley and Sons.

Allen P. and Kovach, M. (2000). The capitalist composition of organic: the potential of markets in fulfilling the promise of organic agriculture. *Agriculture and Human Values* 17, 221–232.

Allen, P., Dusen, D., Lundy, J. and Gliessman, S. (1991). Integrating social, environmental and economic issues in sustainable agriculture. *American Journal of Alternative Agriculture* 6, 34–39.

Allen, P., Fitzsimmons, M., Goodman, M. and Warner, K. (2003). Shifting plates in the agrifood landscape: the tectonics of alternative agrifood initiatives in California. *Journal of Rural Studies* 19, 61–75.

Andrew, J. (2003). Key features of the regional producer network for enabling social learning. *Australian Journal of Experimental Agriculture* 43, 1015–1029.

Barham, E. (1997). Social movements for sustainable agriculture in France: a Polanyian perspective. *Society and Natural Resources* 10, 239–249.

Barry, B. (1999). Sustainability and intergenerational justice. In A. Dobson (Ed.), *Fairness and Futurity: Essays on Sustainability and Social Justice* (pp. 93–117). Oxford: Oxford University Press.

Berry, W. (1977). *The Unsettling of America*. San Francisco: Sierra Club Books.

Bray, D., Sanchez, J. and Murphy, E. (2002). Social dimensions of organic coffee production in Mexico: lessons for eco-labelling initiatives. *Society and Natural Resources* 15, 429–446.

Buttel, F. (1993). The sociology of agricultural sustainability: some observations on the future of sustainable agriculture. *Agriculture, Ecosystems, and Environment* 46, 175–186.

Carson, R (1962). *Silent Spring*. New York: Houghton Mifflin.

Cocklin, C., Bowler, I. and Bryant, C. (2002). Introduction: sustainability and rural systems. In I.R. Bowler, C.R. Bryant and C. Cocklin (Eds), *The Sustainability of Rural Systems: Geographical Interpretations* (pp. 1–12). Dordrecht, Netherlands: Kluwer Academic Publishers.

DeLind, L. (1994). Organic farming and social context: a challenge for us all. *American Journal of Alternative Agriculture* 9, 146–147.

ERS (Economic Research Service) (2001). Structural and Financial Characteristics of US Farms: 2001 Family Farm Report. Available online at: http://www.ers.usda.gov/publications/aib768/(accessed 11/27/05).

ERS (Economic Research Service) (2004). Pennsylvania State Fact Sheets. Available online at: http://www.ers.usda.gov/StateFacts/PA.HTM, United States Department of Agriculture (accessed 1/21/04).

FAO (Food and Agriculture Organization of the United Nations) (2001). *World Markets for Organic Fruits and Vegetables*. Rome: Food and Agriculture Organization of the United Nations and International Trade Centre.

Feenstra, G. (2002). Creating space for sustainable food systems: lessons from the field. *Agriculture and Human Values* 19, 99–106.

Gottleib, R. and Fisher, A. (1996). Community food security and environmental justice: searching for common discourse. *Agriculture and Human Values* 13, 23–32.

Hassanein, N. (1997). Networking knowledge in the sustainable agriculture movement: some implications of the gender dimension. *Society and Natural Resources* 10, 251–257.

Hassanein, N. (1999). *Changing the Way America Farms: Knowledge and Community in the Sustainable Agriculture Movement*. Lincoln, Nebraska: University of Nebraska Press.

Henderson, E., Mandelbaum, R., Mendieta, O. and Sligh, M. (2003). Toward social justice and economic equity in the food system: a call for social standards in sustainable and organic agriculture. Available online at: http://www.cata-farmworkers.org/english%20pages/ socialjusticestandardsOctober2003.doc (accessed 6/10/06).

Hillocks, R.J. (2002). IPM and organic agriculture for smallholders in Africa. *Integrated Pest Management Reviews* 7, 17–27.

Hinrichs, C. (2000). Embeddness and local food systems: notes on two types of direct agricultural market. *Journal of Rural Studies* 16, 295–303.

Jackson, W. (1980). *New Roots for Agriculture*. San Francisco: Friends of the Earth.

Leckie, G. (1996). 'They never trusted me to drive': farm girls and the gender relations of agricultural information transfer. *Gender, Place and Culture* 3, 309–325.

Mariola, M. (2005). Losing ground: farmland preservation, economic utilitarianism, and the erosion of the agrarian ideal. *Agriculture and Human Values* 22, 209–223.

Miller, D. (1999). *Principles of Social Justice*. Cambridge, MA: Harvard University Press.

Moore Lappe, F. (1990). Food, farming and democracy. In R. Clark (Ed.), *Our Sustainable Table* (pp. 143–160). San Francisco: North Point Press.

Morgan, K. and Murdoch, J. (2000). Organic vs. conventional agriculture: knowledge, power and innovation in the food chain. *Geoforum* 31, 159–173.

Murray, D. and Raynolds, L. (2000). Alternative trade in bananas: obstacles and opportunities for progressive social change in the global economy. *Agriculture and Human Values* 17, 65–74.

NEMW (NorthEastMidWest Institute) (2004). Rural population as a percent of state total by state, 2000. Available online at: http://www.nemw.org/poprural.htm (accessed 3/25/04).

Pretty, J. (1995). *Regenerating Agriculture: Policies and Practices for Sustainability and Self-Reliance*. Washington DC: Joseph Henry Press.

Raynolds, L. (2000). Re-embedding global agriculture: the international organic and fair trade movements. *Agriculture and Human Values* 17, 297–309.

Redclift, M. (2000). *Sustainability: Life Chances and Livelihoods*. New York: Routledge

Ritchie, M. (2003). *A Search For 'True Security'*, Keynote address at the Pennsylvania Association for Sustainable Agriculture Annual Conference, 7 February 2003.

Rosset, P. (2000). The multiple functions and benefits of small farm agriculture in the context of global trade negotiations. *Development* 43, 77–82.

Sachs, C. (1983). *The Invisible Farmers*. Totowa, New Jersey: Rowan and Ellanheld.

Sachs, C. (1996). *Gendered fields: rural women, agriculture, and environment*. Boulder: WestviewPress.

Shiva, V. (1999). Monocultures, monopolies, myths and the masculinization of agriculture. *Development* 42, 35–38.

Shrader, R. (2005). Social justice in agriculture forum. *Cooperative Grocer*, 116. Available online at: http://www.cooperativegrocer.coop/(accessed June 2005).

Simpson, C. and Rapone, A. (2000). Community development from the ground up: social-justice coffee. *Research in Human Ecology* 7, 46–57.

Simpson, I.H., Kay, G. and Mason, W.K. (2003). The SGS regional producer network: a successful application of interactive participation. *Australian Journal of Experimental Agriculture* 43, 673–684.

Trauger, A. (2004). 'Because they can do the work': women farmers in sustainable agriculture in Pennsylvania, USA. *Gender, Place and Culture* 11, 289–307.

Tyler, T. (2000). Social justice: outcome and procedure. *International Journal of Psychology* 35, 117–125.

Warner, M. (2006). Wal-Mart eyes organic foods. *New York Times*, 12 May 2006.

Whatmore, S. (1991). *Farming Women: Gender, Work, and Family Enterprise*. Houndmills, Basingstoke, Hampshire: Macmillan Academic and Professional.

Chapter 4

From 'Alternative' to 'Sustainable' Food

Larch Maxey
Department of Geography, Swansea University, UK

Introduction

> Something as simple and basic as food has become the site for manifold and diverse liberations in which every one of us has an opportunity to participate – no matter who we are, no matter where we are.
>
> Shiva 2000: 4

The above quotation introduces four interrelated themes which inform this chapter. First, it points to the importance of food as a contested site through which contemporary processes are shaped. These processes and contestations take place at every temporal and spatial scale; indeed, part of the attraction, complexity and challenge of food is its potential to simultaneously collapse and distil various scales and categories through which we may seek to make sense of it. Secondly, and linked to this, is the growing body of geographical work on food generally and within what have broadly been termed 'Alternative Food Networks' (AFNs) more specifically. Whilst the concept of AFN has developed within English-speaking academic circles, the practices described and analysed within this body of work are located in fields, shops, distribution networks, processing plants, mouths, board rooms and other sites around the world. Thirdly, the quotation points to the fluid, open nature of food as a site, one with which every person and the various networks they co-create can and does engage. Finally, and flowing from this, the quotation points to food's potential to open up and broaden debates and other practices. There is an emerging lacuna, for example, between increasingly nuanced, empirically and theoretically informed academic work on 'alternative' food in Western, or Minority World contexts and 'alternative' food stories in the Majority World, which the above quotation implies.

The concept of sustainability, however, offers ways to engage with all of these themes explicitly and actively. It is somewhat surprising that sustainability has received so little critical and sustained attention within the new geographies of food, given the widespread recognition that geographers are uniquely well placed to contribute to the analysis of sustainability and advocacy of the 'sustainability transition' (Viles 2002; Oldfield and Shaw 2002; O'Riordan 2004). This gap is even more remarkable as the new geographies of food literature so often draws upon sustainability either directly or implicitly. Work on 'alternative' food provisioning in particular deals with issues at the heart of sustainability. This chapter draws upon theoretical and empirical research to consider whether the specific term and wider discursive framework of *sustainability* may be better suited to approaching such food than *alterity*. A brief critique of 'alternative' food provides an introductory pointer to some of the ways sustainability may inform more nuanced approaches to food. Central to the chapter is the elaboration of a critical relational approach to sustainability. Drawing on qualitative research covering three distinct research projects and 15 case studies, the chapter proposes a sixfold model of sustainability. Through this discussion, the chapter considers some of the limitations within the concept of sustainability itself as well as the potential of the notion of 'sustainable food' to inform and open up critical food analysis and action.

New Geographies of Food

The current surge of interest in food amongst geographers points to the diverse and fluid ways it shapes and is shaped by a myriad of social, political and cultural practices. Interests ranging from global–local economics (Jones 2003; Potter and Tilzey 2005), issues of identity (Bell and Valentine 1997) and rural change (Marsden 2003; Renting *et al.* 2003), for example, can all be approached through geographies of food. In this chapter, I consider sustainability's potential to inform these interests. As Whatmore (2002) suggests, there has been a bifurcation between 'cultural' and 'political'/'economic' studies of food. Whilst more recent work, such as that by Cook *et al.* (2004) has begun to address this compartmentalisation, sustainability offers one way of beginning to combine such approaches. Indeed, Cook *et al.* whilst explicitly concerned with the de-fetishisation of commodities, rather than exploring sustainability, conclude by asking what role sustainability may play in such ventures:

> Attempts to de-fetishise commodities raise important but tricky questions. Like, what can any 'radical' and/or 'sustainable' politics of consumption realistically involve? (p. 663).

This chapter offers an initial response to this question, considering the extent to which a critical relational approach to sustainability may provide a useful framework in the exploration of food geographies. I begin this task by looking at a key term through which these 'new geographies' have emerged, namely 'alternative' food.

'Alternative' Food

In one of a series of reviews, Winter (2004: 666) describes "[T]he blossoming of research on alternative food networks". Whilst Winter uses the more common term 'Alternative Food Networks' (hereafter referred to as AFNs), a host of other terms, complete with acronyms, has also appeared, including 'alternative agro-food networks' (AAFNs) (Goodman 2003), 'alternative food initiatives' (AFIs) (Allen et al. 2003), 'short food supply chains' (SFSCs) (Renting et al. 2003) and "market orientated initiatives for environmentally sustainable food production" (MOIs) (Buller and Morris 2004: 1067). There has, indeed, been a wealth of empirical and theoretical work on various aspects of 'alternative' food over the last decade (Venn et al. 2006). Located within the wider genre of the new geographies of food, such work has helped to highlight key contemporary shifts in the growing, processing, selling and eating of food within, for example, Europe and North America. Such shifts include an increase in imported 'luxury' and 'novelty' foods (Barrett et al. 1999), an increase in sales of convenience food and ready meals (Yakovleva and Flynn 2004) and an increasing concentration of these food sales through large multiple supermarket and fast food restaurant chains (Guthman 2004; Smith and Marsden 2004).

Furthermore, there is an emerging consensus within such work that neoliberal economic policies are driving many of these changes. From international bodies, such as the World Trade Organisation (WTO), through corporate executives to national and local governments there is, this work generally suggests, a growing hegemony that places 'the market' as the best, most 'efficient' mechanism through which to organise society generally and food in particular (Watts et al. 2005). The most significant contribution of work on 'alternative' food, however, is not its interrogation of neoliberalism, but its demonstration that even the most influential, most intensely networked actors do not exhibit total control over food networks (Winter 2004). Work on AFNs has demonstrated the capacity of 'ecological entrepreneurs' (Marsden and Smith 2005) and others to shape distinct food networks within a broader homogenising trend.

The notion of 'alternative food', then, has been a useful analytical tool, helping to critique the growing neoliberalism of food and highlighting some of the ways it is being resisted. The limitations of the 'alternativeness' concept are, however, beginning to emerge (Holloway et al. 2005). First and foremost is its ontological root within a binary opposition. Recent food literature has, therefore, generally pitted 'alternative' against 'conventional' food. More recently still, several authors have begun to interrogate this binary framework at the heart of alternative food. Watts et al. (2005), for example, begin to explicitly question the meaning of 'alternative' and propose a framework of weaker–stronger alternatives. This framework begins to render alternative food more nuanced and complex, proposing, at the very least, that there may be a *range* of food practices within the rubric of 'alternative'. Watts et al. (2005: 35) open the way for further interrogation of alternative food suggesting, for example, that ". . . future research could consider. . . whether it may be necessary to begin thinking of alternative systems of food provision as being hybridised when considered at the level of the individual enterprise".

The notion of 'hybridity' appears set to become established as a key concept within the new geographies of food, following Whatmore's (2002: 162) seminal work in which she suggests:

> . . . it is both more interesting and more pressing to engage in a politics of hybridity . . . in which the stakes are thoroughly and promiscuously distributed through the messy attachments, skills and intensities of differently embodied lives whose everyday conduct exceeds and perverts the designs of parliament, corporations and labour.

The approach to sustainability, and through this sustainable food, developed below has considerable potential to contribute to this 'politics of hybridity'. Another contribution of 'hybridity' as an analytical concept stems from its ability to help disrupt taken-for-granted boundaries around, for example, different forms of food provisioning. It thus has considerable potential to help explore the shifting, complex and contested nature of contemporary food systems. Ilbery and Maye (2005) take up the challenge of analysing the extent to which individual alternative food enterprises may be considered hybridised. Their paper is notable in its attempt to explicitly address the 'sustainability' of their case study AFNs. Adopting Sustain's[1] nine point " 'sustainable food' criteria" as a checklist, however, they conclude that none of their 'hybrid' case studies is entirely 'sustainable' but display a mixture of 'alternative' and 'conventional' characteristics.

Whilst advancing the field and opening up grounds for further investigation, work within the new geographies of food to date has yet to fully address the binaries embedded within the notion of 'alternative' and latent within the concept of hybridity.[2] Such binaries suggest the presence of two idealised 'pure' categories that are unlikely to fit the more complex and contingent forms found in practice. Furthermore, such approaches present 'sustainable food' as a given entity which can be measured against a predetermined checklist. Before dealing with this approach to 'sustainable food' more thoroughly in the next section, it is necessary to consider 'alternative food' a little further.

Following Holloway et al. (2005), I suggest this 'hybrid' approach, as it is left by Ilbery and Maye (2005), is in danger of leaving 'alternative' and 'conventional' food intact, as distinct entities with distinct logics and implications, each variously " 'dipped into' by producers in particular instances" (Holloway et al. 2005: 7). Furthermore, as Lockie and Halpin (2005: 305) note, "processes of 'talking about' and 'framing' change actually influence both conceptual and empirical developments". In perpetuating analysis of 'alternative' food, then, we may inadvertently naturalise, normalise and legitimise highly problematic practices and products labelled 'conventional'. One of the reasons for focusing on sustainability here then is to consider whether a sustainability informed framework can radically de-centre the apparently hegemonic and highly problematic term 'conventional' as it pertains to food.

Finally, the term 'alternative' is rarely used by those engaged in the practices academics choose to describe as such. As Seyfang (2006) reports, for example, workers at Eostre Organics, a producer cooperative based in East Anglia in the UK, are only too aware that organic food's continued position as 'alternative' marginalises it and limits their ability to sell products to local hospitals, schools and similar 'mainstream' institutions. Starting with the hypothesis that food practitioners would be similarly ambivalent about the term 'sustainability', my own research over the last decade with a range of

---

[1] Sustain: The alliance for better food and farming is a UK based organisation launched in 1999, see http://www.sustain.org.
[2] This is a moot point which warrants further attention. It could be argued, for example, that a 'hybrid', as the product of two or more discrete and 'incongruous elements' (Brown 1993: 1285), presupposes one or more binaries. The concept of hybridity could thus be seen as implicitly reinforcing discrete divisions whilst explicitly challenging them. This is not my reading of Whatmore's 'politics of hybridity', although this dualistic potential remains if 'hybridity' is used less critically.

food-related case studies suggests a growing willingness amongst growers, sellers and eaters to engage with the term (Maxey 2003, 2006).

'Sustainability'

As Marsden (2003) has shown, sustainability is at the heart of contemporary changes in rural society generally and food geographies more specifically. However, even Marsden's work, so centrally focused on 'sustainability', provides little explicit analysis of 'sustainability' (Kurtz 2004). This reluctance to interrogate sustainability is common to geography generally (Gibbs and Krueger 2005) and recent literature on food specifically. Winter (2003), for example, provides a valuable critique of the local or 'quality turn', unpicking often taken-for-granted assumptions but declines to do this for 'sustainable agriculture'. Whilst it is churlish to single out particular instances in what is a broader trend, and to ignore more sustained appraisals (e.g. Hassanein 2003; Robinson 2004; Watts et al. 2005), this lacuna in much of the literature on food is surprising, given both sustainability's wide ranging salience and the increasingly nuanced and critical nature of work in this area. Buller and Morris (2004: 1070) begin to identify this gap, noting that: "... although the territorial provenance of 'quality' foods and their labelling and marketing have been the subject of a recent wave of writings, much less has been said on their contribution to environmental sustainability".

One reason for the general reluctance to engage explicitly and critically with sustainability may be its uncertain, discursive nature. In common, with several commentators (Dobson 1998; Redclift 2002; Hassanein 2003), Ilbery and Maye (2005) helpfully emphasise that 'sustainable development' is socially and politically constructed. They go on to assert that " 'sustainable agriculture' is an equally slippery and broad ranging term" (p. 333). This raises two potential problems with employing the notion of sustainability, its constructed nature and its indeterminacy.

First, sustainability[3] *is* socially and politically constructed. Whilst this could be regarded as a problem, following Hassanein (2003), I suggest it may be one of its key attractions. In this chapter, I propose that if the broader notion of sustainability is understood as a critical relational process, it can usefully inform understandings and practices associated with the new geographies of food. The starting point of this process is sustainability's potential to offer radical opportunities for reflection. Sustainability, thus construed, encourages everyone to ask, as Engel and Engel (1990) note, the 'big questions' such as: *what is important in my life/our lives?; how do I/we want to live?* Sustainability encourages us all to consider what we want to sustain and to assess the ways we wish to go about this. Furthermore, this approach to sustainability requires that we do this individually and collectively at every scale from households, families and communities to nationally and internationally. Viewed in this way, sustainability's construction is rendered explicit. Sustainability is, at its core, a process of construction and

---

[3] I prefer the term 'sustainability' to 'sustainable development' due to the political and cultural baggage associated with 'development', through which highly unsustainable practices and values have been privileged whilst far more sustainable ones have been undermined (Sachs 1999; Maxey 2006).

negotiation. Therefore, it is not sustainability's *constructed* nature which is problematic, rather it is the way its construction can be *hidden* and manipulated. This is often achieved by presenting sustainability as given, a self-evident 'good' which everyone supports. As Redclift (2002: 194) notes, this is commonly done through a 'scientific' appropriation of sustainability: "the idea of 'sustainability' is invoked in policy discourses as speaking to objective scientific method without the complications of human judgement". Whilst Redclift restricts this critique to policy discourses, I suggest it is pertinent to a broader 'politics of hybridity'. In failing to rigorously scrutinise sustainability, therefore, contemporary food studies not only fail to nurture sustainability's liberatory potential, but also fail to challenge its use as an oppressive tool of surreptitious control.

If the notion of 'alternative' food draws upon and implicitly re-enforces a binary with 'conventional' food, so too, can sustainability be seen as part of an unstated dualism with 'unsustainability'. The surreptitious way in which sustainability is currently employed to represent self-evident 'goods' can be illustrated by the imbalance between the growing number of references to 'sustainable' practices (including food provisioning) versus the scant number of references to 'unsustainable' practices. If the notion of sustainability is to be used, as it so often is, to represent a known or knowable set of things or state of affairs, then this dualistic framework should at least be rendered explicit and the 'unsustainable' named alongside the 'sustainable'. Such analysis, however, assumes that sustainability *can* be defined, quantified and precisely 'known'. The approach I wish to explore in this chapter starts from a very different assumption that sustainability begins with the process of people negotiating what they value and thus what they consider worth sustaining. Within this approach, assumptions as to what sustainable and unsustainable practices may entail can be rendered explicit.

That the unaccountable drive to (pre)circumscribe, (pre)define and claim sustainability can, potentially, be found everywhere is illustrated by attempts to " 'reclaim' sustainability" (Robinson 2002: 186). This process of 'reclaiming' also needs interrogating. Whilst commonly associated with popular attempts to wrest sustainability from corporate neoliberalism, it carries with it implicit binaries such as grass roots-people's sustainability versus 'expert' top-down sustainability. There may well be strong arguments for distinguishing between such different approaches to sustainability. All too often, however, they are left implicit. For those seeking to 'reclaim' sustainability, a helpful, progressive starting point may be to establish who is reclaiming *what* from *whom*? If we are to appeal to sustainability *let us be clear what it is we seek to sustain and why*. Secondly, in addition to its constructed nature, Ilbery and Maye (2005) refer to 'sustainable agriculture' as 'slippery'. Whilst this is a long-standing criticism of sustainability, it can also be seen as one of its strengths. As sustainability has not been restricted to a particular definition, applicable to particular situations, it can be applied more broadly and creatively.

There are, then, tensions at the heart of sustainability. In order to begin considering whether a critically reflexive, process-based approach to sustainability can address such tensions, it is important to note that despite sustainability's fluid, constructed nature: "... a consensus of informed opinion recognises three dimensions in sustainable development – environment, economy and society" (Bowler 2002: 205). As Bowler goes on to note, the key insight provided by this triple bottom line is that sustainability requires all three dimensions to be considered *simultaneously*. Whilst this sounds remarkably simple, it represents a paradigm shift from dominant Western reductionism and contemporary neoliberal privileging of the economic. Amongst those food commentators who do consider sustainability, there is a tendency to bracket it off, referring to 'environmental sustainability', for example. This is often done in the interests of clarity and precision, a form of 'reclaiming' sustainability's environmental component, perhaps. However, I argue that bracketing sustainability off in this way risks undermining it. If we are to begin addressing contemporary environmental, social and economic crises, which is one goal sustainability offers, then human activities and decisions need to consider all three aspects together.

Whilst this minimal definition of sustainability is potentially useful in pointing towards the kind of radical shifts required, it leaves much open and unclear. Bowler's (2002: 205) reference to "a consensus of informed opinion", for example, could lead to the privileging of 'expert' sustainability and obscure the central importance of participation and enabling everyone to co-negotiate sustainability. Furthermore, Bowler (2002) implies sustainability inevitably involves economic growth. This may be a reasonable assumption, given the apparent hegemony of the growth paradigm within contemporary political and economic discourses. Indeed, Yakovleva and Flynn (2004) make similar assumptions, as do a wide variety of commentators. Leaving such an assumption hidden and unaccountable, however, fails to grasp the full potential of sustainability as a tool to guide and broaden analyses. Robinson (2002: 186), for example, criticises those who have equated sustainability with "the long term maintenance of economic growth... thereby severely devaluing the term". Indeed, there are signs that even Western governments, wedded to economic growth since their inception, are beginning to appreciate that it may conflict with, or at least not be central to sustainability (see DEFRA 2005).

Relational Sustainability and Sustainable Food

The tension between 'alternative' and 'conventional' food mirrors that between 'alternative' and 'mainstream' approaches to sustainability (Daly and Cobb 1989; Pearce *et al.* 1989; Beckerman 1994, 1995; Daly 1995, 1996; Serafy 1996; Krueger and Agyeman 2005; Seyfang 2006). Whilst each context is unique, such binary framing presents similar tensions in each case, as we have seen above, offering the promise of clarity and a basis for (re)action, whilst running the risk of essentialising and obscuring complex contingencies. In an attempt to begin elucidating a more dynamic, relational approach to sustainability as it pertains to food, therefore, I propose the six key components model of sustainability set out in Figure 4.1. This has been developed from the four principles of physical limits, futurity, participation and equity adopted by Bhatti *et al.* (1994) and my own research into activism, sustainable communities and sustainable food systems (Maxey 2003, 2004a,b, 2005a,b, 2006).

Research Case Studies and Methodology

The sixfold model of sustainability outlined in detail below has been developed in response to primary research spanning 11 years (1995–2006) and three distinct research projects. In each project, the research methodology involved in-depth interviews and participant observation with three-six case studies (see Figure 4.2). Every willing participant was interviewed in each case study. Most interviews were conducted with just one

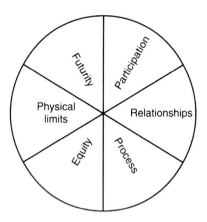

Figure 4.1: Sixfold model of sustainability.

respondent; however, group interviews were also held where participants' time commitments necessitated. The number of interviews carried out at each case study varied from two in the case of the smallest food operations to 18 in the case of the larger communities. Around 67 interviews were held in total across the 15 case studies. At the behest of participants, the case study names have been kept, whilst the names of individual participants have been coded to provide anonymity.[4]

Research Projects

(I) From 1995 to date, I have been working with three small-scale communities in England and Wales. The work grew out of a PhD thesis on intentional communities' potential contribution to sustainability (Maxey 1999, 2003, 2005a). This research has since broadened to include research into Low Impact Development within the UK (Maxey et al. 2006; Maxey and Pickerill 2007). Each of these case studies raises a range of interesting questions and insights into the ideas and practices of 'sustainable food'.

(II) From 1998 to date, I have been undertaking a comparative research project working with small-scale vegetable producers. Three of these case study producers were located in Southern Ontario, Canada, and three were based in South Wales, UK. These producers were all involved in direct distribution schemes where their products were sold directly to customers through farm shops and direct delivery box/bag schemes. Between 3 and 22 acres were cultivated for vegetable production in the various case studies (see Everdale 2005; Plan B 2005; Maxey 2006; OTG 2006 for further details).

(III) From 2004 to date, I have been working with six Welsh case studies all involved in small-scale pig farming. This work explores the shifting and contested

---

[4] Thus 'PB 2: 678' after a quotation in this chapter refers to a comment made by the second member of the 'Plan B' case study to be interviewed, and the final number refers to the interview tape counter number.

From 'Alternative' to 'Sustainable' Food 63

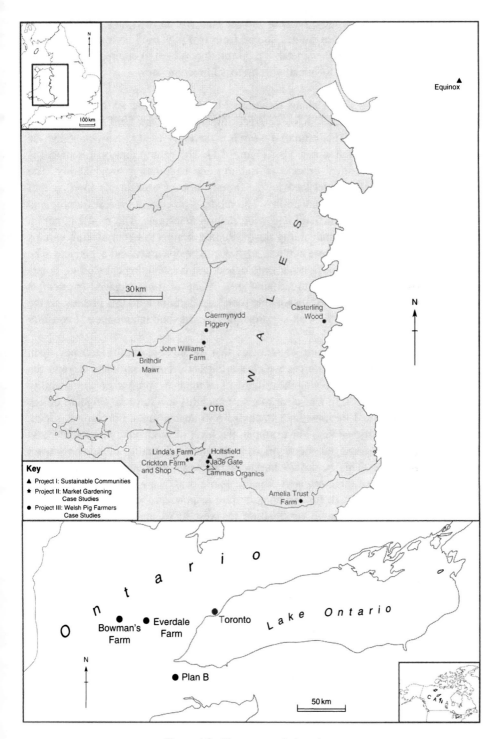

Figure 4.2: The case study locations.

relationships between these farmers and the land and animals they work with (Maxey 2005b). The case studies ranged from two pigs used on a seven-acre farm as part of an organic vegetable crop rotation system to commercial concerns focused on pig rearing for meat with up to 13 breeding sows.

Following an abductive approach to research (Mason 2002), the sixfold model and the notion of 'sustainable food' have emerged directly from the field data (Pain and Francis 2003; Pain 2004). This primary research is therefore used to help elaborate the sixfold model explained below (see also Figure 4.1). The central aim of this model is to encourage an open, critical process of framing and negotiating sustainability. The framework offered by the model leads to the introduction of 'sustainable food' as part of a wider engagement with sustainability. The model's process of critical framing and negotiation involves asking difficult and wide ranging questions. The model is not an attempt to definitively pin sustainability down. Rather, it aims to begin making various components of sustainability more explicit, so they can be discussed and negotiated. The model is a tool to help us question our ontological and epistemological starting points and to see how/where these fit into sustainability. Many of those engaged in attempts to produce, supply and consume 'sustainable food' are familiar with this process, as the case-study material attests. Whilst all six principles overlap and interconnect, I consider each briefly in turn.

The principle of *physical limits* emphasises that all human activity depends upon a complex range of natural resources which are currently being exploited beyond the Earth's carrying capacity. Following the critique of attempts to appropriate sustainability (Harvey 1996; Redclift 2002), it is clear that sustainability can have no singular, objective 'limit'. The physical limits we acknowledge will depend upon our various social, cultural and other positionalities. For example, the level of biodiversity and wilderness we wish to see will shape our views on what volume and quality of human impact we deem appropriate. This begins to highlight the ways food 'choices' shape food sustainability. Across all three research projects, participants were concerned to encourage biodiversity. In Project II, for example, despite respondents' concern regarding the cost and bureaucracy of organic certification (Maxey 2006), most supported organic certification bodies' stringent measures to encourage on-farm biodiversity.

The process of producing food rendered the implications of food-related choices stark within all case studies. Drawing on participants' experiences, I suggest that engaging critically with sustainability can highlight that the *types* of food and food provisioning we choose are crucial. This insight can be used to question assertions such as Bowler's (2002: 210) statement that productivist agriculture is inevitably needed to "produce the volumes of food necessary to support the urban-industrial population". Whilst globally meat consumption is escalating, it takes, on average, four times less resources to supply a vegetable-based diet than a meat-based diet, for example (Roberts *et al.* 1999; Pretty *et al.* 2001). Indeed, Cowell and Parkinson (2003) suggest even densely populated areas such as the UK could be self-sufficient in food, and this would be facilitated by a shift towards more vegetable-based diets. This reflects the findings of a Dutch transdisciplinary research programme, aiming to develop more sustainable food systems (see Jongen and Meerdink 2001).

Arguments for a more vegetable-based diet have been well rehearsed within the literature on global environmental and animal health for the last 30 years. Indeed, those approaching sustainability in search of a fixed and clearly knowable agenda and end point often present vegan and vegetarian diets as more sustainable (Strachan 2004). However, the case studies highlighted the more complex and textured nature of sustainable food choices. This female resident of Brithdir Mawr, for example, suggested:

> ... veganism works if you're living in a city, you know, buying your food from supermarkets and the like. But here we really want to feed ourselves from the land and so we have to work with what the land wants to do (BDM 2: 128).

In this example, 'working with the land' won official recognition for their role in re-establishing traditional native pasture. At Casterling Wood, it involves sophisticated and sensitive management in which 20 acres of woodland is integrated with the raising of traditional and rare-breed wieners for slaughter. Through in-depth knowledge and careful monitoring, biodiversity has increased within the woodland.[5]

The principle of *futurity* within the sixfold model means that our levels and type of activity should be maintainable into the future. Despite widespread recognition of this futurity component (Gibbs and Krueger 2005), most literature is vague as to how *far* into the future sustainability should extend. Within the case studies, each resident tended to develop a shifting series of temporal scales to negotiate concerns over the future. These scales were grounded in respondents' every day practices and the choices, values and ideas which sprang from them. For a female organic vegetable grower on Gower, for example, the future would at various times mean the impact her work would have the next day, for the next weeks' delivery, the next season's growing and the people who took on the farm after her. As a mother she often considered the impacts her actions would have upon her daughter as she grew up, but like most respondents, her concern for the future did not start or end with her daughter but covered a range of socio-cultural issues across wide ranging temporal and spatial scales.

Equity is clearly linked to futurity and the principle that our actions now should not adversely affect future generations. However, in addition to inter-generational equity, intra-generational equity is a key component of sustainability, as the Brundtland Report (WCED 1987: 43) noted: "Even the narrow notion of physical sustainability implies a concern for social equity between generations, a concern which must logically be extended to equity within each generation". Whilst intra-generational equity has been underplayed in much of the literature (Dovers and Handmer 1993), an increase in unequal wealth distribution has been one of the defining characteristics of recent economic globalisation and neoliberalism (Beck 2000; Redclift 2002). Food is a site that renders rising global inequality stark. The Food and Agriculture Organisation (2002), for example, estimates that 798 million people suffer from chronic hunger as their daily intake of calories is insufficient for them to lead active and healthy lives. The UN estimate that $16 billion could eradicate all hunger and fund basic healthcare programmes, compared

---

[5]Originally, the participants hoped to establish a coppicing business on the site. These plans evolved into the pork business upon learning that planning permission for a family dwelling required them to demonstrate a 'functional need' to be on the land. Under current UK planning law, raising pigs provided this functional need, but coppicing would not have.

with the $17 billion the United States and European Union spend on pet food each year and the $40 billion spent on food advertising each year (Millstone and Lang 2003). Residents in each case study throughout all three research projects felt strongly that access to healthy food is a basic human need which everyone should enjoy. Whilst several of the pork and vegetable producers were aware that some of their products may have been regarded as luxury or 'quality' food, none were comfortable with this and all sought to produce healthy, affordable food, as this comment from a female smallholder illustrates:

> I don't think food should be considered a luxury just because it hasn't got chemicals and antibiotics in it! I want everyone to be able to buy my pork; we all deserve food that's delicious and nutritious! (LF 2: 319).

This commitment to accessible, affordable 'nutritious and delicious' food throughout the case studies contributed to the notion of 'sustainable food' and critique of 'quality' food outlined below.

Participation is crucial to sustainability both ethically and practically. The changes which sustainability implies include potentially disruptive processes, particularly in the Minority World, from reducing consumption levels to superficially mundane changes in daily routines such as reusing, repairing and recycling, from redesigning our built environment to reshaping our careers. In order, both practically, to draw upon our existing knowledge and experiences, and ethically to be able to promote and expect such changes, sustainability requires a high degree of participation at every scale and stage (UNCED 1993).

There are strong links, as yet largely unexplored, between food democracy (Shiva 2000; Hassanein 2003), food citizenship (Wilkins 2005; Seyfang 2006) and sustainability. Drawing on sustainability's central principle of participation, I suggest 'sustainable food' as with 'food democracy' "... rests on the belief that every citizen has a contribution to make to the solution of our common problems" (Hassanein 2003: 85). This may go even further than DuPuis and Goodman's (2005: 361) appeal for "a broadly representative group to explore and discuss ways of changing their society". A more participatory approach may involve one of a plethora of forms of direct democracy (Madron and Jopling 2003; Monbiot 2004). Detailed critical engagement with participation is largely missing from the 'new geographies of food'. Participation will be shaped by individual and shared resources, including power, time and access to information. Several commentators have shown, for example, how powerful actors such as supermarkets and food transnationals are able to exert asymmetrical influence on current food policies (Monbiot 2000; Shiva 2000; Cahill 2001; ETC Group 2006). Such analysis could feed into a wider and more sustained analysis of participation within new food geographies.

The three research projects clearly demonstrate the value of active participation in the process of sustainable food production, distribution and consumption. Each case study was laden with examples of the wisdom, ingenuity and creativity people bring to bear on this process. At Equinox Housing Co-operative in Manchester, for example, residents with little prior experience of food growing took a small urban concrete back garden and turned it into a highly productive green oasis. At Plan B, Ontario residents with little initial knowledge established a successful and innovative community-supported agriculture (CSA) within three years (Plan B 2005). From a similar starting point, Jade Gate Organic Farm, South Wales, developed a unique system of cultivation and delivery using a single working Cobb.[6]

Residents' direct and meaningful participation was integral to the high levels of innovation and achievement found throughout the case studies, as this comment from a woman working at Casterling Wood suggests:

> I don't think any amount of training with animals could really prepare you for what it's like to actually have them on your land and be responsible for them, because when they are yours and your livelihood, everything that happens to them matters. You really don't know what its like until you do it (CWP 2: 352).

Active participation with sustainable food often opened up wider engagements with sustainability. Interests, questions and concerns across all case studies tended to progressively broaden over time through this process. An initial concern with self-sufficiency for several residents at Brithdir Mawr, for example, led eventually to non-violent direct action to prevent local trials of genetically modified food. For several other residents, including those at Plan B, Organics To Go and Jade Gate, an initial concern with neoliberal globalisation and the politico-ethics of food led to participants establishing their own food-growing enterprises. This process was not limited to the 'middle class intellectuals' within the case studies, but included case studies run by third-generation farmers: Bowman's Farm, Crickton Farm Shop and John Williams' Farm. In each case, the process of engaging with sustainable food opened up a plethora of insights and challenges which were not discretely bounded around their food production but spiralled off into many other aspects of their lives and their relationships. Actively participating in the process of sustainable food, therefore, holds considerable potential to become part of a more deeply transformatory process and to inform a broader engagement with sustainability.

As the above discussion begins to suggest, the component of *relationships* connects all the others within the sixfold model as each involves a shift in or acknowledgement of our relationships, to ourselves, with other humans, other species and the wider world. Furthermore, these relationships are dynamic, and thus we are encouraged, through sustainability, to consider how these may change in the future. Explicitly addressing relationships helps us to focus on the qualitative as well as the quantitative within sustainability generally and sustainable food specifically. Approaching sustainability is a radical, challenging task for most people, particularly in the Minority World. The principle of acknowledging and working on relationships within a sustainability framework also means disrupting boundaries such as those between 'humans' and 'nature', between ourselves now, as we have been socialised within a largely unsustainable society, and ourselves as we may become.

The importance of relationships within sustainability came out very strongly in the case studies, where engaging with sustainable food radically shifted various forms of relationships. Project III's work with pig farmers, for example, originally sought to

---

[6]Thorn, the animal in question, was technically a large pony, although to the lay person he would generally be regarded as a powerful horse.

explore the different ways pigs are raised within a rural Western nation (Wales). The aim was to draw out lessons that can be applied to sustainable food systems more generally. What emerged strongly from the research, though, was the plethora of ways engaging with pigs as part of the 'sustainable food' process led to the radical and ongoing (re)negotiation of relationships. Pigs performed a range of functions across the case studies, including providing income, food, manure, woodland management and helping one set of respondents to acquire planning permission for their family home (see Footnote 5). However, in every case study these relationships extended beyond a purely functional one. One respondent, who had run his own building firm for over 20 years before establishing Caermynydd Piggery, for example, described his bottom line: "I would pack it in if I lost my sense of playing with the pigs, for me its all about enjoying the animals" (CP 1: 631). The relationships participants developed with each other, the land, animals and wider cosmos were challenged and transformed by the process of engaging with sustainable food in all 15 case studies. Several participants made comments similar to this one from a Holtsfield resident:

> Our culture increasingly wants to be immortal, we forget that our own bodies are just bits of clay, instead they become things we get attached to (HF 8: 622).

This destabilising of boundaries, identities and relationships resonates with recent work on hybrid and transgressive food geographies, such as that pioneered by Whatmore (2002). It also helps to highlight the rapidly changing relationships within food provisioning more generally. Whilst this is often presented as the growing formality of productivist provisioning versus the intimacy and 'human connection' of AFNs (Hinrichs 2003: 295), the approach to sustainability explored here suggests our relationships with food and food provisioning are heterogeneous and dynamic. Rather than slipping into reductionist binaries, a process-based sustainability analysis of food emphasises the various relationships at play and our ability to shape and be shaped by them.

Finally, the principle of sustainability as a *process* explicitly acknowledges that sustainability is not a fixed, ideal end point. It is an active, dynamic process in which we are all engaged (Warburton 2000; Gibbs and Krueger 2005). In the literature on food, as elsewhere, there is a tendency to view sustainable development as a process and sustainability as a particular state or condition, an (idealised) end point (Krueger and Agyeman 2005). Robinson (2002: 188), for example, suggests "sustainable development is a process not a single attainable outcome". Given the problems associated with the notion of 'development' (e.g. see Escobar 1995; Sachs 1999; Simon 2006), the relational approach to sustainability proposed here emphasises that it, too, is a process, as Hassanein (2003: 85) suggests:

> ...achieving sustainability involves conflicts over values and there is no independent authority, such as science or religion, to which we can appeal for resolution of these conflicts. Therefore sustainability must be defined socially and politically, and our collective understanding of it will evolve over time as conditions change.

This approach resonates with Krueger and Agyeman's (2005) analysis of 'actually existing sustainabilities'. Such approaches reduce the tendency to prescribe sustainability and for particular groups to appropriate and foreclose it, though the case studies demonstrated that even small, grass-roots food projects may attempt this. Approaching sustainability as a process, then, emphasises not which particular point has been reached, but that everyone, individually and collectively, be allowed, empowered and encouraged to actively and critically engage in the process of moving towards sustainability. It is through this focus on process and the other principles of sustainability outlined above that we may consider the value of shifting the focus of analysis from 'alternative food' to 'sustainable food'. The notion of 'sustainable food' is proposed not because it offers an objective judgement on the state of our food, but because it helps to focus attention on this creative, participatory and partial process, part of a broader sustainability transition.

Although this relational approach to sustainability may appear weak and uncertain, it is potentially a most liberatory force. Instead of re-enforcing the adversarial, binary patterns which currently dominate Western, if not global, food discourses, this approach has the potential to embrace conflict and difference as "grounds for respectful - and even productive - disagreement" (DuPuis and Goodman 2005: 361), an approach food commentators are increasingly calling for (see Hassanein 2003). Furthermore, this approach may contribute to what Redclift (2002: 190) terms "new sustainability discourses".

Conclusion: Towards 'Sustainable Food'?

A critical relational approach to sustainability along the lines sketched out above may be capable of informing food geographies in a number of ways. Recently, for example, there have been a growing number of calls for research which does not focus solely on producers but explicitly considers consumer–producer relations (Goodman and DuPuis 2002; Goodman 2003; Holloway *et al.* 2005). Sustainability explicitly encourages broad, multidimensional approaches, exploring various aspects of food networks from a range of perspectives. Furthermore, a *relational* approach to sustainability is particularly well suited to begin interrogating precisely the kind of food relations such authors call for. Whereas research on food geographies, if it considers sustainability at all, tends to look at 'sustainable agriculture', I suggest a focus on 'sustainable food' can draw upon insights from recent relational approaches to food and related geographies (Whatmore 2002; Lee and Leyshon 2003; Massey 2004). In addition to facilitating analysis of consumer–producer relations, for example, focusing on 'sustainable food' rather than 'sustainable agriculture' helps to avoid the privileged status of productivism and Minority World perspectives.

As this chapter has suggested with the help of empirical research from a range of case studies, there are several advantages to approaching food from the analytical frame of 'sustainable food' rather than 'alternative food'. These arguments tend to build upon each other. First, practitioners themselves are increasingly happy to engage with the notions and associated practices of sustainability and sustainable food, whereas this is not the case with 'alternative food'. As the case studies unanimously reported and as other studies are beginning to show (e.g. Seyfang 2006), many producers are uncomfortable with the inherently marginal position 'alternative food' often entails. In contrast, sustainability increasingly occupies centre stage within popular, media and governmental discourses. Secondly, and linked to the first, sustainability and the notion of sustainable food advocated here offers the opportunity for practitioners to simultaneously appeal to mainstream food provisioning and subvert it. For example, the ideas and practices associated with sustainable food offer a radical critique of 'conventional food', highlighting how unconventional it is in historical perspective and how unsustainable. Conversely, 'alternative food' implicitly endorses 'conventional food' and the multifarious practices it entails. The very notion of 'alternative food' implicitly normalises and validates 'conventional food'.

Thirdly, and again building on the last point, 'sustainable food' offers greater scope for disrupting dualistic framing of food provisioning. 'Alternative food' is built upon a binary relationship with 'conventional' or 'mainstream' food. As is so often the case with binaries, this operates to privilege one part of the pair at the expense of the other and to encourage reductionist thinking and practice. Whilst 'sustainable food' is certainly capable of endorsing its own implicit binary (with 'unsustainable food'), the worst excesses of a dualistic framing can be avoided if a more open, relational and process-based approach to sustainability and sustainable food is adopted. Finally, approaching 'sustainable food' in this relational way allows its critical perspective to be applied to many areas of the new geographies of food. One particular area which the notion of 'sustainable food' could inform is 'organic' food and debates around the 'conventionalisation' thesis (Guthman 2004; Lockie and Halpin 2005). Given the dramatic shifts in the organic sector and the heated nature of current debates (Smith and Marsden 2004; Guardian Leader 2006; Melchett 2006), such analysis is urgently needed. The framework of sustainability and sustainable food outlined here is inherently partial and formative. In many respects this chapter can only begin to flag up areas where future attention might be focused. This may include notions and practices such as those surrounding 'alternative', 'local', 'quality' and 'community' food, climate change and food democracy (Carlsson-Kanyama 1998; Jones 2001; Hassanein 2003). A brief look at 'quality food' illustrates how a sustainable food perspective can inform such work.

The considerable attention paid by academics to 'quality' food has recently given rise to more critical, nuanced understandings. Whilst he does not make this connection, Winter's (2003) critique, for example, begins pointing towards the value of sustainability as an analytical framework within the new geographies of food. Whereas the 'quality turn' may not be as radical as many claim, sustainability may offer a more rigorous and radical framework through which to approach food. As with the binary of 'alternative' versus 'conventional' food, the language of 'quality' has questionable implications, normalising 'non-quality' food and marginalising, or rendering exceptional, food which people actually desire and enjoy. The notion of 'sustainable food', with its central commitment to equity, presents more progressive potential than 'quality'. This was supported by the case studies where participants were uncomfortable with the idea of supplying luxury food and emphasising that 'nutritious and delicious' food is a right for all. Whereas the 'turn' to quality has been described as a "contested process of transition" (Goodman 2003: 1; Miele and Murdoch 2002), attention could perhaps be more productively turned to the sustainability transition (see also O'Riordan 2004).

Viewing sustainability as a relational process may cause alarm for those who seek certainty. 'Sustainable food' as a process cannot be neatly defined, prescribed or quantified beyond suggesting that it requires a popular and ongoing negotiation of debate and action as part of a broader approach to sustainability. This process can be guided by models such as the sixfold model and by ongoing reflection on our shifting circumstances. As the case studies highlighted, engaging in this process is itself often transformatory. The research reported here emphasises grass-roots, small-scale producer–supplier–consumer engagement with sustainable food. At this level, the research suggests, the value of the relational approach to sustainability is clear as it draws upon and encourages individual and collective ingenuity, creativity and cooperation. Whilst participation remains an under-explored component of sustainability, sustainability is not restricted to operating at the grass-roots level. Governments, corporations and other actors at all scales are increasingly engaged with and committed to sustainability. If these actors, as well as practitioners on the ground, begin to engage with sustainability as a process, then its full transformatory potential may be released.

Acknowledgements

The research on which this paper is based was made possible by the commitment and cooperation of case studies' participants. It has benefited considerably from feedback provided at two ESRC seminars in the 'Approaches to Sustainable Farmland Management' series, at which research from Projects II and III was initially reported. The paper has also benefited from feedback following the presentation of an earlier version of this paper at the Royal Geographical Society's annual conference in London, September 2005. Comments from the anonymous review process, as well as from Damian Maye and Lewis Holloway, have also been highly valuable. Anna Ratcliffe, Department of Geography, School of the Environment and Society, Swansea University helped reproduce the figures.

References

Allen, P., Fitzsimmons, M., Goodman, M. and Warner, K. (2003). Shifting plates in the agrifood landscape: the tectonics of alternative agrifood initiatives in California. *Journal of Rural Studies* 19, 61–75.
Barrett, H., Ilbery, B., Browne, A. and Binns, T. (1999). Globalization and the changing networks of food supply: the importation of fresh horticultural produce from Kenya into the UK. *Transactions of the Institute of British Geographers* 24, 159–174.
Beck, U. (2000). *World Risk Society*. Malden, MA: Blackwell.
Beckerman, W. (1994). 'Sustainable development': is it a useful concept? *Environmental Values* 3, 191–209.
Beckerman, W. (1995). How would you like your 'sustainability', Sir? Weak or strong? A reply to my critics. *Environmental Values* 4, 169–179.
Bell, D. and Valentine, G. (1997). *Consuming Geographies: We are Where We Eat*. London: Routledge.
Bhatti, M., Brooke, J. and Gibson, M. (Eds) (1994). Housing and the new environmental agenda: an introduction. *Housing and the Environment: A New Agenda* (pp. 1–24). Coventry: Chartered Institute of Architecture.
Bowler, I. (2002). Developing sustainable agriculture. *Geography* 87, 205–212.
Brown, L. (Ed) (1993). *The New Shorter Oxford Engilsh Dictionary of Historical Principles*, Oxford: Oxford University Press.
Buller, H. and Morris, C. (2004). Growing the goods: the market, the state, and sustainable food production. *Environment and Planning A* 36, 1065–1084.
Cahill, K. (2001). *Who Owns Britain?* London: Canongate.

Carlsson-Kanyama, A. (1998). Climate change and dietary choices – how can emissions of greenhouse gases from food consumption be reduced? *Food Policy* 23, 277–293.

Cook, I. *et al.* (2004). Follow the thing: Papaya. *Antipode* 36, 642–664.

Cowell, S. and Parkinson, S. (2003). Localisation of UK food production: an analysis using land area and energy as indicators. *Agriculture, Ecosystems and Environment* 94, 221–236.

Daly, H. (1995). On Wilfred Beckerman's critique of sustainable development. *Environmental Values* 4, 49–55.

Daly, H. (1996). *Beyond Growth: The Economics of Sustainable Development*. Boston: Beacon Press.

Daly, H. and Cobb, J. (1989). *For the Common Good: Redirecting the Economy Towards Community, Environment and a Sustainable Future*. London: Green Print.

DEFRA (2005). *Securing the Future – UK Government Sustainable Development Strategy*. London: HMSO. Available online at: http://www.sustainable-development.gov.uk/publications/uk-strategy/uk-strategy-2005.htm (accessed: 11/12/05).

Dobson, A. (1998). *Justice and the Environment* Oxford: Oxford University Press.

Dovers, S. and Handmer, J. (1993). Contradictions in sustainability. *Environmental Conservation* 20, 217–222.

DuPuis, E. and Goodman, D. (2005). Shall we go 'home' to eat?: towards a reflexive politics of localism. *Journal of Rural Studies* 21, 359–371.

Engel, J.R. and Engel, J.G. (Eds) (1990). *Ethics of Environment and Development: Global Challenge, International Response*. Tucson: The University of Arizona Press.

Escobar, A. (1995). *Encountering Development: The Making and Unmaking of the Third World*. Princeton, NJ: Princeton University Press.

ETC Group (2006). *Monsanto Acquires Delta & Pine Land and Terminator*. Ottawa: ETC Group. Available online at: http://www.etcgroup.org (accessed 21/9/06).

Everdale (2005). *Everdale Organic Farm and Environmental Learning Centre* Available online at: http://www.everdale.org/(accessed 11/12/05).

Food and Agriculture Organisation (2002). *Reducing Poverty and Hunger: The Critical Role of Financing for Food, Agricultural and Rural Development*. Paper prepared for The International Conference on Financing for Development Monterrey, Mexico, 18–22 March 2002. Rome: FAO. Available online at: ftp://ftp.fao.org/docrep/fao/003/y6265E/Y6265E.pdf (accessed: 10/12/05).

Gibbs, D. and Krueger, R. (2005) Exploring local capacities for sustainable development. *Geoforum* 36 (4), 407–409.

Goodman, D. (2003). Editorial: the quality 'turn' and alternative food practices: reflections and agenda. *Journal of Rural Studies* 19, 1–7.

Goodman, D. and DuPuis, E. (2002). Knowing food and growing food: beyond the consumption-production debate in the sociology of agriculture. *Sociologia Ruralis* 42, 5–22.

Guardian Leader (2006). Green gauges. *The Guardian*, 7/10/06. Available online at: http://www.guardian.co.uk/food/Story/0,1889715,00.html#article_continue (accessed: 7/10/06).

Guthman, J. (2004). The trouble with 'Organic Lite' in California: a rejoinder to the 'conventionalisation' debate. *Sociologia Ruralis* 44, 301–316.

Harvey, D. (1996). *Justice, Nature and the Geography of Difference*. Oxford: Blackwell.

Hassanein, N. (2003). Practicing food democracy: a pragmatic politics of transformation. *Journal of Rural Studies* 19, 77–86.

Hinrichs, C. (2003). The practice and politics of food system localization. *Journal of Rural Studies* 20, 33–45.

Holloway, L., Kneafsey, M., Venn, L., Cox, R., Dowler, E. and Tuomainen, H. (2005). Possible food economies: food production-consumption arrangements and the meaning of 'alternative'. Cultures of Consumption Working Paper Series No. 25. Available online at: http://www.consume.bbk.ac.uk/publications.html (accessed: 10/12/05).

Ilbery, B. and Maye, D. (2005). Food supply chains and sustainability: evidence from specialist food producers in the Scottish/English borders. *Land Use Policy* 22, 331–344.

Jones, A. (2001). *Eating Oil: Food Supply in a Changing Climate*. London: Sustain and Elm Research Centre.

Jones, A. (2003). 'Power in place': viticultural spatialities of globalization and community empowerment in the Languedoc. *Transactions of the Institute of British Geographers* 28, 367–382.

Jongen, W. and Meerdink, G. (2001) Pea proteins based food products as meat replacers: The Profetas concept. *Nahrung-Food* 45, 402–404.

Krueger, R. and Agyeman, J. (2005). Sustainability schizophrenia or "actually existing sustainabilities?" Toward a broader understanding of the politics and promise of local sustainability in the US. *Geoforum* 36 (4), 410–417.

Kurtz, M. (2004). Review: the condition of rural sustainability. *Journal of Rural Studies* 20, 257–258.

Lee, R. and Leyshon, A. (2003). Conclusions: re-making geographies and the construction of spaces of hope. In A. Leyshon, R. Lee and C. Williams (Eds) *Alternative Economic Spaces* (pp. 193–198). London: Sage.

Lockie, S. and Halpin, D. (2005). The 'conventionalisation' thesis reconsidered: structural and ideological transformation of Australian Organic agriculture. *Sociologia Ruralis* 45, 285–307.

Madron, R. and Jopling, J. (2003). *Gaian Democracies: Redefining Globalisation and People Power*. Totnes, Devon: Green Books.

Marsden, T. (2003). *The Condition of Rural Sustainability*. Assen, Netherlands: Royal Van Gorcum.

Marsden, T. and Smith, E. (2005). Ecological entrepreneurship: sustainable development in local communities through quality food production and local branding *Geoforum* 36 (4), 441–451.

Mason, J. (2002). *Qualitative Researching*. London: Sage.

Massey, D. (2004). *For Space*. London: Sage.

Maxey, L. (1999). Beyond boundaries? Activism, academia, reflexivity and research. *Area* 31, 199–208.

Maxey, L. (2003). *One path forward? Three sustainable communities in England and Wales*. Unpublished doctoral thesis, Swansea University.

Maxey, L. (2004a). Moving beyond from within: reflexive activism and critical geographies. In D. Fuller and R. Kitchin (Eds), *Radical Theory/Critical Praxis: Making a Difference Beyond the Academy?* (pp. 159–171). Canada: Praxis (e) Press, Critical Topographies Series, Okanagan University College, Vernon and the University of Victoria, Victoria, BC. Available online at: http://www.praxis-epress.org/rtcp/ljm.pdf. (accessed: 7/10/06)

Maxey, L. (2004b). The participation of younger people within intentional communities: evidence from two case studies. *Children's Geographies* 2, 29–48.

Maxey, L. (2005a). La construccíon del desarrollo sustentable en comunidades regionales: La experiencia de Holtsfield en Gales. In R. Bernal-Meza and S. Saha (Eds), *Economía Mundial y Desarrollo Regional* (pp. 349–380). Buenos Aires: Nuevohacer Grupo Latinoamericano.

Maxey, L. (2005b). Pigs: from functional unit to family friends. Paper presented at the ESRC Transdisciplinary Seminar Series: *Approaches to Sustainable Farmland Management Seminar 4: Ethical Production and Protection for Sustainable Farmland Management*. London: Royal Geographical Society (18/01/05).

Maxey, L. (2006). Can we sustain sustainable agriculture? A comparative study of small scale producers and suppliers in Canada and the UK. *The Geographical Journal* 172 (3), 230–244.

Maxey, L. and Pickerill, J. (2007). Lammas: land and liberty. *The Land* 3, 35–36.

Maxey, L, Pickerill, J. and Wimbush, P. (2006). New planning opportunities for low impact settlements. *Permaculture Magazine* 50, 32.

Melchett, P. (2006). We're not blurring our principles on organic food: We are campaigning to change our food culture so that cheap means nasty. *The Guardian*, 17/10/06. Available online at: http://environment.guardian.co.uk/food/story/0,,1924192,00.html (accessed: 17/10/06).

Miele, M. and Murdoch, J. (2002). The practical aesthetics of traditonal cuisines: slow food in Tuscany. *Sociologia Ruralis* 42(4): 312–23.

Millstone, E. and Lang, T. (2003). *The Atlas of Food: Who Eats What Where And Why*. London: Earthscan.

Monbiot, G. (2000). *Captive State: The Corporate Takeover of Britain*. London: Macmillan.

Monbiot, G. (2004). *Age of Consent: A Manifesto for New World Order*. New York: New Press.

Oldfield, J. and Shaw, D. (2002). Revisiting sustainable development: Russian cultural and scientific traditions and the concept of sustainable development. *Area* 34, 391–400.

O'Riordan, T. (2004). Environmental science, sustainability and politics. *Transactions of the Institute of British Geographers* 29, 234–247.

OTG (2006). Welcome to Organics To Go. Available online at: http://www.organicstogo.info/ (accessed 5/6/06).

Pain, R. (2004). Social geography: participatory research. *Progress in Human Geography* 28 (5), 652–663.

Pain, R. and Francis, P. (2003). Reflections on participatory research. *Area* 35 (1), 46–54.

Pearce, D., Markanyda, A. and Barbier, E. (1989). *Blueprint for a Green Economy*. London: Earthscan.

Plan B (2005). What is a CSA? Available online at: http://www.angelfire.com/ca/planBcsa/page2.html (accessed 21/12/05).

Potter, C. and Tilzey, M. (2005). Agricultural policy discourses in the European post-Fordist transition: neoliberalism, neomercantilism and multi-functionality. *Progress in Human Geography* 29, 581–600.

Pretty, J., Brett, C., Gee, D., Hine, R.C., Mason, C., Morsion, J., Raymen, M., van der Bijl, G. and Dobbs, T.L. (2001). Policy challenges and priorities for internalizing the externalities of modern agriculture. *Journal of Environmental Planning and Management* 44, 263–283.

Redclift, M. (2002). Pathways to sustainability. *Geography* 87, 189–190.

Renting, H., Marsden, T. and Banks, G. (2003). Understanding alternative food networks: exploring the role of short food supply chains in rural development. *Environment and Planning A* 35, 393–411.

Roberts, W., MacRae, R. and Stahlbrand, L. (1999). *Real Food for a Change: How the Simple Act of Eating Can: Boost Your Health and Energy, Knock Out Stress, Revive Your Community, and Clean Up Your Planet*. Vancouver: Random House.

Robinson, G. (2002). Sustainable development – from Rio to Johannesburg. *Geography* 87, 185–188.

Robinson, G. (2004). *Geographies Of Agriculture: Globalisation, Restructuring and Sustainability*. Harlow: Pearson.

Sachs, W. (1999). *Planet Dialectics: Explorations in Environment and Development*. London: Zed Books.

Serafy, S. (1996). In defence of weak sustainability: a response to Beckerman. *Environmental Values* 5, 75–81.

Seyfang, G. (2006). Ecological citizenship and sustainable consumption: examining local organic food networks. *Journal of Rural Studies* 22 (4), 383–395.

Shiva, V. (2000). *Stolen Harvest: The Hijacking of the Global Food Supply*. Cambridge, MA: South End Press.

Simon, D. (2006). Separated by common ground? Bringing (post)development and (post)colonialism together. *The Geographical Journal* 172 (1), 10–21.

Smith, E. and Marsden, T. (2004). Exploring the 'limits to growth' in UK organics: beyond the statistical image. *Journal of Rural Studies* 20, 345–357.

Strachan, G. (2004). *Sustainable Lifestyle Quiz*. Unpublished document available from the author. Available online at: http://www.esd-wales.org.uk (accessed: 7/10/06)
UNCED (1993). *Agenda 21: Programme of Action for Sustainable Development*. New York: United Nations.
Venn, L., Kneafsey, M., Holloway, L., Cox, R., Dowler, E. and Tuomainen H. (2006). Researching European 'alternative' food networks: some methodological considerations. *Area* 38 (3), 248–258.
Viles, H. (2002). Editorial. *Area* 34 (4), 339.
Warburton, D. (2000). A passionate dialogue: Community and sustainable development. In D. Warburton (Ed), *Community and Sustainable Development: Participation in the Future* (pp. 1–39). London: Earthscan.
Watts, D., Ilbery, B. and Maye, D. (2005). Making reconnections in agro-food geography: alternative systems of food provision. *Progress in Human Geography* 29, 22–40.
WCED (1987). *Our Common Future*. Oxford: Oxford University Press.
Whatmore, S. (2002). *Hybrid Geographies: Natures, Cultures And Spaces*. London: Sage.
Wilkins, J. (2005). Eating right here: moving from consumer to food citizen. *Agriculture and Human Values* 22, 269–273.
Winter, M. (2003). Embeddedness, the new food economy and defensive localism. *Journal of Rural Studies* 19, 23–32.
Winter, M. (2004). Geographies of food: agro-food geographies – farming, food and politics. *Progress in Human Geography* 28, 664–670.
Yakovleva, N. and Flynn, A. (2004). Innovation and sustainability in the food system: A case of chicken production and consumption in the UK. *Journal of Environmental Policy and Planning* 6, 227–250.

Chapter 5

Beyond the 'Alternative'–'Conventional' Divide? Thinking Differently About Food Production–Consumption Relationships

Lewis Holloway[1], Moya Kneafsey[2], Rosie Cox[3], Laura Venn[4], Elizabeth Dowler[5] and Helena Tuomainen[5]
[1]Department of Geography, University of Hull, UK
[2]Department of Geography, Environment and Disaster Management, Coventry University, UK
[3]The School of Continuing Education, Birkbeck, University of London, London, UK
[4]West Midlands Regional Observatory, Birmingham, UK
[5]Department of Sociology, University of Warwick, Coventry, UK

Introduction

Considerable research effort is being directed at examining food systems which are regarded in one way or another as 'alternative' to 'conventional' ways of food provisioning. This chapter stems from research which explored such 'alternatives' from the perspectives of producers and the consumers who obtain some or all of their food from them. It examined six very different, relatively small-scale food projects which aim to establish closer connections (in both physical and social senses) between consumers and the people, places and processes associated with the food they eat. The projects include quite mundane examples, such as a farm shop and box scheme, as well as more unusual examples, including community-supported agriculture, an urban market garden and an Internet-based sheep-adoption scheme which enables consumers from almost anywhere in the world to receive sheep's milk cheese from Italy by airmail.

Our work on these food projects has increasingly made the term 'alternative', as used in relation to particular sorts of food network, seem conceptually problematic. The idea of separate, if intertwined, 'conventional' and 'alternative' food networks is developed by several authors (e.g. Ilbery and Maye 2005; Watts *et al.* 2005; Morgan *et al.* 2006) in ways which draw attention to their different economic, geographical and sociological structures. Morgan *et al.* (2006), for instance, discuss this in terms of a comparison between 'deterritorialised' conventional food networks and 'reterritorialised' alternative food networks. Such distinctions are useful and informative, but, as their authors acknowledge, included within either of the 'conventional' or 'alternative' categories are many different examples of food production–consumption, relating to different foods, cuisines (Murdoch and Miele 2004) and social, economic and geographical circumstances. Methodologically, then, we have encountered in our research a need to formulate a strategy for working through very different examples of food production–consumption, in order to move towards a fuller understanding of how they are assembled and how they function in their specific local contexts.

In this chapter, then, we suggest that we need to move beyond a prevalent 'alternative'–'conventional' dualism in order to describe the constitution of food production–consumption in more complex and relational ways. Rather than categorising heterogeneous modes of food provisioning as 'alternative', we explore how particular food projects can be understood as arranged across a series of interrelated 'analytical fields' in ways which make their operation possible and which, following the calls by Lockie and Kitto (2000), Goodman (2002) and Goodman and DuPuis (2002), explicitly relate food production and consumption in ways which have received relatively little attention (although see, e.g., Kirwan 2004). Yet, although arrangements are specific to individual projects, we can nevertheless examine them in relation to a common notion of a struggle to take control and develop visions of possible food systems which, for those involved, seem to improve on existing arrangements. We recognise that there is a wide range of actors (e.g. in retail, storage, processing, transport, policy and regulation) involved in food production–consumption networks, whether regarded as 'alternative' or 'conventional'. However, the nature of the particular food projects which form the basis of our research means that we attend here to the perspectives of consumers and producers engaged in contemporary, heterogeneous food production–consumption systems (see Halkier 2004), and to the relational nature of production and consumption (e.g. Goodman and DuPuis 2002; Guthman 2002, 2003).

We begin by briefly reviewing how the term 'alternative' has been used in relation to food production–consumption, before discussing the structure and value of our analytical framework. We then apply this framework to three of the food projects which we examined in our research. We end by emphasising the significance of our approach for studying food production–consumption arrangements, and by acknowledging the further questions it raises.

Thinking Beyond the 'Alternative'–'Conventional' Divide

Goodman (2003) suggests that writing on 'alternative food networks' can conveniently be understood as falling into either 'European' or 'North American' schools. While there is not space here for a full review of these literatures, we outline their main points and suggest that the use of the term 'alternative' is problematic in both cases. In the 'European' tradition, 'alternative' food networks tend to have been discussed in terms of their potential to contribute to the survival of small rural businesses (particularly farms), and more widely to processes of rural development through processes of 'adding-value' in various ways to farm outputs (e.g. Marsden *et al.* 2000, 2001, 2002; Marsden and Smith 2005). Such businesses are envisaged as being able to survive around the margins of a food supply system heavily biased towards large-scale food processing and retailing corporations, responding to consumer anxieties about the effects of 'industrial' food supply chains and the possible risks to themselves (Goodman 2003, 2004; Stassart and Whatmore 2003; Whatmore *et al.* 2003). Entrepreneurially minded individuals are seen as able to carve out niches for speciality food enterprises in a demanding business environment by, for example, inscribing their products with ecological credentials (Marsden and Smith 2005). In this context, debates have centred around a series of characteristics which have become associated with food produced by such small businesses. These include ideas of 'quality' (e.g. Gilg and Battershill 1998), 'localness' (e.g. Norberg-Hodge *et al.* 2002), 'embeddedness' (e.g. Murdoch *et al.* 2000; Sage 2003) and 'reconnection' (including ideas of Short Food Supply Chains) (e.g. Renting *et al.* 2003; Ilbery and Maye 2005), and much effort has been expended on trying to define these terms and explore how they are played out in particular instances (e.g. Renting *et al.* 2003; Winter 2003; DuPuis and Goodman 2005; Ilbery and Maye 2005; Watts *et al.* 2005). While recognising the usefulness of these debates, we find that the term 'alternative' which is tangled up with them does not become any clearer in association with this parallel terminology. Indeed, it becomes rather more opaque, as it seems to have to represent a collection of other terms and senses, and is used in a polarised manner as part of a conventional–alternative dualism.

In the 'North American' literature, according to Goodman (2003), 'alternative' has tended to be used in a rather more politicised discourse of oppositional activism (Allen *et al.* 2003). As with the 'European' literature, 'alternative' is associated with a range of descriptors and is presented as one half of a dualism (e.g. Kloppenburg *et al.* 1996; Grey 2000). But in addition, North American authors have in many cases defined a political agenda related to food production–consumption, which emphasises environmental and social justice, and 'alternative' economic relationships, rather than focusing simply on rural development (e.g. Hendrickson and Heffernan 2002; Allen *et al.* 2003; Hassanein 2003). Again, we find these ideas stimulating and valuable. However, alongside the earlier issues of opacity in the use of 'alternative', in this literature we are wary of the risks of romanticising the radicalised 'alternatives' in such a way that they are not subject to the same degree of critical reflection which is currently being applied to 'mainstream' food supply systems (see Massey 2000, in relation to romanticisation of the alternative more widely).

The problems which we have identified in the conceptualisation of 'alternative' in these literatures concerning food networks, in particular, suggest that there should be other ways of thinking about food networks which retain a sense of the diversity and particularity of different food networks, but which also allow us to say something useful about them in terms of relations of power and struggles over how food production and consumption should be arranged in a society. Here, our discussion about food networks begins to follow other arguments about how power, resistance and 'alternativeness' can be thought about more widely. Leyshon and Lee (2003) outline the commonly understood thesis of a hegemonic neoliberal global capitalism which is opposed in a whole variety of different ways. Oppositional discourses have led to "the efforts of individual and collective actors to imagine and, more importantly, to perform, economic activities in a way that marks them out differently from the dictates and conventions of the mainstream economy" (Leyshon and Lee 2003: 5). While this seems to persist with the representation of 'alternatives' opposing an established and dominant way of doing things, Leyshon and Lee draw on Gibson-Graham (1996) to point to a more complex understanding of these relationships, arguing that within what they present as a fragile and susceptible capitalism there are possibilities for a proliferation of economic spaces and practices which are centred less around capital accumulation and more around social, ecological and ethical concerns. Examples of "performing the economy otherwise" (Leyshon and Lee 2003: 16) include Local Exchange Trading Systems (LETS), in a situation where global capitalism is open to challenge and not necessarily as hegemonic as is sometimes suggested. What is important, though, is to avoid reverting to a binary opposition of 'alternative' and 'mainstream', and to recognise the relational contingency of what is regarded as 'alternative' at any one time and in any one place (Crewe et al. 2003; Lee and Leyshon 2003). Although discourses of 'alternativeness' might be powerful in stimulating challenges to what are felt to be, or experienced as, unjust economic relations, 'the alternative' itself is a slippery concept, resisting definition and shifting as soon as attempts are made to tie it down.

In this context, we introduce our interrelated analytical 'fields' as a heuristic device allowing us to describe and explore, in multidimensional terms, what makes the ongoing production of particular food networks possible.

A Heuristic Framework for Exploring Food Projects

Our identification of a series of analytical fields derives from an initial analysis of over 100 food projects, all with UK consumers and almost all actually based in the UK. These projects were identified via an extensive and systematic search of Internet and other sources. The fields emerged from qualitative assessment of how different projects defined themselves, how they were represented (e.g. in websites and leaflets) and how they actually worked in practice (see Venn et al. 2006 for more details).

Six food projects were selected as case studies for detailed research. They were not intended to act as representatives of different types of 'alternative' food network, but to be illustrative of the heterogeneity of modes of food production–consumption encountered during the initial analysis. The research included interviewing producers and consumers and focusing in more detail on consumers through workshops and household observation. After describing the fields, we discuss how three examples from our case studies can be arranged across them in different ways and suggest that it is these arrangements which make them possible and which make possible their engagement in entangled relations of power. While some of our case studies are explicitly involved in attempts to challenge relations of domination in food supply systems, others appear to exist in ways more similar to powerful actors in those systems, although their existence and practices might still be understood as oppositional to 'powerful' food networks. This diversity confirms our sense that an 'alternative'–'conventional' dualism setting our case studies against a monolithic dominant food system is inappropriate; instead, we see our projects as arranged across our heuristic fields, in a way that preserves their specificity and diversity. Our emphasis on arrangements across fields also allows us to see our case study schemes as spatially and temporally dynamic and involved in the production of particular spatialities.

Table 5.1: Analytical fields for describing food projects.

Heuristic 'analytical field'	Examples from sample food projects
Site of food production	Community garden, school grounds, urban 'brownfield' sites, farm, rented field, allotments...
Food production methods	Organic, biodynamic, consumer participation, horse ploughing...
Supply chain	Local selling/procurement, Internet marketing...
Arena of exchange	Farm shops, farmers markets, home delivery, mobile shops, PYO...
Producer–consumer interaction	Direct selling, email, newsletters, cooking demonstrations, food growing work (e.g. weeding parties), farm walks, share/subscription membership schemes...
Motivations for participation	Business success, making food accessible, social/environmental concerns, anxiety avoidance, sensory pleasure...
Constitution of individual and group identities	Customers, participants, stakeholders, supporters groups, children's groups, disability groups, women's groups...

Table 5.1 lists the fields which we have used for characterising a matrix of relationships across which the particularities of individual projects can be distributed. The table gives some examples of what might be included within each field in the cases of our particular database of food projects, emphasising the diversity in both the fields and the particular arrangements across fields that constitute specific food projects. Examples could, however, be drawn from *any* food network, including those where food is produced and consumed within globalised or industrialised systems. We emphasise too the importance of the material and symbolic presence of food itself across each of the fields, which implies that description of food networks should trace how food is implicated in the 'holding together' of particular sets of relationships and spatio-temporal arrangements (see Cook 2004). Each field is briefly discussed below before looking in more detail at three of our case studies.

- *Site of food production.* This refers to the place where food is grown and/or processed. It encompasses a range of contrasting sites, including some which have a degree of permanence and are established within traditions of food production in the UK (e.g. farms) and others which are more ephemeral and makeshift, occupying 'new' or temporary sites like ex-industrial 'brownfield' sites or small areas of rented farmland. Spatial scale and location is also important here. For our projects, their relatively small size, in terms of food production area, and their position in relation to, for example, particular groups of consumers, other producers or institutions which support their operations, is important to how they operate.
- *Food production methods.* Many projects emphasise the ways food is grown and prepared, in particular where these are thought to challenge the prevalence of industrial methods in agriculture. The emphasis placed on these methods, such as

'organic' production, may also be interpreted as demonstrating producers' assessments of consumers' motivations to consume food produced in these ways, for example, in assuming consumer anxiety in relation to food produced using pesticides. Food growing is thus seen as the result of producer–consumer relationships, and in some cases (especially the cooperative or community-supported projects) negotiations, rather than simply a product of growers' decision-making.

- *Supply chain.* We use this field to indicate a sense in which food literally moves between different arenas via different technologies and organisations of movement. Our case study projects use supply chain mechanisms ranging from 'low-tech' methods of local supply to the use of the Internet and air freight to supply an international consumer base. Again, producer–consumer relationships are key to the understanding of these food chains, as they are mediated by the particular mechanisms as they operate.

- *Arena of exchange.* This refers to the concrete and meaningful spaces in which food is exchanged. At one level, it refers to the site in which exchange occurs, such as a shop or market stall. At another level, it also refers to what the exchange is actually *of*, both materially and symbolically. Food is usually exchanged for money, but in other cases for work, communal activity, LETS, longer-term financial commitments (such as in share or subscription projects), and also, at least in part, a sense of 'regard' (Sage 2003) and other intersubjective aspects of producer–consumer relationship (see below).

- *Producer–consumer interaction.* The material and symbolic, formal and informal 'meeting points' of consumers and producers in food networks are emphasised in this field. Interaction might thus be face-to-face or involve communications at a distance through various technologies. The importance of this interaction is in the establishment of particular sorts of intersubjective and spatio-temporal relationship which influences the ways food projects emerge and change over time. These interactions are thus important processes of effecting change, contributing to the changing spatial relationships involved in particular examples of our food projects, and in the ways they are explicitly or implicitly engaged in challenges to dominant food systems.

- *Motivations for participation.* This describes the reasons people have for participating in particular food networks as consumers or producers and relates these reasons to particular forms of behaviour. Motivations clearly are subject to change, and there are likely to be shifts within and between producers' and consumers' motivations and processes of negotiation between differently motivated actors. Motivations and behaviours are thus seen as 'becoming', rather than as a fixed part of stable identities. Motivation is important here as it allows us to examine how participants in food networks describe and explain their own participation. It also points to the importance of producer–consumer relationships in the sense that each has an understanding of the motivations of the other that in turn influences their own behaviours.

- *Constitution of individual and group identities.* This final field attempts to account for the ways in which particular food networks, first, depend on or assume particular subject positions or identities and second, actually produce or reproduce particular subjectivities. For example, ideas of the 'ethical consumer' are important in some food networks, alongside others which produce particular group identities, such as food projects centring round women, ethnic minority groups or people with disabilities. More widely, this is part of an argument that a food network or project is never simply external to the producers and consumers but is in relationship to the people involved. This field, then, allows a sense of subjective 'becoming' in relation to food production–consumption. We use this field to account for the co-constitutive relationships between human identity and the shifting spatial and social formations making up the heterogeneous food networks which people participate in.

Exploring Possible Food Economies: Three Case Studies

We now examine three case studies, illustrating first, how they are arranged across the fields of our heuristic model, and second, how these arrangements enable them in different ways to offer possibilities for resistance to centres of power in food systems – whether or not this is an explicit objective of a project. The case studies discussed here are, first, Earthshare, a community-supported agriculture project near Forres, Scotland; second, Salop Drive market garden, located in the urban West Midlands of England and third, the 'adopt a sheep' scheme, in the Abruzzo region of Italy (We only have space here to briefly outline the projects in relation to the fields, but for further details about these and some of our other case studies see Holloway 2002; Holloway and Kneafsey 2002, 2004; Holloway et al. 2006, 2007).

Earthshare

Earthshare identifies itself as a community-supported agriculture project, with a team of six producers supplying soft fruits and vegetables to around 200 families in its local area via a box delivery scheme. Table 5.2 outlines some of the key aspects of the project in relation to our analytical fields.

Food is grown using organic principles, and although there is some mechanisation, some cultivation work is done using horses. The diversity of crops produced on the site contrasts with the surrounding larger-scale farming practices in the area. The main site is located close to the Findhorn Community (an organisation dedicated to holistic and spiritual relationships between people and non-human nature), and there have been exchanges between Earthshare and Findhorn which have supported the project's development. Land is rented from a livestock farm, and the vegetable production rotates around a series of fields. There is a subsidiary site at Cullerne, where delicate crops are grown. This means that there is a sense that the project is ephemeral in space and time, having no fixed occupancy or structures, so that there is little sense of a physical farming establishment despite the strength of the project as a social structure which has allowed it to persist for many years. Although the producers feel secure in having a 7-year renewable tenancy agreement, some consumers have expressed concerns about what they feel is an insecure situation and have argued that the scheme should attempt to acquire land to be held in joint ownership as a land trust. Consumers subscribe to the project on an annual basis, paying either a whole year in advance or by monthly direct debits in exchange for the weekly vegetable box. Their involvement

Table 5.2: Earthshare community-supported agriculture, near Forres, Scotland.

Heuristic 'analytical field'	Brief description
Site of food production	Around 9 ha rented land. Close to Findhorn Foundation
Food production methods	Organic; mechanical and horse power; consumer involvement in growing
Supply chain	Box delivery
Arena of exchange	Money, work and LETS exchange; food embedded with ethical value
Producer–consumer interaction	Website, notes in boxes, working parties, social events; emphasis on 'closeness', connection and commitment
Motivations for participation	Social, economic and environmental concerns; desire for fresh, seasonal produce
Constitution of individual and group identities	Consumers are participants, sharers, subscribers to values; group growing work

in the project is extended, however, by expectations that they participate in working groups ('weeding parties') on the production site, three times a year. For those that do, the money cost of their boxes is reduced, so that the arena of exchange is defined by an exchange of labour for food, as well as money for food. Further widening this arena, part payment for food can be made using the local LETS scheme. The food itself is clearly highly meaningful within this arena of exchange; as the Earthshare website says, "Earthshare does a lot more than just provide organic vegetables!" (Earthshare 2005). On its website, the project's food is contrasted with supermarket food, in terms of both its money cost to the consumer, as well as in terms of the relative economic, social and environmental costs. So, for example, Earthshare argues that its operation circulates money within the local economy, provides jobs and stimulates a sense of local community and mutual support. This food is, too, represented as different to supermarket food in its freshness, taste and seasonality. Earthshare produces very direct forms of producer–consumer interaction through the working groups and through social events such as garden parties. In terms of motivations to participate, our research with consumers showed that, although some initially joined for a range of personal reasons, involvement with the project leads them to increasingly identify with some of the wider social, economic and environmental values which are important to the project overall.

The field of motivation, then, through relationships and negotiations between producers and consumers, becomes dominated for both producers and consumers by the opportunity to be involved in a mode of food production–consumption which accords with a particular set of ethical concerns for how economies and communities should function. This requires them to have effected an arrangement of production–consumption relationships which in several ways operates counter to dominant relations evident in much of the food system. As well as being evidenced in the site and mode of food production, and in the inclusion of non-monetary forms of exchange, this is particularly apparent in the identities supported and produced by Earthshare. Thus, for example, consumers are represented as subscribers, indicating a sense of both long-term commitment and, beyond a money payment, a subscription to particular sets of values and beliefs. These are captured in the name of the project, which implies a notion of shared production and shared relationship with a particular place and its people. Subscribers are thus more actively participating in food production, in direct and indirect ways, though their support of Earthshare, than they would be in buying from a supermarket. Indeed, in Earthshare, the conventional distinction between producers and consumers is to an extent blurred, as through the participative 'ownership' conferred by subscription and the physical engagement of subscribers in farm work, food consumers become, at least in part, food producers. Our analysis thus begins to show that it is particular aspects of the project's arrangements across our heuristic analytical fields which make possible this specific way of producing-consuming food, and which effect its potential to produce changes in food production–consumption relationships more widely through the example it provides of organisational and ethical principles. Our second case study bears some similarities to this in that it blurs the boundaries between producers and consumers, but it operates in an urban context and also locates itself within broader health, inclusion and lifestyle agendas.

Salop Drive Market Garden

Salop Drive market garden is described as "a three-acre site which provides a working market garden for people to learn at first hand about food production, undertake healthy exercise and learn about healthy eating and lifestyles" (http://www.idealforall.co.uk). Salop Drive is located in Sandwell, an economically depressed urban borough in the West Midlands. Sandwell is characterised by ill health, premature death and deprivation. Evidence from a number of sources indicates that access to healthy food within walking distance is a significant issue for many low-income households (Dowler *et al.* 2001). This local context is important for understanding the principal aims and ethos of the project, which are discussed in more detail below. Table 5.3 outlines some of the key aspects of the project in relation to our analytical fields.

The site of food production was formerly an allotment site which had fallen almost derelict over a period of about 15 years. Much work was needed to reclaim the overgrown site and then improve the soil, which was severely compacted once the clearance had finished. Green waste from Herefordshire council has been used to improve fertility. Access roads have been installed and wooden huts erected which serve as meeting and resting places for people working on the site. Six commercial sized poly-tunnels are now in place and the produce is grown 'sustainably'. (In reality the production is organic, but interestingly, this term is not used in the publicity materials for the garden). The site of food production – unlike many sites of industrial food production – is thus highly visible because it is surrounded by houses and flats and is also very accessible. In 2003, 74 bags of fruits and vegetables were delivered each week to three drop-off points at local residents' homes and at an Independent Living Centre in the borough. If necessary, the bags are bulked up with produce bought from a local supplier. The project also offers leisure, social and learning activities which include basic and intermediate skills training, community events and celebrations.

Table 5.3: Salop Drive market garden, West Midlands, United Kingdom.

Heuristic 'analytical field'	Brief description
Site of food production	2.7 acres in the heart of a deprived inner city housing estate
Food production methods	Organic
Supply chain	Fruit and vegetable bag deliveries
Arena of exchange	Consumers pay a fixed amount per bag, depending on size of bag required. Consumers can also volunteer to work on the site and will often receive free produce in return
Producer–consumer interaction	Consumers are frequently consulted by project workers. Use of leaflets, newsletters, questionnaires, workshops and organised social/educational events
Motivations for participation	The scheme was established to encourage healthy living and healthy eating, with a specific focus on involving disabled people. Consumer participation driven by desire for fresh, accessible, good quality and cheap vegetables
Constitution of individual and group identities	Consumers are local residents, community members, users and participants; project workers (growers) try to work in consultation with the users of the scheme

At the time of research, the project was staffed by a horticultural site worker, a development worker and a support worker. However, volunteers are central to its survival, and from its very beginnings one of the key aims of the project has been to enable people with disabilities to participate in the work. In terms of producer–consumer interaction, therefore, the Salop Drive market garden is represented – and organised – as a 'user-led' scheme, with participatory processes at its heart. The scheme was originally driven by local and regional food policy initiatives dating back to 1995 and is now managed by Ideal for All, an organisation set up in 1996 to offer services for disabled people living in Sandwell and beyond. Local residents and users of the scheme have regularly been consulted, and the garden is portrayed as an outcome of the vision and hard work of many people. Questionnaires, workshops and social events such as community lunches and visits to other gardens have all been used to encourage community interaction with the scheme. Leaflets, recipes and newsletters are included in the vegetable bags, and the garden has also supplied vegetables to Healthy Living Network Community Cafes and regular cook, grow and eat workshops (Ideal for All 2004). A small number of mini-allotments are also available on site, and these are tended by local residents and people referred from the Independent Living Centre. There are plans to develop a wildlife garden, build more seating and create herb gardens and raised beds to further increase accessibility for those with disabilities.

Regarding the motives for participation, the scheme is first and foremost concerned with encouraging healthy eating and healthy living, with a specific focus on involving disabled people. The project aims to contribute to the government's national agenda for encouraging people to eat five pieces of fruit and vegetable a day. The scheme is not, therefore, driven by a profit-making motive, although it does seek to become self-supporting and decrease its reliance on public funding. From the perspective of the residents and consumers involved, our research indicates that the key motives for participation are the freshness and quality of the produce, along with the convenience of the delivery system. Consumers also reported that they have tried new vegetables and learned new cooking skills as a result of being in the scheme. They also appreciated the positive contribution that the garden makes to the community in terms of providing a social contact point for people who would otherwise be isolated and generally encouraging people to 'get involved'. For those involved in volunteering at the garden, there were many benefits including the enjoyment of getting out of the house, meeting new people, getting some physical exercise, relaxation, stress-reduction and gaining a sense of pride and satisfaction from seeing the results of their work. In relation to the final analytical field in our table, Salop Drive presents a clear case of the co-constitution of individual and group identities in the sense that the project workers are highly attuned to the needs of the local community and the disabled volunteers in particular. As such, growing decisions are made in response to these needs rather than in response to commercial pressures. The project is very much a partnership between different actors and consumers tended to identify with the scheme itself rather than any individual growers or staff members. They recognise that the scheme is about much more than simply the production of food; it has community and environmental dimensions which were greatly appreciated by the consumers involved. Many of the consumers have also been able to develop new food-related skills and knowledge.

This project then reveals much about the diversity of activities and agendas which can be pursued within differently structured producer–consumer relationships. Importantly, this case is about so much more than reconnection in the sense of supplying local food for local people. It has a broader agenda of widening participation, promoting wildlife conservation, environmental sustainability and crucially, improving public health not only by increasing access to fresh fruits and vegetables but also by enabling people to enjoy the physical and mental health benefits associated with growing their own food. Whilst resistance to the dominant food system is by no means an overt aim, this project can be thought of as a form of resistance for those marginalised by such systems – those who cannot easily access supermarkets because of physical disability, lack of private transport, low income and so on. Examining the case in terms of our analytical fields allows us to identify just how this particular scheme can empower groups and individuals to participate more fully in the production of their food and also exert more control over their own health and well-being.

Adopt a Sheep

This case study concerns a scheme in Abruzzo, Italy which allows people located anywhere in the world to 'adopt' a milking sheep on an Italian mountain farm and receive the products of the farm (cheese and salami) by post. There are currently around 1100 adopters, who receive a certificate of adoption, a photo of 'their' sheep and are able to give it a name. For other accounts of the scheme, see Holloway (2002), Holloway

Table 5.4: Adopt a sheep scheme, Anversa, Abruzzo National Park, Italy.

Heuristic 'analytical field'	Brief description
Site of food production	High mountain pasture and farmstead located in a depopulating village in the Abruzzo national park; farmstead also has processing plant, shop, accommodation and restaurant
Food production methods	'biological' (organic); emphasis on 'traditional' breeds, farming and processing methods; use of local 'wild' herbs. Produce is high money value speciality cheese and meat
Supply chain	Regional: speciality food shops. Global: air freight
Arena of exchange	Adoption relationship; money transaction; food highly embedded with (1) qualities derived from locality and (2) ethical value
Producer–consumer interaction	Internet technologies – 'closeness at a distance'; farm visits and holidays.
Motivations for participation	Producer: develop a sustainable business associated with concern for local economy, community and environment and for 'traditional' farming and processing. Consumers: similar concerns and/or 'conspicuous consumption'
Constitution of individual and group identities	Consumers represented as concerned individuals and connoisseurs. Producers as upholders of 'traditional' practices and values

and Kneafsey (2004) and Holloway *et al.* (2006). Table 5.4 shows schematically some of the important features of the scheme in relation to our analytical fields.

The scheme was started by someone from outside the region, 'M', who was inspired to take steps to preserve what she saw as traditional ways of life and rural landscapes by establishing a mode of enrolling consumers into a food production and consumption network. This particular scheme can be understood at different scales. On the farm itself, the adopt a sheep project is part of a series of linked enterprises aiming to add value to the farm produce, including a farm shop, accommodation for visitors and a restaurant. Secondly, within the locality, it is part of a wider cooperative network of farms supplying high value sheep's cheese and meat to specialist retailers and restaurants, together with agri-tourism projects. At the European scale, the cooperative has received funding from the EU LEADER programme, and is thus linked into EU rural development ideology. Finally, this set of on-farm, local and European relationships allows the adopt a sheep scheme specifically to be part of an international network of communications and transportation technologies allowing farm produce to be mailed anywhere globally. Adopters are located, for example, in Japan, the USA, Australia and the UK. These sets of local and international relations reproduce an arena of exchange which deploys the trope of 'adoption' to establish and attempt to sustain relations of close connection and care between the farm and consumers of cheese and meat. The food itself is represented and experienced as meaningful in different ways in this arena.

For example, as the product of 'your sheep' it carries the significance of the 'adoption' trope. It is also strongly identified by the producer with the physical and ecological characteristics of the locality, ideas of tradition, and the importance of sustaining local communities and economies. In this sense, we can identify the motivations of M. as the producer and founder of the scheme as a key analytical field for this scheme. Her motivations, as she stated to us and as evidenced on the scheme website, are to preserve a particular type of rural economy, community and environment, characterised by what are represented as traditional lifestyles, foods, relations of production and human–livestock–natural relationships. These are, for her, contrasted favourably against urban lifestyles and industrialised foods, and people's loss of a connection with rurality and food production. Her scheme thus attributes special values to the food itself, but in association with attempts to challenge what she sees as the relationships of power within society and its food supply system which threaten her ideal types of rural existence.

A central mechanism for these attempts to attribute value to food and challenge dominant power relations is the website through which the scheme's relationships with adopters and potential adopters are mediated, so that the analytical field of producer–consumer relationships is also key to our understanding of this particular project. The website uses photographs and text to do three things. First, it draws attention to the sensual qualities of the local environment and the farming experience, evoking in viewers a vicarious experience of tastes, sights, smells, sounds and textures. In doing this, attempts are made to 'virtually' establish sensual connections between viewers and place, thus enrolling viewers into a representation of the place as special, but vulnerable and worthy of protection. Second, it makes an argument about the special value of 'traditional' rural lifestyles, communities, economies and environments, and urges that these are worthy of viewers' concern and support. Included within that are representations of farm workers and local people as having the authentic knowledge and practices needed to reproduce this specific rural environment and 'traditional' community. Third, it constructs a picture of the viewer as both a connoisseur of particular foods and sensual experiences and as an ethically concerned consumer. In this way, a particular identity is constructed for consumers to align themselves with. This perhaps contrasts with a sense we get from consumers that their participation in the scheme is more to do with its novelty, as something to be conspicuously consumed *in itself* alongside the consumption of cheese and the vicarious consumption of distant lifestyles and environments. This has meant that, despite the attempt to enrol consumers into lasting relationships with the scheme, many fail to renew their adoption subscriptions after the first year. As a result, what seems more important in making this scheme possible is the motivation of the producer, and her agency in developing a representation of consumer subjectivity and the pleasures of consumption that, for whatever reasons, people will actually buy into.

This outline description of the arrangement of this particular project across our analytical fields again points to the particular aspects of the project which make its operation possible, and allow it in some ways to challenge established power relations in the food supply system by the way specialist cheeses are produced. We emphasise in this case the importance of the producer's motivations, the representation of potential consumers as connoisseurs and ethically motivated individuals, the deployment of the 'adoption' trope in attempts to enrol consumers into relations of care, the spacing of the scheme's local and global connectivities and the way those connectivities are effected through the use of communications and transport technologies.

Conclusions

These brief descriptions of three of our case study food projects begin to demonstrate how our analytical fields can be used as a heuristic device for thinking about food provisioning. Although the projects are clearly very different – in terms of their structures, producer–consumer relationships and locations – describing the arrangements of particular projects across the fields allows us to assess how the projects work in their different ways, and to begin to find out exactly where in the projects the potential is found for countering prevailing power relations in food supply systems.

It is difficult to assess the effectiveness of such countering, but we can note two things. First, actors' behaviours in these projects seem to have effects of engaging with and resisting prevailing power relations. Second, in many instances such resistance is 'unintentional' in that many actors (as producers or consumers) do not set out to deliberately challenge structures of power in food supply but nevertheless contribute to a practical critique of those structures through their actions and discourse (of course, for others, the resistance *is* intentional). The value of our approach is that, for a particular project, even though the evidence within some of the fields is for practices which are not counter to dominant power relations, in other fields things are happening which do resist or challenge the status quo. The plurality of relationships between schemes and power relations in an overall food supply system is emphasised by the sort of multidimensional analysis advocated in this chapter. It allows us to retain a sense of the diversity and specificity of projects whilst also locating where, in which fields and in what relationships between fields, is the capacity to effect change and challenge established power relations.

Finally, there is the question of how, despite our desire to emphasise difference and particularity, some sort of broader political project might be established which seeks to change the overall food supply system for the better, without reverting to a story of 'alternatives' positioned against a monolithic 'conventional' food system. One way of starting to answer this may be to focus on the effects of combinations of different projects, with their different ways, wheres and whys of arranging food production–consumption, in the struggle for what might be seen as more 'progressive' food systems. Despite their diversity, there are nevertheless at least some commonalities in the discourses and visions for a 'better' food system which are drawn upon by those involved in projects as producers and consumers. Although we can suggest as a result of our approach that there is no such thing as a singular 'alternative' food economy or system, there are important discourses surrounding being different and doing things differently. Drawing on these in varying ways makes possible a politics of food which goes beyond simply labelling a collection of different practices 'alternative', and instead examines how the specific ordering and spatiality of particular projects can effectively challenge centres of power in food supply.

Acknowledgements

Research for this chapter was funded by the ESRC/AHRC Cultures of Consumption research programme, project reference: RES-143-25-0005. A version of this chapter appears in *Sociologia Ruralis* (Holloway *et al.* 2007), and we are grateful to the editor of that journal for permission to present this version of the paper here. We gratefully acknowledge the help provided by producers and consumers associated with our case study food projects.

References

Allen, P., Fitzsimmons, M., Goodman, M. and Warner, K. (2003). Shifting plates in the agrifood landscape: the tectonics of alternative agrifood initiatives in California. *Journal of Rural Studies* 19, 61–75.
Cook, I. (2004). Follow the thing: papaya. *Antipode* 36, 642–664.
Crewe, L., Gregson, N. and Brooks, K. (2003). Alternative retail spaces. In A. Leyshon, R. Lee and C. Williams (Eds), *Alternative Economic Spaces* (pp. 74–106). London: Sage.
Dowler, E., Blair, A., Donkin, A., Rex, D., Grundy, C. (2001). *Measuring access to healthy food in Sandwell*. Final report, copies available from E. Dowler, University of Warwick, Department of Sociology.
DuPuis, E. and Goodman, D. (2005). Shall we go 'home' to eat?: towards a reflexive politics of localism. *Journal of Rural Studies* 21, 359–371.
Earthshare (2005). Available online at: http://www.earthshare.co.uk.
Gibson-Graham, J.K. (1996). *The End of Capitalism (As We Knew It): A Feminist Critique of Political Economy*. Oxford: Blackwell.
Gilg, A. and Battershill, M. (1998). Quality farm food in Europe: a possible alternative to the industrialised food market and to current agri-environmental policies: lessons from France. *Food Policy* 23, 25–40.
Goodman, D. (2002). Rethinking food production-consumption: integrative perspectives. *Sociologia Ruralis* 42, 271–277.
Goodman, D. (2003). Editorial: the quality 'turn' and alternative food practices: reflections and agenda. *Journal of Rural Studies* 19, 1–7.
Goodman, D. (2004). Rural Europe redux? Reflections on alternative agro-food networks and paradigm change. *Sociologia Ruralis* 44, 3–16.
Goodman, D. and DuPuis, E. (2002). Knowing food and growing food: beyond the production-consumption debate in the sociology of agriculture. *Sociologia Ruralis* 42, 5–22.
Grey, M. (2000). The industrial food stream and its alternatives in the United States: an introduction. *Human Organization* 59, 143–150.
Guthman, J. (2002). Commodified meanings, meaningful commodities: re-thinking producer-consumer links through the organic system of provision. *Sociologia Ruralis* 42, 295–311.
Guthman, J. (2003). Fast food/organic food: reflexive tastes and the making of 'yuppie chow'. *Social and Cultural Geography* 4, 45–58.
Halkier, B. (2004). Handling food-related risks: political agency and governmentality. In M. Lien and B. Nerlich (Eds), *The Politics of Food* (pp. 21–38). Oxford, Berg.
Hassanein, N. (2003). Practicing food democracy: a pragmatic politics of transformation. *Journal of Rural Studies* 19, 77–86.
Hendrickson, M. and Heffernan, W. (2002). Opening spaces through relocalisation: locating potential resistance in the weaknesses of the global food system. *Sociologia Ruralis* 42, 347–369.
Holloway, L. (2002). Virtual vegetables and adopted sheep: ethical relation, authenticity and internet-mediated food production technologies. *Area* 34, 70–81.
Holloway, L. and Kneafsey, M. (2000). Reading the space of the farmers' market: a preliminary investigation from the UK. *Sociologia Ruralis* 40, 285–299.
Holloway, L. and Kneafsey, M. (2004). Producing-consuming food: closeness, connectedness and rurality in four 'alternative' food networks. In L. Holloway and M. Kneafsey (Eds), *Geographies of Rural Cultures and Societies* (pp. 257–277). London, Ashgate.

Holloway, L., Venn, L., Cox, R., Kneafsey, M., Dowler, E. and Tuomainen, H. (2006). Managing sustainable farmed landscape through 'alternative' food networks: a case study from Italy. *Geographical Journal* 172, 219–229.

Holloway, L. Kneafsey, M., Venn, L., Cox, R., Dowler, E. and Tuomainen, H. (2007). Possible food economies: a methodological framework for exploring food production-consumption relationships. *Sociologia Ruralis* (in press).

Ideal for All (2004). Available online at: http://www.idealforall.co.uk (accessed 18.12.06).

Ilbery, B. and Maye, D. (2005). Alternative (shorter) food supply chains and specialist livestock products in the Scottish-English border. *Environment and Planning A* 37, 823–844.

Kirwan, J. (2004). Alternative strategies in the UK agro-food system: interrogating the alterity of Farmers Markets. *Sociologia Ruralis* 44, 395–415.

Kloppenburg, J., Hendrickson, J. and Stevenson, G. (1996). Coming in to the foodshed. *Agriculture and Human Values* 13, 33–42.

Lee, R. and Leyshon, A. (2003). Conclusions: re-making geographies and the construction of spaces of hope. In A. Leyshon, R. Lee and C. Williams (Eds), *Alternative Economic Spaces* (pp. 193–198). London: Sage.

Leyshon, A. and Lee, R. (2003). Introduction: alternative economic geographies. In A. Leyshon, R. Lee and C. Williams (Eds), *Alternative Economic Spaces* (pp. 1–26). London: Sage.

Lockie, S. and Kitto, S. (2000). Beyond the farm gate: production-consumption networks and agri-food research. *Sociologia Ruralis* 40, 3–19.

Marsden, T. and Smith, E. (2005). Ecological entrepreneurship: sustainable development in local communities through quality food production and local branding. *Geoforum* 36, 441–451.

Marsden, T., Banks, J. and Bristow, G. (2000). Food supply chain approaches: exploring their role in rural development. *Sociologia Ruralis* 40, 424–438.

Marsden, T., Banks, J., Renting, H. and van der Ploeg, J.D. (2001). The road towards sustainable rural development: issues of theory, policy and research practice. *Journal of Environmental Policy and Planning* 3, 75–84.

Marsden, T., Banks, J. and Bristow, G. (2002). The social management of rural nature: understanding agrarian-based rural development. *Environment and Planning A* 34, 809–826.

Massey, D. (2000). Entanglements of power: reflections. In J. Sharp, P. Routledge, C. Philo and R. Paddison (Eds), *Entanglements of Power: Geographies of Domination/Resistance* (pp. 279–286). London: Routledge.

Morgan, K., Marsden, T. and Murdoch, J. (2006). *Worlds of Food: Place, Power and Provenance in the Food Chain*. Oxford: Oxford University Press.

Murdoch, J. and Miele, M. (2004). Culinary networks and cultural connections: a conventions perspective. In: A. Hughes and S. Reimer (Eds), *Geographies of Commodity Chains* (pp. 102–119). London: Routledge.

Murdoch, J., Marsden, T. and Banks, J. (2000). Quality, nature and embeddedness: some theoretical considerations in the context of the food sector. *Economic Geography* 76, 107–125.

Norberg-Hodge, H., Merrifield, T. and Gorelick, S. (2002). *Bringing the Food Economy Home: Local Alternatives to Global Agribusiness*.London: Zed Books.

Renting, H., Marsden, T. and Banks, G. (2003). Understanding alternative food networks: exploring the role of Short Food Supply Chains in rural development. *Environment and Planning A* 35, 393–411.

Sage, C. (2003). Social embeddedness and relations of regard: alternative 'good food' networks in south-west Ireland. *Journal of Rural Studies* 19, 47–60.

Stassart, P. and Whatmore, S. (2003). Metabolising beef: food scares and the un/remaking of Belgian beef. *Environment and Planning A* 35, 449–462.

Venn, L., Kneafsey, M., Holloway, L., Cox, R., Dowler, E. and Tuomainen, H. (2006). Researching European 'alternative' food networks: some methodological considerations. *Area* 38, 248–258.

Watts, D., Ilbery, B. and Maye, D. (2005). Making reconnections in agro-food geography: alternative systems of food provision. *Progress in Human Geography* 29, 22–40.

Whatmore, S., Stassart, P. and Renting, H. (2003). Guest editorial: What's alternative about alternative food networks? *Environment and Planning A* 35, 389–391.

Winter, M. (2003). Embeddedness, the new food economy and defensive localism. *Journal of Rural Studies* 19, 23–32.

Chapter 6

Globally Useful Conceptions of Alternative Food Networks in the Developing South: The Case of Johannesburg's Urban Food Supply System

Caryn Abrahams
School of Geography, Archaeology and Environmental Studies,
University of the Witwatersrand, Johannesburg, South Africa

Introduction

Literature on alternative food networks (AFN) has hitherto included multifaceted foci such as short food supply chains (Renting *et al.* 2003; Ilbery and Maye 2005), local food supply systems (Hinrichs 2000; Winter 2003) and local supply chain sourcing (Ilbery and Maye 2006). Other related literatures have focused on the 'quality turn' in food supply (Goodman and DuPuis 2002; Weatherell *et al.* 2003), culturally embedded food systems (Hinrichs 2000), direct farm retail (Brown 2001; Renting *et al.* 2003; Weatherell *et al.* 2003), community-supported agriculture (Allen *et al.* 2003), 'good food'/specialist production (Ilbery and Kneafsey 1999; Sage 2003) and hybrid food networks that include 'alternative' and 'conventional' elements (Ilbery and Maye 2005; Ilbery and Maye 2006).

AFN have been presented either as a new evolution of agro-food systems emerging as a response to crisis-ridden conventional agribusiness, as a "popular mobilisation against US cultural and corporate food imperialism" (Whatmore *et al.* 2003: 389), or a transitionary move towards some kind of alternative or post-productivist era (see Ilbery and Bowler 1998). AFN have been characterised by a different phase of trade relations, sourcing practices or era of production and consumption, as compared to globalised agri-food processes (Goodman and DuPuis 2002), exhibiting defining characteristics that are succinctly reflected in Ilbery and Maye (2005). These characteristics include food that is "fresh", "diverse", "organic", "slow" and/or "quality" based (Ilbery and Maye 2005: 824), and networks or supply systems that are "small-scale", "short", "traditional"

Alternative Food Geographies
D. Maye, L. Holloway & M. Kneafsey (Editors)
ISBN: 978-0-08-045018-6

Copyright © 2007 by Elsevier Ltd.
All rights of reproduction in any form reserved.

"local", environmentally "sustainable" and "embedded" (Ilbery and Maye 2005: 824). The common factor is that all these characteristics are oppositional to characteristics of conventional food supply systems that can be defined as "processed", "mass (large-scale) production", "long food supply chains", formal retailing – "hypermarkets" and "disembedded" (Ilbery and Maye 2005: 824).

Existing literature acknowledges that the emerging research on AFN has widely centred on northern, predominantly European and North American, contexts (Whatmore *et al.* 2003; Goodman 2004; Hughes 2005; Ilbery and Maye 2005; Watts *et al.* 2005). It is thus not surprising (and arguably necessary) that AFN evident in the north are markedly different from AFN in a developing world context. At a broader conceptual level, there is also the emergence of alternative discourses (Ilbery and Holloway 1997) which are both driven by and fundamentally linked to spatial policy developments (Watts *et al.* 2005), including agricultural policy. The emergence of AFN in the European experience is situated within ideas about food safety, and within rigorous quality and safety regulations developed as a result of policy and institutional change within a transitional European economy (Ilbery and Holloway 1997; Evans *et al.* 2002; Goodman and DuPuis 2002; Goodman 2004). The European-wide paradigm shift in quality consumption (Goodman and DuPuis 2002) has been facilitated by alternative transitions within the 1999 CAP reforms (Goodman 2004). Naturally, agricultural transformation linked to these, as in the post-1992 reforms – like any policy response – "[is] necessarily embedded in recent historical, political, economic and cultural contexts which, while enabling or encouraging certain types of (in)activity, proscribe or constrain others" (Ilbery and Holloway 1997: 185). In the case of AFN, this means that policy reforms, driven by institutional structures (albeit signalled by 'on the ground' experience), create and enable the conditions necessary for AFN to emerge and exist (Campbell and Coombes 1999). It follows that respective national, regional or institutional priorities will determine the nature of food networks, amongst other things, and that a particular articulation of AFN will emerge from its contextual space, be it institutional, regional, political or cultural.

The emergence of AFN in North America is directly attributed to oppositional social movements within activist circles (Goodman 2004). The emergence of AFN and its concomitant literature gained impetus primarily from "[its] oppositional status and [the] socio-political transformative potential of alternative agro food networks" (Goodman 2004: 4). The North American move to embrace and encourage AFN is as a result of a strong commitment to social justice movements (Allen *et al.* 2003) and increasing antagonism towards the hegemony of the productivist complex (Grey 2000a,b). AFN are conceptualised as a quasi-revolutionary movement which "return[s] insistently on the central question of [the] capacity [of AFN] to wrest control from corporate agribusiness and create a domestic, sustainable and egalitarian food system" (Goodman 2004: 4).

Furthermore, for those adhering to it, a North American articulation of AFN or 'alternative agri-food initiatives' represents a lifestyle statement that for instance reflects an avant-garde, socially reflexive transition for the young, 'yuppie type' who do it because it is fashionable and the 'right[1] thing to do'. Allen *et al.* (2003: 61/2) accurately question whether alternative food initiatives (AFI) are "significantly *oppositional* or primarily *alternative*" (original emphasis), and whether there is a potentially transformative process of consumption choice suggested by the rising popularity of AFI. The premium

---

[1](Read left/oppositional).

paid for local, organic, environmentally sustainable and ethically sourced food is set up as the price to pay for having a social conscience and opposing a global regime that is wrought with opacity, and which contributes to exploitation and environmental degradation. These initiatives can arguably be perceived then as being embedded within either a socially reflexive frame of reference or an elitist one. Irrespective, it is clear that particular articulations of AFN emerge from differing contexts, even though the observable characteristics may be similar.

The European and North American contextualisations of AFN have therefore emerged predominantly from policy agendas concerning food quality and an effort to *re*-value rural (and economically declining) spaces, and also from an activist impetus linking economic justice and food safety. I argue in this chapter that even though AFN in the south are oppositional to conventional food supply systems,[2] they are fundamentally different from AFN in the north. AFN in the south are defined here as the entire food supply system that, in part or fully, contests or opposes the dominance of conventional food networks within urban areas of the developing south. I show that AFN in the south have emerged as a response to the inability of communities to access conventional food supply systems, and should not be understood primarily as a marker of remnant informal food systems. These AFN represent a grass-roots endeavour for culturally diverse communities to consume culturally specific food and for poorer communities to make use of accessible food networks within the emergent context of supermarket dominance. While I argue that AFN should be understood contextually, the larger point is that 'contextually based' AFN in the south have globally useful significance to an emergent alternative agro-food theory and could potentially become universalised beyond the south.

A conception of AFN in the south offers a timely contribution to the emergent theoretical construction of AFN by arguing that issues that are perceived to have greater significance in a developing world context are in fact integral to the formation of an inclusive worldwide knowledge base of AFN. The significance of extending the theory on *Alternative Food Geographies* also suggests a correlation with critical research. Emergent literatures are too often implicated in exclusionary academic practises (Yeung and Lin 2003; Berg 2004; Hughes 2005), which exclude knowledge that emerges from non-UK or US contexts (Paasi 2005) and, more particularly, from the developing south. Theory – which often becomes universalised within academic space (Yeung and Lin 2003) in general – needs to include empirical evidence from arenas other than the dominant European and North American contexts (Berg 2004). It also needs to "incorporate a 'theorizing back' from South to North" (Hughes 2005: 502) to be found estimable. Researchers both within and outside the south have recognised boundaries to the advancement of theory concerning issues affecting a 'southern'/developing world category such as poverty or cultural diversity (Braun 2003; Minco 2003; Vaiou 2003; Yeung and Lin 2003; Berg 2004; Milbourne 2004).

To bring this argument back to the emergent theory on *Alternative Food Geographies*, the discursive space suggests that there may be "a dearth of space and time among the

---

[2] A conventional food network is defined in this chapter as the food supply system that is linked to large-scale, productivist agriculture and supermarket retail, and the highly industrialised food supply chains associated with industrialised, large-scale production and consumption.

'disenfranchised' for democratic participation and imagining alternatives" (Samers 2005: 882), which in this context plays out as the exclusion of globally useful south-based agri-food debates in agri-food research. The focus of south-based agri-food debates is not the celebration of (alternative) consumption (Goss 2004) but rather issues of food poverty and insecurity and the potential of agricultural development for alleviation (Wrigley 2002; Allen *et al.* 2003; Allen 2004; Coen-Flynn 2005). Since they are perceived as having more to do with survival rather than consumption, these arguments have hitherto been ignored by those contributing to the emergent AFN literature.

Even though Goodman and DuPuis propose a research agenda of alternative geographies of food which stretches "beyond the production-consumption debate" (Goodman and DuPuis 2002: 5), these debates have failed to transcend the celebration of alternative consumption evident in much of the northern literature (Goss 2004). It is for this reason that within the literature on alternative geographies of food, issues that are typically classified as 'developing world issues' (e.g. poverty, food security and cultural diversity) do not occupy dominant space within the nascent northern consumption-framed AFN literature. This chapter thus argues that, beyond the consumption debate, emergent themes in a study on AFN in the south have relevance to agro-food studies in an era of urban poverty and cultural diversity in the south *and* the north. AFN articulated in the south do not simply offer a developing world *perspective* on AFN but should challenge a hitherto northern and exclusionary conception of AFN and propose a globally usefully conception of alternative geographies of food.

Methodology

This study is based on research undertaken between March 2004 and September 2005. The broader focus of the research is part of an ongoing investigation around urban food supply systems in the developing south. The study consisted of ethnographies of a small sample of farms, and over 40 semi-structured, in-depth interviews with agricultural practitioners, vendors and a number of random consumers in the Johannesburg area, in the Gauteng Province, South Africa (see Figure 6.1). Lenasia, in the southwest of Johannesburg, is a formally racialised Indian township community which now is more diverse and is also home to a growing informal settlement population. A number of interviews with farmers in the south of Johannesburg and more affluent northern areas of Johannesburg were also undertaken, as well as interviews with relevant stakeholders in the Department of Social Development and the agricultural community. During a brief visit to France, Germany and England, a number of ethnographic observations and short interviews were also undertaken at farmers' markets and allotment gardens.

The rest of the chapter is divided into four main sections. The first section describes and contextualises AFN in the south, the second reflects on the themes which an analysis in the south highlights, and the third argues that a 'south' conception of AFN has implications for northern AFN theory. The conclusion reflects on the point that a potentially universalised AFN theory needs to go beyond geographically and socially exclusive arguments by presenting a case in 'defence' of why AFN in the south should be considered alternative at all.

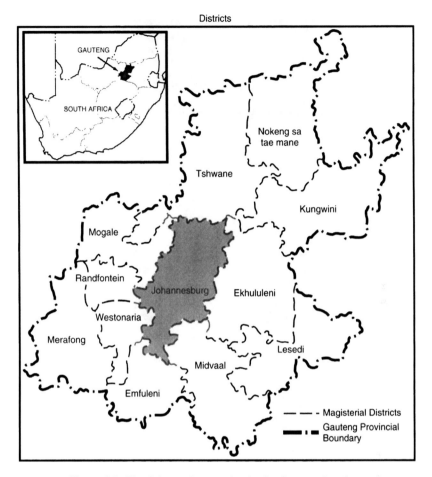

Figure 6.1: The Johannesburg region in the Gauteng Province.

An Empirical and Contextual Description of AFN in the South

Johannesburg's alternative food supply system appears to be marked by stark contrasts with Lenasia's peri-urban farms juxtaposing large supermarket complexes. Young families walk alongside the busy freeway and carry their fresh food purchases – anything from leafy vegetables to squawking chickens – to where they live. While there is a supermarket complex, within close driving distance to where people live, that sells fresh vegetables and pre-packed chickens at a fraction of the price of live chickens, the formal retail supply system is generally inaccessible for those without private transport. Scores of fruit and vegetable vendors in Lenasia's main trading area make fresh produce more accessible to the majority of the community. This produce is either sourced from surrounding farms or from the Johannesburg Fresh Produce Market. Other foodstuff is sold directly by farmers at car boot sales, in makeshift market stalls, or from head-balanced buckets (see Figures 6.2 to 6.5). Women sit in front of large metal drums, *braai* (barbeque) chicken feet, heads and giblets, and sell them as snacks in the main

Figure 6.2: The road-side sale of live farm chickens.

Figure 6.3: Cultural chicken snacks.

Figure 6.4: Traditional nuts sourced directly from a farm.

Figure 6.5: Traditional African vegetables.

shopping area. A person nearby can buy half a boiled sheep's head (still in the skull) to eat with a spoon during a break from shopping.

Other food distribution systems, which are local and against the logic of conventional retail, include the sale of *maas* (cultured sour milk) from local small producers and un-sliced quarter loaves of bread that are sold to schoolchildren during breaks. Alternative food supply systems for poor people are particularly noteworthy; surrounding farms sell poultry and beef off-cuts to residents in the township who re-sell them, while others give away the offal, head, skin and feet of their slaughtered livestock to the poorer residents at no cost. Informal alternative food supply systems that cater to poorer members of the community are preferred by poorer people even though the food may not always be cheaper or safer. Face-to-face interaction of producers and consumers are typical, and relationships of trust are increasingly the *modus operandi* of small peri-urban producers to gain access to speciality markets, even in formal retailing structures. The farms that facilitate direct sale are legitimate landholdings, and the enterprise conducted is regulated.[3] The following four sections explain the food provisioning system in the south: direct farm retail, local cultural food provisioning networks, cultural and religious food networks and food supply chains for the urban poor.

Direct Farm Retail

More than 25 of the farms that fringe the peri-urban region have begun over the past decade to engage in direct sale to the local community, both in formal residential areas and in informal settlements. Neither the cultural food supply system nor food networks catering for the urban poor have *intentionally* developed as a community food-security specific or culturally specific food networks, although they can now be perceived as such. The community market relationships rely heavily on local knowledge of religious regulation or ceremonial expertise. As one producer/retailer put it: "when I sell to the Indians I leave the giblets in a little packet with the Cornish [rock chicken] and I keep the feet separately" (Interview, 05/2005). Again, these networks are not part of the informal economy although the end sale sometimes does not occur at a formal store. The owner of one of the largest formal farm enterprises, which has four home-outlet stores said: "just because we don't sell to Spar [a local supermarket] doesn't mean we're a fly-by-night joint" (Interview, 03/2004). A larger percentage of local food supply systems are hybrid and make cultural food more accessible to urban communities.

Local Cultural Food Provisioning Networks

In areas of Johannesburg, cultural modes of production and regulation, which intercept formal, more conventional retail networks, are familiar and are increasingly becoming the norm. One farmer/retailer explains "the customers want to know that I am Halaal[4] certified. The other HACCP[5] one is already met" (Interview, 04/2005). Another informant added: "it's safe, it's cheap and people can get what they want. Checkers [a discount supermarket] is cheap but it's not Halaal" so "people trust my food" (Interview, 04/2005).

The success of the cultural food network, in particular from the production side, lies in the community relationships it makes use of. Some farmers cite the fact that they can speak the language of the consumer is an added bonus, which helps to build

---

[3]This is with reference to the notion that African food supply systems should be classified as part of the informal economy. The classification of the economic systems as informal is based on whether or not they are economically regulated (see Rogerson and Preston-Whyte 1991).
[4]Halaal is a religious certification ensuring that food products do not come into contact with any forbidden substance under strict religious sanction (The Muslim Food Board).
[5]Hazard Analysis Critical Control Point (a standardised international food regulatory certification framework).

cultural trust. Home deliveries, drive-through vendors, telephone orders and outlet stalls outside mosques, radio stations and banks, all function as a unique community 'service'. While hardly any of the small farms have formal marketing, the advertisements of one large poultry farm in community newspapers under the slogan "we only sell what we slaughter" (Interview, 09/2005) highlight the importance of consumer trust that is linked to religious Halaal certification.

Many of the farmers in this area do not have the capacity to enter the formal market, but there are a considerable percentage of farmers who are relatively well-off, and whose businesses are thriving and growing, despite the lack of government support. These farms thus present a fundamentally different conception of typical smallholder agriculture in the south that engages in subsistence agriculture or informal trade. These farms potentially have access to formal retail supply chains because of the quantity and quality of their output but choose not to engage in formal chains. As one farmer explains: "I can sell to Spar, but I have enough people who buy from me and I don't even advertise. Someone tells someone else and that's how the word gets around" (Interview, 04/2004). The dominant motivation for direct farm retail is the customers' perception that farmers provide a service to members of the community who would otherwise have no other channel to purchase some of the cultural foodstuff and ceremonial livestock they offer. One consumer said: "it's so convenient, and even if it wasn't, you can't get stuff like this in the shops. It stays fresher for longer and we know it's good and healthy. Plus it's cheaper. I wish I could buy all my fruit and vegetables here" (Interview, 08/2005). Traditional food networks are not only strongly visible and accessible because of consumer food choice; another driving factor is religious-sanctioned foodstuff under the Islamic Halaal regulation.

Cultural and Religious Food Networks

Cultural or religious food networks that are specifically located in Indian and Black townships facilitate cultural hubs in the racially diverse south of Johannesburg. These cultural zones have emerged 'from below' and are not always voluntary. They have developed as an alternative method of cultural survival, and they have formed on the basis of local knowledge (cf. Sage's (2003) point about 'defensive localism'). Culturally driven AFN in the south suggests a 'necessary' and niche opportunity, which caters to enclaves of cultural groupings. As one dairy farmer put it: "we sell the real thing [cultured sour-milk]... we can say 'this is stuff you grew up with'" (Interview, 07/2004).

Since cultural supply systems continue to exist in a post-apartheid era, it is clear that cultural patterns of consumption are persistent and go beyond local AFN debates. Ex/non-residents of Black and Indian townships commute to these 'speciality' localities for culturally specific food items. Indian vegetables, Halaal poultry and products and livestock, used in traditional African and/or Islamic ceremonies and festivals, are increasingly produced by peri-urban farms in the area. With the alternative of travelling up to 600–1000 km for cultural religious food, the peri-urban fringe of many black townships is fast becoming the core of alternative food provisioning.

The religious Halaal regulation governing the production of chicken and poultry products is a fascinating element of cultural food networks that intercepts 'conventional' regulatory frameworks. Communities in the developing world, which are increasingly under the jurisdiction of Islamic food regulation, do not purchase Halaal products predominantly because they want healthier lifestyles, although one farm retailer does suggest that the Halaal certification is so stringent that it meets all of the international quality controls anyway. The 'choice' of exclusively buying Halaal is a religious imperative. The process of production is heavily regulated by The Muslim Food Board (2005) (noted earlier) and governs the growth, storage and processing of all meat products. In Lenasia and its surrounding areas, the largest market for these products is the Indian Islamic community. The cultural enclave of Johannesburg now functions as the only place where some cultural/religious food networks are available. As an aside, since most food enterprises – retail, wholesale and restaurants – are owned and managed by members of the Islamic population in Lenasia and other majority Muslim communities, the Halaal marker can now be seen as an alternative regulatory standard within African cities. While Halaal provisioning practices have been in place for as long as the Islamic religion has existed, that it is now facilitated by formal regulatory frameworks is a recent occurrence much like the French AOC regulation. The final type of food provisioning network in the south facilitates accessible food supply for the urban poor.

AFN for the Urban Poor

AFN for the urban poor is a highly significant aspect of food supply in the south, even though cultural food supply systems are dominant. While supermarkets provide cheaper and more accessible food, the urban poor (mainly from informal settlements and low-income areas) do not have refrigeration facilities to store fresh food and meat products nor do they own private vehicles with which to transport large quantities of foodstuff. While the urban poor have to access some non-food items in supermarkets, most food products are procured from surrounding farms that are within walking distance and where they can purchase an un-plucked live chicken that is unavailable in formal retail outlets. While this chicken may be close to double the price of a pre-packed frozen supermarket chicken, every part of the chicken is consumed by these populations. These chickens are mainly purchased by women from informal settlements, in and around the farm area, who will carry one back home by foot, then pluck, singe the hairs and cook every part of the animal, including the head, feet and beak with the exception of the rectal bags.

Poultry AFN are different to free-range/quality/organic AFN as portrayed in a European context for at least two reasons: first, the kind of product that communities purchase from urban and peri-urban farms are not available in a conventional food supply system and secondly, although consumers pay more for supposedly more accessible food, they sometimes get an inferior quality product. Pre-packed chicken in a supermarket is cleaner and at times almost half the price of a live chicken, but cultural modes of consumption and accessibility, in this sense, take precedence over conventional food networks that according to an agro-industrial logic offer safer, cheaper food. The poultry network only briefly illustrates the point that within a multifaceted developing world context, alternative modes of food provisioning play an important role for both poorer and culturally diverse communities, who may not have access to formal retail outlets or who require speciality foodstuffs that are not available in large supermarkets.

No local government, farmer unions or formal marketing bodies support these enterprises through such initiatives as farmers' markets. A farm may hold an open day where there are pony rides for kids and *braais* for the family, using foodstuff from the farm.

With these exceptions, however, the success of the network depends solely on enterprise innovation and the demand for cultural food. This, together with cultural customer familiarity and communal ties, promotes a 'by us, for us' ethos. One farmer succinctly sums it up: "it's our own product for our people" (Interview, 07/2004). Food supply systems in the south thus cater for culturally diverse urban communities whose food requirements are not available in supermarkets. AFN in this context *are* enterprise/business-driven but are also increasingly based on cultural preservation articulated through market relations of trust. These are forged through personal marketing strategies, the ability to speak the language of the consumer, the possibility of making other products according to customer demand and the initiative on the part of farmers to set up car boot farm stores or makeshift stalls outside primary schools where parents frequent. As discussed, the most notable AFN is the culturally specific Halaal and traditional food network for urban residents.

The point of this largely descriptive section is that the kind of food supply systems represented in a city of the south cannot be classified as either informal or traditionally remnant. These alternative supply chains occur in a context of increased access to conventionally available (supermarket) food, and yet, through direct farm sale, are geared to meet the consumption needs of the urban poor and culturally diverse that are not met by conventional food supply systems. Much like culturally specific and 'traditionally remnant' food consumption practises in the north, which are now facilitated by AFN, alternative and hybrid food supply systems in the south have become dominant in the food provisioning choices of south-based consumers. Unlike food supply systems represented in the literature around AFN, however, in the alternative food supply systems in the developing world context there is no romanticised return to the local and no quest of the idyllic countryside lifestyle. Although there are consumers who are concerned with greenness, fresh-from-the-earth produce, highly regulated quality measures and who do want to make a socio-political statement by what and where they obtain food, for poorer consumers these are not always the primary motivations. Thus, while these food networks may be perceived as informal or remnant, they represent an alternative consumption space for the urban poor and culturally diverse communities in the south.

The next section presents a brief review of the literature regarding the increasing dominance of supermarket retail in the south, as the conventional food supply context to which food supply systems in the south present a valid alternative. Even though consumers' access to cheaper retailed (supermarket) food is increasingly greater given the proximity and proliferation of supermarkets, the real access poorer people in urban areas could enjoy is counteracted by rising inflation and stagnant (if not decreasing) real incomes, higher public transportation costs and lack of refrigeration. In *this* context AFN, in the south have emerged both as a result of the unavailability of certain cultural food stuffs in supermarkets and as an alternative means of food provisioning for poorer people.

AFN in the Context of Increasing Formal Retail

Food supply systems in the south may, in the past, have been characterised entirely by informal networks or subsistence, but from the above description it may appear that these food supply systems are survivalist or are merely contemporary reflections of traditional food consumption. More recent research, however, helps to build the argument that these direct, cultural and accessible food supply systems occur in a context of increased accessibility of formally retailed foodstuffs in developing world cities. Over the past four decades, formal retail networks, and supermarkets in particular, have become more dominant in food provisioning. Research on Southern Africa (Weatherspoon and Reardon 2003) and Latin America, including Brazil (Farina 2002), Chile (Faiguenbaum et al. 2002) and Argentina, agrees that decentralised supermarkets replace traditional stores which are directed towards poorer members of local communities as the "nature of the domestic market changes in general with development" (Weatherspoon and Reardon 2003: 8). While food supply purely within a myopic sense of informal trade is still evident, informalised food outlets linked to formal farm sale cater for the 'poor consumer segment' and culturally diverse communities (Rogerson 2003).

Furthermore, it is argued that those developing world countries, where agro-industrial activity is on the increase, exhibit the growth that is necessary for their further economic development (Reardon and Barrett 2000). Nevertheless, within the specific contexts mentioned above, there is a reactionary politic not in opposition to retail-led hegemony, agro-industrialisation or institutionally supported anti-globalisation food networks, as in northern cases of AFN. Instead, AFN in the developing world are interpreted here as fundamentally survivalist in nature (see also Rosset 1998).

Consumption or Survival? Cultural and Accessible Food Supply Systems

In the context of the dominance of formal, supermarket-driven food networks, AFN emerge as a food provisioning mechanism that supply culturally diverse communities and poor people with accessible, culturally specific food. More generally, AFN in the south are linked to urban and peri-urban agriculture (UPA) and have become a strategic, alternative income-generating enterprise for some, as well as a coping/survivalist mechanism (Rogerson 2003; Porter et al. 2004; Coen-Flynn 2005). These particular contexts enable and explain the existence of AFN linked to local, direct, relational, fresh produce, organic and culturally specialist supply chains, which occur in response to inadequate conventional supermarket food supply. When it comes to the speciality cultural foodstuffs like Indian vegetables, dairy products, religiously sanctioned foodstuffs and ceremonial livestock, the food networks are noteworthy not because they are perceived as being informally available but because they are unavailable through other conventional retailing systems.

AFN in the South

AFN in the south have not emerged primarily from a top-down policy proviso, neither are they predominantly driven by NGOs in search of economic justice. They emerge as a grass-roots development imperative which includes survivalist enterprise, accessible food networks for the urban poor and cultural food networks for diverse communities. While AFN in the south exhibit certain elements similarly articulated in the north, the driving factors for these different AFN are contextual and have emerged from fundamentally different politico-economic spaces. AFN in the south exhibit short food chain, quality food, slow food, local/speciality food, cultural speciality food and direct farm sale characteristics. However, the economic context within the agro-food complex in South Africa is not lagging but is in a dynamic phase of growth (Mather 2005). The distinction of "advanced economies" (Renting *et al.* 2003: 394) in reference to the context from which AFN emerge, suggests that other (less 'advanced') contexts do not have the same driving factors. In the Johannesburg case, the driving factors include inadequate provisioning of supermarkets for traditional, religiously sanctioned foodstuffs and accessible foodstuffs for the urban poor.

AFN in the South exhibit two dominant characteristics: cultural food networks and food networks for the poor. The local context includes culturally diverse communities, a large percentage of whom are Islamic, and poorer communities who do not have access to supermarkets. It has been suggested, in light of the empirical description, that these food networks be classified as being part of a more established research tradition on African food systems (cf. Guyer 1987) that examines either *informal* food networks or that they should be framed within a literature on the broader informal economy. It is clear that all kinds of consumption that make use of informal networks in the developing south continue to occur. However, the AFN in the south cannot be equated with informal food networks entirely, even though they may exhibit 'informal' characteristics such as catering for the poor or bypassing formal retail structures. Formal retailing structures do exist, and the fact that these AFN are not simply remnant from some pre-conventional era suggests that they cannot be classified as merely being part of survivalist informal economic enterprise. For much the same reason as AFN in the north are not considered as part of the informal economy despite the fact that they too make use of car boot sales (Crewe and Gregson 1998; Fisher 1999), marginalised consumption spaces (Crewe and Gregson 1998), less technological production or word-of-mouth marketing, or are based on remnant cultural food needs, food networks in the south occupy an alternative space. They do so because they occur (and have recently emerged) within a context of the dominance and increased accessibility of conventional food networks. They are reactionary to inadequate supermarket food provisioning, and the AFN materialise as a result of the demand for cultural food that was previously home-grown or unavailable.

Samers (2005) argues that the myopia of classing certain systems as either informal or formal is linked to the imaginaries associated with what the term 'informal' conjures up; in Africa, for example, poor, illicit, ethnic, exotic and/or chaotic street trade. Daniels (2004: 507) is critical of the notion that economies "graduat[e] from the informal to the formal economy" as a reflection of their social progress. To suggest that food supply systems are informal based on their context is problematic. Whatmore *et al.* (2003: 389 argue that AFN in Europe are "[f]ar from disappearing... [and are] diverse and dynamic food networks that had been cast as remnant or marginal in the shadow of productivism [but] have strengthened and proliferated". It is strange to note that what is considered 'alternative' in a developed world context is often classed as being part of the 'informal economy' in a developing world context – with both having similar characteristics. The premise of a recent work by Coen-Flynn is reflective of a conception of African food supply as still occupying a pre-modern, non-industrialised space that asks "[w]ithout grocery stores, supermarkets, food delivery services, or convenience stores, how do people acquire food in Africa?" (Coen-Flynn 2005: 1). On the contrary, in a liberalised and deregulated developing world context, an increased dominance of formal retailing marked by the increase in large retail supermarkets is evident. While current debates suggest the increase in formal retail, and others argue the persistence of informal (pre-formal) retailing in the developing world (cf. Samers 2005), there is evidence of an emerging countermovement of AFN in the developing world; speciality food networks, cultural food supply chains and direct selling of food grown on urban and peri-urban farms in the vicinity as a response to inadequate supermarket supply.

Within an increasingly culturally diverse milieu that is particularly pervasive in developing world cities, alternative types of food provisioning and consumption behaviours that contest the conventional, formal and retail-led food supply system are necessary for survival. Porter et al. (2004: 31) argue that "in urban areas [in the developing world] there is a diverse range of consumption behaviours shaped by ethnicity, household structure and poverty". In comparison to AFN emerging from the north, these AFN do not necessarily represent lifestyle alternatives to conventional food systems, which are arguably elitist and geographically exclusionary. Issues of culturally diverse and accessible food for the urban poor, which have fallen out of view in northern literature, are crucial to an understanding of useful AFN in the developing world. I suggest the term 'useful AFN' because, in a southern context, AFN has the potential to challenge and extend a theoretical politico-economic construct that will not only inform agri-food debates and policy but, as I argue below, will also be of actual value to poorer and culturally diverse people by addressing their access to food. Theory which includes food systems of the poor in the north, however, has been to some extent advanced by northern researchers.

Food Insecurity in the North

Where poorer UK residents do not have access to transport or to supermarkets, a few theorists have seen the importance in addressing "issues such as how low-income families living two bus rides from a grocery store get access to fresh fruit and vegetables" (Wrigley 2002: 2032). As is evident in a developed country where access to retailed food is assumed to be higher, certain members of the community are marginalised. Facilitating AFN that cater for the urban poor in developed countries may prove not only to be a necessary policy interest in regard to quality control but also a growing social solution to food poverty and concerns of cultural food marginalisation.

Wrigley (2002) notes decreasing access to food for poorer parts of the UK population and argues that this type of social exclusion is becoming increasingly important to policy makers. In North America, it is argued that "without access to supermarkets, which offer a wide variety of foods at lower prices, poor and minority communities may not have equal access to the variety of healthy food choices available to non-minority and wealthy communities" (Morland et al. 2001: 23). Although there is a need for a policy focus on access to food for poorer communities, in the north this kind of research does not emanate predominantly from agro-food studies but hails from concerns around welfare and national longitudinal nutritional studies (Morland et al. 2001; Cummins and Macintyre 2002). The increase in concerned consumerism is not argued in relation to supporting survivalist economies but buying local foods (Weatherell et al. 2003). If increasing poverty and access to food is indeed such a hugely urgent future policy agenda (Cummings and Macintyre 2002; Pearson et al. 2005), it is increasingly necessary for alternative food studies to engage with these issues in agri-food debates (and also surprising that it has not done so already). Food networks for the urban poor and for marginalised and/or multicultural communities have evaded the perception of contemporary research agendas within agro-food studies. The poor are regular features in literature around both food security and poverty alleviation (Coen-Flynn 2005; D'Haese and Van Huylenbroeck 2005). The failure to include poor and culturally diverse communities in a 'broader' theory on alternative agro-food systems, however, with the exception of Wrigley (2002), Allen *et al.* (2003) and Allen (2004), implicitly excludes a crucial segment of society. A powerful argument in relation to this is made by Rigg and Ritchie (2002: 360) who assert that the reason "why post-productivist rural scholarship on the developing world has yet to make much of a mark *is* because it is out of step with [other] realities in the poor world" (original emphasis).

A Globally Useful AFN Conception: Implications for AFN Theory

A developing world perspective on AFN is useful to countries in the north since rural and urban poverty is an increasing trend across the globe. A developing world articulation of AFN includes survivalist strategies, poverty alleviation and community development initiatives, and does not merely fulfil the hankering for the rural idyllic nor the often elitist quest for conscience-quenching food. Like promoting economically declining regions in Europe and North America, AFN in these contexts have the potential to reflect broader agricultural transformation with respect to food security. More recent AFI (Allen *et al.* 2003) in some parts of North America recognise the need to cater for the urban poor – particularly low-income Black and Hispanic groups – and tackle issues of empowerment and education; homeless peoples' gardens, school garden projects and skills-based training (Allen *et al.* 2003). In New Jersey, for example, youth empowerment projects include using urban food systems as a tool to teach responsibility, financial independence and to increase food security of at-risk youth (Hamm and Baron 1999), while in Paris, farmers' markets function as an accessible source of fresh produce. In Sandwell, in the United Kingdom, food banks accumulated through urban agriculture are used to provide for poorer people, and allotment gardens in the urban and peri-urban areas of the country could arguably facilitate culturally driven types of AFN by catering for large and growing migrant communities. The potential is far greater than just community-*supported* agriculture (Grey 2000a; Hinrichs 2000), which is a formal project-based initiative supported by USDA and other farmer organisations (Wilkinson 2005).

There is the possibility of making use of contextual drivers of AFN to further the kind of AFN that would benefit the urban poor or culturally diverse communities. Using the European AFN paradigm that is based on policy-driven agricultural transformation, and through understanding the North American impetus for reflective social AFN as the "critical impulse of social resistance in everyday life" (Harvey 1996 cited in Allen *et al.* 2003: 62), AFN could be a transformative tool with which to guide agricultural policy, social development, poverty alleviation and cultural modes of provisioning. More importantly, since AFN may also be understood as a value-based reaction (Grey 2000a), AFN has the potential to be more than just a theoretical construct.

It is significant that developing world conceptions of AFN are not necessarily only valid within a developing world context. AFN in the south illustrate that while it is crucial to examine the geographic, socio-economic and institutional contexts from where particular types of AFN emerge, issues like cultural food networks and accessible, alternative food provisioning for poorer populations are increasingly applicable to other contexts. In this sense, it is vital within North American and European conceptions (and markets) to broaden the paradigm of AFN to include urban and peri-urban agriculture, food networks that cater for the urban poor and cultural modes of provisioning. Of course, evidence of larger multicultural communities and increasing poverty in other communities in the north is now more visible than ever before. For example, the German Ministry of Social Development is proactively involved with the holistic integration of their Turkish communities (Halm and Sauer 2004), and the large Indian, Polish and Caribbean populations in the United Kingdom (amongst other ethnic groups) have increasingly popular cultural food supply systems in various urban areas. Again, the allotment garden offers potential for cultural food supplies to migrant communities. The point is that AFN must be contextually relevant to diverse communities. Certainly, holistic research into AFN – that by definition is not conventional and *cannot* therefore assume a conventional customer – must include these marginalised communities.

Conclusion: Beyond Geographically and Socially Exclusive AFN

Were I to sketch the profile of the imaginary customer envisaged by a fictitious AFN 'practitioner' – as suggested by AFN literature based in the north – she may look something like this: White, upper-middle class, with a sophisticated sensibility, middle to late-middle age, professional, academic, with access to transport and credit. She frequents farm stores and chique produce markets to buy organic, fresh, fairly traded food which is "not only good to eat, but good to think" (Lockie and Halpin 2005: 284) since it is healthy, good for the environment *and* donates 80 pence to start a school in the rural area in the developing world from which the produce comes. She drives a huge SUV a longer distance to the next rural village where speciality foodstuffs are locally produced. She takes pleasure in consuming the rural countryside and returns home to fine wine and prepares her ethically traded, local, safe and high-quality organic food.

The more conventional customer, whose class precludes her from being a consumer of alternative food, would then have to be one of two types of people. The first is a person who lives from hand-to-mouth and makes use of informal food networks because there are no other available sources of food. Of course, she is not perceived as consuming at all, merely surviving. The second type of consumer is a lower-middle class, public transport-using person, who wields a large shopping cart through an overcrowded discount supermarket looking for store brand goods in bulk. She stops for lunch, with her tirade of children, at McDonalds, and returns home fat, unhealthy and swarming with untraceable bacteria and hungry children.

AFN catering to people fitting the first profile, and a conception of AFN engendering this class of consumer, does not only exclude two-thirds of the world's population but overlooks the presence of socio-economic classes who do not meet the profile of the kind of AFN consumer suggested in the literature. Surely this kind of notion cannot be considered the forerunner of conceptually mapping alternative geographies of food, without it also being implicated as a geographically and socially exclusive, *elitist* geography of food. Similarly, one of the vanguards of northern AFN, the Slow Food Movement, is critiqued as being "[a] movement [that] encourages careful, 'reflexive', spare-no-expense food production both on the farm and in the kitchen – a social and political ideal that... has now become a statement of elite class structure" (Freidberg 2003: 5). In a rather scathing account, Goss (2004) argues that agro-food systems and research which has an established exclusive market in mind are symptomatic of the celebration of a consumption culture. He shows how the celebration of consumption within northern agri-food studies set the paradigm for all other kinds of contemporary research. Of course, taking advantage of the 'festivity' is not possible for most of the planet, making the celebration, like the preferred AFN profile, one to which the large majority of the world's population has no access.

Since emergent literature on AFN was hitherto based within European and North American contexts, it is natural that the theory is eschewed to these geographical domains and reflects issues perceived to be relevant in the north. However, this chapter has argued that a contextual focus that includes AFN perspectives from the south – food systems for the poor and for culturally diverse communities – are progressively more relevant in a developed world context of racial diversity and urban poverty. A south-based articulation of AFN is equally oppositional to the conventional food system but includes AFN for poor and culturally diverse people, for reasons not altogether different to European and American contexts: accessibility and cultural food provisioning. While evidence of high-quality, local, organic networks does exist, dominant indicators of AFN emerge from grass roots, survivalist enterprise that link production and consumption so that these networks cater for the urban poor and culturally diverse communities. Allen (2004) similarly argues that AFN play an important role in the ability of poorer members of the community to access safe food. Within global agri-food studies and the literature on AFN emerging from the north, arguments of this nature, highlighted most evidently in southern contexts, have not yet been adequately engaged with.

Korf and Oughton (2006) argue that developing world methodologies and theory can benefit transforming and emerging conceptions of the European countryside. If this does not occur, the growing body of agri-food knowledge may obscure diverse social realties in contexts where a more inclusive theory of alternative agro-food geographies would be useful in serving real social agendas (not just in knowledge production). While critical researchers may be reflexive about this danger, an emergent theoretical paradigm on *Alternative Food Geographies* developed by researchers primarily from the north has the potential to become universalised. Without a broadening of the parameters which define the theory, it also has the potential to become an exclusive, elitist academic discussion. As such, a south-based conception of AFN has a much greater global usefulness beyond the (albeit important) critical geography contention that research from the south should 'theorise-back' to the north.

Acknowledgements

This chapter has been made possible by research funded by Wits University and the Mellon Award in 2004. Earlier versions of this chapter were presented at the RGS-IBG Conference, London in August–September 2005 and the Society of South African Geographers Conference in September 2005. Thanks to Ms W. Phillips in the Wits Cartography Unit for the preparation of area maps. Special thanks to my mentor Charlie Mather from the University of the Witwatersrand and to Tom Molony from the University of Edinburgh for their helpful comments and close critique. I am also grateful to the editors of this collection and to the anonymous reviewer of this chapter. Their constructive comments and suggestions are greatly valued.

References

Allen, P. (2004). *Together at the Table: Sustainability and Sustenance in the American Agrifood System*. Pennsylvania: Pennsylvania State University Press.
Allen, P., FitzSimmons, M., Goodman, M. and Warner, K. (2003). Shifting plates in the agrifood landscape: the tectonics of alternative agrifood initiatives in California. *Journal of Rural Studies* 19, 61–75.
Berg, L.D. (2004). Scaling knowledge: towards a *critical geography* of critical geographies. *Geoforum* 35(5), 553–558.
Braun, B. (2003). Introduction: tracking the power geometries of international critical geography. *Environment and planning D* 21(2), 131–133.
Brown, A. (2001). Counting farmers' markets. *The Geographical Review* 91 (4), 655–674.
Campbell, H. and Coombes, B. (1999). Green protectionism and organic food exporting from New Zealand. *Rural Sociology* 64(2), 302–319.
Coen-Flynn, K. (2005). *Food Culture and Survival in an African City*. Manchester: Palgrave/Macmillan.
Crewe, L. and Gregson, N. (1998). Tales of the unexpected: exploring car boot sales as marginal spaces of contemporary consumption. *Transactions of the Institute of British Geographers* 23 (1), 39–53.
Cummins, S. and Macintyre, S. (2002). 'Food deserts' – evidence and assumption in health policy making. *British Medical Journal* 325 (7361), 436–438.
Daniels, P. W. (2004). Urban challenges: the formal and informal economies in mega-cities. *Cities* 21 (6), 501–511.
D'Haese, M. and van Huylenbroeck, G. (2005). The rise of supermarkets and changing expenditure patterns of poor rural households: a case study is the Transkei area, South Africa. *Food policy* 30, 97–113.
Evans, N., Morris, C. and Winter, M. (2002). Conceptualizing agriculture: a critique of post-productivism as the new orthodoxy. *Progress in Human Geography* 26 (6), 313–332.
Faiguenbaum, S., Berdegué, J. A., Reardon, T. (2002). The rapid rise of supermarkets in Chile: effects on dairy, vegetable and beef chains. *Development Policy Review* 20 (4), 459–471.
Farina, E. (2002). Consolidation, multinationalisation and competition in Brazil: impacts on horticulture and dairy production systems. *Development Policy Review* 20 (4), 441–457.
Fisher, A. (1999). Hot peppers and parking lot peaches: evaluating farmers' markets in low income areas. *Community Food Security Coalition*. Available online at: http://www.foodsecurity.org/HotPeppersPeaches.pdf (accessed: May 2006).
Freidberg, S. (2003). Editorial – Not all sweetness and light: new cultural geographies of food. *Social and Cultural Geography* 4(1), 3–6.
Goodman, D. (2004). Rural Europe redux? Reflections on alternative agro-food networks and paradigm change. *Sociologia Ruralis* 44 (1), 3–16.
Goodman, D. and DuPuis, M. (2002). Knowing food and growing food: beyond the production-consumption debate in the sociology of agriculture. *Sociologia Ruralis* 42 (1), 5–22.
Goss, J. (2004). Geography of consumption I. *Progress in Human Geography* 28 (3), 369–380.
Grey, M. (2000a). The industrial food stream and its alternatives in the United States: an introduction. *Human Organization* 59 (2), 143–150.
Grey, M. (2000b). "Those bastards can go to hell!" Small-farmer resistance to vertical integration and concentration in the pork industry. *Human Organization* 59 (2), 169–176.

Guyer, J. I. (Ed.) (1987) *Feeding African Cities: Studies in Regional Social History*. Manchester: Manchester University Press.

Halm, D. and Sauer, M. (2004). Freiwilliges engagement von Türkinnen und Türken in Deutschland, Project of the Stiftung Zentrum für Türkeistudien im Auftrag des Bumdesministeriums füur Familie, Snioren, Frauen und Jugend. Available online at: http://www.bmfsfj.de/Publikationen.engagementtuerkisch/zusammenfassung.html (accessed: April 2006).

Hamm, M.W. and Baron, M. (1999). Strategies for developing an integrated, sustainable urban food system: a case study from New Jersey. In M. Koc, R. MacRae, L.J.A. Mougeot and J. Welsh (Eds), *For Hunger-Proof Cities: Sustainable Urban Food Systems* (pp. 54–59). Ottawa: IDRC Press.

Hinrichs, C.C. (2000). Embeddedness and local food systems: notes on two types of direct agricultural market. *Journal of Rural Studies* 16, 295–303.

Hughes, A. (2005). Geographies of exchange and circulation: alternative trading spaces. *Progress in Human Geography* 29 (4), 496–504.

Ilbery, B. and Bowler, I. (1998). From agricultural productivism to post-productivism. In B. Ilbery (Ed), *The Geography of Rural Change* (pp. 57–84). Brian. Essex: Addison Wesley Longman Ltd.

Ilbery, B. and Holloway, L. (1997). Responses to the challenge of productivist agriculture. *Built Environment* 23 (3), 184–295.

Ilbery, B. and Kneafsey, M. (1999). Niche market and regional speciality food products in Europe: towards a research agenda. *Environment and Planning A* 31, 2207–2222.

Ilbery, B. and Maye, D. (2005). Alternative (shorter) food supply chains and specialist livestock products in the Scottish-English borders. *Environment and Planning A* 37, 823–844.

Ilbery, B. and Maye, D. (2006). Retailing local food in the Scottish-English borders: a supply chain perspective. *Geoforum* 37 (3), 307–440.

Korf, B. and Oughton, E. (2006). Rethinking the European countryside – can we learn from the South? *Journal of Rural Studies* 22 (3), 278–289.

Lockie, S. and Halpin, D. (2005). The conventionalisation thesis reconsidered: structural and ideological tranformation of Australian organic agriculture. *Sociologia Ruralis* 45(4), 284–307.

Mather, C. (2005). The growth challenges of small medium enterprises (SMEs) in South Africa's food processing complex. *Development South Africa* 22 (5), 607–622.

Milbourne, P. (2004). The local geographies of poverty: a rural case study. *Geoforum* 35(5), 559–575.

Minco, C. (2003). Critical geographies. *Environment and Planning D* 21(2), 160–168.

Morland, K., Wing, S., Diez Roux, A., Poole, C. (2001). Neighborhood characteristics associated with the location of food store and food service places. *American Journal of Preventative Medicine* 22 (1), 23–29.

Muslim Food Board (The) (2005). Available online at: http://www.tmfb.net/js/whatishl.html (accessed: October 2005).

Paasi, A. (2005). Globalisation, academic capitalism, and the uneven geographies of international journal publishing spaces. *Environment and Planning A* 37, 679–789.

Pearson, T., Russell, J., Campbell, M. J. and Barker, M. E. (2005). Do 'food deserts' influence fruit and vegetable consumption? – A cross-sectional study. *Appetite* 45, 195–197.

Porter, G., Lyon, F., Potts., D. and Bowyer-Bower, T. (2004). *Improving Market Institutions and Urban Food Supplies for the Urban Poor: A Comparative Study of Nigeria and Zambia: Scoping Phase.* London: DFID.

Reardon, T. and Barrett, C. B. (2000). Agroindustrialization, globalization, and international development: an overview of issues, patterns and determinants. *Agricultural Economics* 23, 195–205.

Renting, H., Marsden, T. and Banks, J. (2003). Understanding alternative food networks: exploring the role of short food supply chains in rural development. *Environment and Planning A* 35 (3), 393–411.

Rigg, J. and Ritchie, M. (2002). Production, consumption and imagination in rural Thailand. *Journal of Rural Studies* 18, 359–371.

Rogerson, C.M. (2003). Towards "pro-poor" urban development in South Africa: the case of urban agriculture. *Acta Academica Supplementum* 1, 130–158.

Rogerson, C. M. and Preston-Whyte, E. (Eds) (1991). *South Africa's Informal Economy*. Cape Town: Oxford University Press.

Rosset, P. M. (1998). Alternative agriculture works: the case of Cuba. *Monthly Review* 30 (3), 137–146.

Sage, C. (2003). Social embeddedness and relations of regard: alternative 'good food' networks in south-west Ireland. *Journal of Rural Studies* 19, 47–60.

Samers, M. (2005). The myopia of "diverse economies", or a critique of the "informal economy". *Antipode* 37 (5), 875–886.

Vaiou, D. (2003). Radical debate between 'local' and 'international': a view from the periphery. *Environment and Planning D* 21(1), 133–137.

Watts, D.C.H., Ilbery, B. and Maye, D. (2005). Making reconnection in agro-food geography: alternative systems of food provision. *Progress in Human Geography* 29 (1), 22–40.

Weatherell, C., Tregear, A. and Allinson, J. (2003). In search of the concerned consumer: UK public perceptions of food, farming and buying local. *Journal of Rural Studies* 19, 233–244.

Weatherspoon, D.D. and Reardon, T. (2003). The rise of supermarkets in Africa: implications for agrifood systems and the rural poor. *Development Policy Review* 21 (3), 333–355.

Whatmore, S., Stassart, P. and Renting, H. (2003). Guest editorial: What's alternative about alternative food networks? *Environment and Planning A* 35, 389–391.

Wilkinson, J. (2005). *Community supported agriculture. OCD Technote 20*. Washington: Office of Community Development, U.S. Department of Agriculture, Rural Development. Available online at: http://www.rurdev.usda.gov/ocd/tn/tn20.pdf (accessed: April 2006).

Winter, M. (2003). Embeddedness, the new food economy and defensive localism. *Journal of Rural Studies* 19, 23–32.

Wrigley, N. (2002). 'Food deserts' in British cities: policy context and research priorities. *Urban Studies* 39 (11), 2029–2040.

Yeung, H.W. and Lin, G.C.S. (2003). Theorizing economic geographies of Asia. *Economic Geography* 79 (2), 107–128.

Chapter 7

Justifying the 'Alternative': Renegotiating Conventions in the Yerba Mate Network, Brazil

Christopher Rosin
Centre for the Study of Agriculture, Food and the Environment, University of Otago, Dunedin, New Zealand

Introduction

Since the publication of Whatmore and Thorne's (1997) frequently cited chapter, alternative food geographies, both as real features of contemporary agriculture and as a concept, have received much attention among scholars of agri-food networks. In this field, the identification and examination of the food networks associated with such geographies is particularly attractive because they involve functioning, and apparently viable, alternatives to a 'conventional' food network that is increasingly associated with unsustainable social and environmental outcomes. Thus, alternative food networks (AFNs) are recognised in such features as organic management practices, the creation of short commodity chains, fair trade branding and geographical designations. These practices are seen as means through which producers, and in many cases consumers, can exert some control over food networks that are dominated by corporate interests. As alternatives to the 'conventional' food network, AFNs are upheld for their potential to contribute to more equitable and/or sustainable food provision. The value of research on AFNs, as well as their promise as beneficial alternatives, is evident in the remaining works in this volume and the prominent position of AFNs in reviews of progress in rural (Winter 2004) and economic geography (Hughes 2005).

In spite of the attraction of AFNs, several limitations to the concept have emerged. The most common challenge involves the potential for the strategies employed in the construction of AFNs to be incorporated by the 'conventional' food network (see Guthman 2004 for an example from the organic sector). In addition, there is some question about the strength of the distinction between 'alternative' and 'conventional' networks. In a review of the literature on AFNs, Watts *et al.* (2005) argue that the potential for alternative strategies to remain viable outside conventional networks depends on the latter's capacity to imitate the alternative. Thus, they suggest that AFNs defined by a set of accepted practices are likely to be mimicked; whereas short supply chains relying on more direct relationships between producer and consumer are less easily replicated by conventional marketing practices. The authors' assessment of the viability of types of AFNs, while not critical of the concept, does indicate the potentially ephemeral nature of such networks. A final limitation is more directly related to the concept of AFNs and disputes the extent to which such networks can be generalised. More specifically, the fact that the consumption nodes of a majority of AFNs are located in First World markets suggests that Third World producers, despite realising significant benefits from Fair Trade networks, may have limited capacity to construct and determine the parameters of the AFNs. Cidell and Alberts (2006) address this point from the perspective of chocolate as a processed commodity. These authors argue that, because the quality of chocolate (however it is defined) largely reflects distinct *processing* methods, alternative sourcing of raw material in the tropics limits farmers' capacity to claim the benefits associated with AFNs.

In this chapter, an alternative approach to AFNs is proposed that, rather than examining the limitations outlined above, attempts to offer a more universal perspective on emerging trends in the global food system of which AFNs are a part. This approach is grounded in French convention theory (CT) and views AFNs as routine features of the global food system which rely on the strategic promotion of a product's unique characteristics or its geographical origin in order to differentiate it from more uniformly designated products (e.g. Champagne from sparkling wine, or Fair Trade coffee from coffee more generally). This viewpoint does not challenge the significance of existing literature on AFNs and, in fact, extends existing analyses of the construction of food qualities within AFNs using a CT perspective (see Murdoch *et al.* 2000 and Barham 2002). It does, however, suggest the value of examining 'marginally' alternative situations to gain insight into the processes through which AFNs become established. I argue that this is especially true of food networks that are only marginally defined as 'alternative' relative to the ideal examples of AFNs often presented in the literature. Here, the example of the yerba mate food network (a tea that is an important cultural and nutritional staple for populations in Argentina, southern Brazil, Paraguay and Uruguay) is used (see Figure 7.1 for examples of yerba mate packaging) to demonstrate the potential of CT to inform the analysis of AFNs more generally.

Convertion Theory (CT) and Alternative Food Geographies

CT, at a general level, posits that social relations, including economic production and exchange, are facilitated by the negotiation of conventions that diminish the uncertainties inherent to social interaction. These conventions may involve anything from unarticulated expectations of another's actions based on the understanding of that person's 'rules' of engagement (see Wilkinson 1997a for a discussion of the relative merits of CT and game theory approaches) to the formalised rules of business contracts or international trade treaties. To some extent, the relative formality of conventions can be claimed to increase in relationships that involve recurrent interactions

Figure 7.1: Examples of packaging used by Santa Catarina yerba mate processors.

among individuals. Thus, from a CT perspective, what is commonly referred to as the 'conventional' food network in AFN research would be addressed as a temporal consolidation of conventions facilitating the provision of food.[1] While it would be impossible to argue that this 'conventional' network is advantageous to all participants in the network, its conventions do establish a set of expectations that allow

---

[1] The labelling of an established set of economic conventions in this manner can be compared to that undertaken by Storper and Salais (1997) in their designation of 'worlds of production', albeit the designation of a 'conventional' food network involves a further level of generalisation.

actors to develop strategic means of operating within it. In other words, because actors develop an 'understanding' of the network's operation and expectations of the actions of other participants, they are able to better negotiate some of the uncertainties associated with participation in the network. AFNs, by contrast, challenge the predominant conventions of the 'conventional' network and attempt to establish alternative sets of expectations governed by a distinct group of conventions. These alternative conventions will persist to the extent that they are accepted by network actors as they negotiate the social relations involved in the movement of food products from field to table.

The application of CT developed by Luc Boltanski and Laurent Thévenot, in particular, provides an excellent vehicle with which to address existing considerations within the examination of AFNs. A central emphasis of this approach to CT has involved the examination of challenges to (or denunciations of) established conventions in public forums (Boltanski and Thévenot 1999, 2006 [1991]). In an early work, Boltanski and Thévenot identified six regimes of justification (civic, market, industrial, domestic, fame and inspiration, which are presented in more detail in Table 7.1) that were strategically employed by actors seeking to either challenge or defend existing conventions. As *public* justifications, each of these regimes was associated with a political philosophy that proposed a means of ordering interactions with reference to the good of a common humanity. As such, the application of a given justification involves the valorisation and ordering (or an 'order of worth' in CT terms) of humans and objects which interact in a given network. The focus on relative values in this approach contributes to its frequent application in the analysis of the qualities attached to commodities (e.g. Favereau *et al.* 2002; Levy 2002). This aspect of CT is of particular salience to the analysis of AFNs, which often reflect attempts to distinguish the product of alternative networks from that of the 'conventional' one. Whereas this representation of CT has been critiqued for its reliance on idealised polities for structuring society (see discussion in Latour 1998 and Wagner 1999 for a critique specific to a subsequently identified environmental regime), the relevance of the regimes of justification to the negotiation of real world social relations is demonstrated in investigations of AFNs and in the case of the Brazilian yerba mate network examined in this chapter.

Agri-food systems analysis has already proven a fertile ground for insights drawn from a CT approach. Perhaps the most radical argument for its application is presented in Wilkinson's (1997a/b) case for integrating CT with actor-network theory and neo-Schumpeterian economic theory as a new and improved perspective on economic relations in the agricultural sector. To date, however, its application has predominantly focused on two distinct aspects of agri-food analysis: the classification of firm behaviour based in Storper and Salais' 'worlds of production' and analyses of quality designations. The former approach is most commonly identified with the work of Murdoch and Miele (1999) that examined the developments of an egg processing firm and an organic food retailing firm in Italy. These firms provide ready examples of movement between 'worlds of production' that are mapped on a conceptual Cartesian space defined by axes extending from standardised to specialised and dedicated to generic production. Stræte (2004) employs a similar approach to compare dairy processing firms in Norway, one of which focuses on bulk milk supplies to a nearby urban market while another has developed specialised processed milk products in order to retain the firm's viability. More recently, several authors have argued for an expansion of CT approaches

Table 7.1: Regimes of justification[a]

	Civic	Market	Industrial	Domestic	Fame	Inspiration	Environmental
Representative author	Rousseau	Smith, Adam	Saint-Simon	Bossuet	Hobbes	Augustine	N/A[b]
Common principle	Pre-eminence of collective	Competition	Efficiency	Engenderment according to tradition	Public opinion, renown	Inspiration	Inherent value of nature
Order of worth							
Humans	Civic rules and representation	Wealth	Efficiency	Hierarchy; superiority	Fame; reputation	Achieving perfection or happiness	Relation to nature
Objects	Legal forms; rights	Luxury; value	Standards; tools/methods	Association with rank or title	Brand names	Relation to ideas and dreams	Inherent value as part of nature
Relationships	Solidarity; membership	Possession	Control; function	Etiquette; respect	Recognition; identification	Uniqueness	Systems; humans as part of nature
Challenge to other regimes	Demonstration of just cause	Price or value	Effectiveness	Authority and tradition	Public opinion	Abandoning habits	Sustainability

[a] Each regime of justification is characterised by a political philosophy associated with a specific designation of a common humanity (except, according to Latour 1998, the environmental regime). Based on that philosophy, it is possible to identify a regime's underlying principle, its means of establishing an order of worth for humans and objects, its basis for relationships and the rationale for challenging contrasting regimes.
[b] The environmental regime was not part of Boltanski and Thévenot's (2006 [1991]) initial framework, but was added by the latter author to account for justifications developed by Green and environmental political movements. As such, the regime is not associated with a single author.

that more closely approximates that employed here. Stræte and Marsden (2006), for instance, attempt to move CT beyond categorisation to analysis of the processes through which firms occupy a given 'world'. Morgan *et al.* (2006: 23; their emphasis) similarly argue for an expansion of CT and 'worlds of production' from *'economic . . .* to *cultural, ecological,* and *political/institutional* logics'.

As a vehicle for the examination of quality designation, CT approaches incorporating Boltanski and Thévenot's regimes of justification have also addressed AFNs. Employing a perspective similar to that of Busch (2000) and Mansfield (2003), this literature examines the social construction of quality. In an important early application of CT in the analysis of AFNs, Murdoch *et al.* (2000) argued that the explanation and promotion of qualities in foods require reference to nature as an active element of production. These authors claim that, whereas both CT and actor-network theory more successfully incorporate nature within analyses than does a political economy approach, concepts of strategic justification of worth in CT facilitate a more developed analysis of quality. Marsden *et al.* (2000) similarly propose CT as a means of assessing the ordering of quality designations in AFNs that emphasise the naturalness of a product. The authors follow the movement of participants in such networks between emphases on quality or price and general and interpersonal interactions. Raynolds has used applications of CT to explain the potential for 'fair trade' (Raynolds 2002) and organically certified (Raynolds 2004) products to compete with more 'conventional' products in international markets. She argues that the success of the designated products stems from their ability to appeal to consumers who value civic and green worlds of justification. Similar arguments are used to examine 'fair trade' (Renard 2003) and geographic (Barham 2003) quality designations for agricultural products. More recently, Ponte and Gibbon (2005) have employed regimes of justification in global value chains to explain the dominance of buyer or producer driven chains in specific sectors of the international economy. The authors compare the global clothing and coffee value chains, arguing that the first relies on inter-firm relationships and compatibilities (industrial justifications), whereas the latter is more subject to emphasis on the geographical origin (domestic justifications) of the coffee beans. However, by reducing the analysis to competing regimes of justification (civic vs. market, domestic vs. industrial) which conform to the dichotomy between a conventional and an alternative form of marketing, each of these applications of CT fails to acknowledge the broader negotiation of justifications among all participants in a given food network.

In several recent works, authors have developed approaches that focus more specifically on regimes of justification as an *explanatory mechanism* as opposed to classificatory scheme. Kirwan (2006), for example, examines the interactions between producers and consumers that occur at farmers' markets in the UK. He argues that the objectives of buyers and sellers at these markets diverge from the civic or domestic justifications with which they are readily associated in AFN analyses. The case study thus shows how the potential to realise higher sales volume and profit is the primary motivation of sellers, whereas buyers are looking for greater variety and freshness and will compare products on offer with those in supermarkets. Furthermore, Kirwan identifies additional values attached to the market experience which he attributes to a *regard* regime of justification. While this entails a more developed application of regimes of justification, normative values are applied to specific regimes. Thus, rather than operating as appeals to public acceptance of a form of production, the regimes are presented as individual rationalisations of action. In contrast, Murdoch and Miele (2004) demonstrate the manner in which regimes of justification are employed in public appeals to assign value to two distinctive food types: fast food exemplified by McDonalds' and slow food as represented by the slow food movement. In both cases, the authors examine the strategic use of regimes as a means to facilitate network possibilities and environments of action. This latter approach, which avoids restricting analysis to the normative differentiation between fast and slow foods, most closely resembles that pursued in this chapter.

The following analysis of the yerba mate food network in Brazil is an effort to develop a more dynamic examination of alternative food geographies through an application of CT that moves beyond the dichotomy of alternative and conventional (see Stræte and Marsden 2006 for a similar argument). The approach used in the chapter proceeds from the assertion that CT sees uncertainty and incompleteness as inherent features of market exchange, and social interaction more generally (Storper 2000). Certainty is only arrived at as actors refer to conventions, which may rely on compromises in the form of objects or tests. Because conventions become infused with certainty through reference to a given regime of justification and its associated valorisation of humans and objects, they are temporal solutions to uncertainty. As conditions change (e.g. externally as a network is expanded or internally as the existing order of worth becomes unacceptable or is not fairly employed), the aggrieved actors can challenge and upset this temporal stability. This acknowledgement of and interest in the emergent nature of conventions facilitates an approach to 'the procedures of qualifying the [world of things as] so many occasions . . . for quarrels between varied and contradictory logics' (Dosse 1999: 108). Wagner (1999) attributes similar advantages to CT to the extent that, as a critical approach to sociology, it examines the justifications operating in social interactions without committing to a normative implication; that is, no regime of justification is inherently better than another. As such, regimes of justification are viewed as tools employed to secure public 'support' for positions of negotiation. Conventions are, therefore, not exclusively imposed by the powerful. Rather they require the collaboration of other actors and the potential compromise between competing orders of worth. In the case examined below, the efforts of a relatively empowered set of participants (i.e. yerba mate processors) can, thus, be interpreted as initial (strategic) attempts to renegotiate the conventions operating within Brazil's yerba mate network.

A Brief Narrative of Competition in Brazil's Yerba Mate Market

South America's yerba mate market is largely defined by a domestic orientation in which consumers generally have a selection of only domestic, and often regional, brands. This orientation is due, in part, to historical tariff protection on the order of 20 per cent or more. A further result has been the emergence of quality standards reflecting the relative strengths of the respective processing industries. However, the implementation of MERCOSUL[2]. in 1995 initiated a period of heightened competition

---

[2]MERCOSUL is the Portuguese acronym for the Common Market of the South, a regional free trade agreement (more accurately classified as a customs union) between Argentina, Brazil, Paraguay and Uruguay founded with the Treaty of Asunción (1991) and enacted in 1995. Since 1995, Venezuela has joined the customs union and Bolivia, Chile, Columbia, Ecuador and Peru are associate members.

throughout the yerba mate market. Prior to the free-trade agreement, processors in the principal producing countries (Argentina, Brazil and Paraguay) engaged primarily in competition with domestic counterparts. Most importantly, relative to competition in the MERCOSUL market, the Argentine processors benefited from a strong centre of demand in Buenos Aires as well as from an extensive state support for a product perceived to be an important element of national food security. Their position had been further reinforced through continued protection (reduced tariffs over 5 years, reaching 0 per cent in 2000) as a designated 'threatened product' under the MERCOSUL agreement and by a well-established research and extension infrastructure. By comparison, the Brazilian and Paraguayan industries were relatively unimportant to their national economies and lagged in the development of intensive production and processing technologies. Thus, the elimination of protective tariffs combined with a large supply surplus in Argentina initiated a period of uncertainty for the Brazilian and Paraguayan industries.

During the first 5 years of the MERCOSUL agreement, the impact of heightened competition was strongest in Brazil. Argentina's processors strategically situated themselves as active players in Brazil's domestic market by offering low-cost, already processed yerba mate at enticing financial terms such as repayment over several months. In this manner, the Argentine processors (emboldened by their economic and political strengths domestically) attempted to establish their product as the most technologically advanced and, by their estimation, highest quality within the region. Evidence of the success of this strategy is demonstrated in the growth of official imports of Argentine yerba mate to Brazil from none to an estimated US$10 million in 1995 (Rücker 1995). The imported yerba mate was characterised, however, by a distinctively bitter taste for the Brazilian palate. Thus, it was primarily used either by large firms as an input to prepared beverages (iced tea or soft drinks) or by firms in less advantageous locations relative to the regions of production. The latter firms would blend imported and domestic product in their loose leaf tea, compensating for the bitterness by adding sugar. As a result, competing firms were confronted with stiffer price competition.

The threat of cheap imported yerba mate and its impact on domestic markets in Brazil elicited a multi-faceted response among that country's processors. Perceived by many as a feature of globalisation associated with MERCOSUL, the heightened competition challenged the economic viability of what had become traditional processing practices. Furthermore, the emerging economic policy environment in Brazil touted industrial justifications of free trade, claiming that competition would promote higher quality and more efficiently produced goods. As such, processors were pressured to adapt by means of mechanisation or alterations of existing conventions of production. However, the majority of processors in Brazil are medium-to-small sized firms, and few were sufficiently well capitalised to invest in retooling. As a result, Brazilian processors proposed an alternative network founded in shorter (domestic) supply chains and a distinctly Brazilian quality standard (both commonly recognised features of AFNs) in opposition to the conventional network associated with MERCOSUL and imported yerba mate. The following section examines the response of processors in Brazil's Santa Catarina State (see Figure 7.2) through the application of Boltanski and Thévenot's regimes of justification.

Figure 7.2: Map of the study area: Santa Catarina, Brazil.

Responding to MERCOSUL Competition

By exerting their competitive advantage within the economic structure of MERCOSUL, the Argentine yerba mate industry initiated a situation of perceived crisis for processors in Santa Catarina. From a CT perspective, this 'critical moment' within an existing set of conventions provides the context for challenges to network structures and attempts to alter the conventions through which they operate (Boltanski and Thévenot 1999). In the case of yerba mate processors in Santa Catarina, the initial response of individuals referred to personal justifications (see Wagner 1999: 346–347) based on localised experiences in the industry and in their own capacity to engage with a more competitive environment. In addition, processors attempted to improve the efficiency of the network by renegotiating its conventions. However, this renegotiation required other network actors (including producers, the Brazilian state and consumers) to coordinate actions and compromise with the processors (Thévenot 2001). Such coordination and compromise among the various justifications employed by these actors are facilitated by reference to a mutually recognised object (Thévenot 2002) – that is pure green, domestically sourced yerba mate. The emerging conventions reflect the competing objectives of participants in the network with regard to the value of (and price for) raw material, the relative anonymity of yerba mate relative to other agricultural commodities and the diverse beverage options on retail shelves.

Personal Justifications

The personal justifications employed by yerba mate processors were strongly influenced by the underlying political justifications of MERCOSUL. The Brazilian government has generally employed neoliberal arguments of improved quality and efficiency through competition to demonstrate the value of a regional free-trade agreement. These arguments have percolated into understandings of economic exchange in Brazil as production emphases changed from productivity to quality differentiation. This underlying industrial regime of justification often conflicted with existing exchange relations that relied on a processor's ability to foster personal relations. As such, the decision to adopt industrial perspectives and revise personal approaches to the processes and exchanges in the yerba mate network was not without difficulties.

The attempt by some of the processors to meet the challenge of imported Argentine yerba mate through capital investment corresponds closely to industrial justifications. Investments designed to increase the efficiency of processing included natural gas and steel drums, which allowed for more consistent drying of the raw material. Such changes generally did not create a need for public justification as they contributed to increasing the throughput of firms as opposed to the replacement of manual labour. Furthermore, the processors' actions were reinforced by the conviction that the flavour characteristics associated with 'improved' processing methods appealed to the taste preferences of the domestic market. An additional effect of this investment was to raise demand for raw material harvested throughout the year (as opposed to the traditional June–August harvest period) in order to allow continuous utilisation of the equipment. In this sense, processors able to make investments in machinery had incorporated the industrial justifications underlying MERCOSUL within their personal justifications. While the explicit intent of such investment was to compete with an Argentine industry, it also provided efficiency advantages relative to domestic competitors.

Efforts to raise industrial justifications to a public level in the yerba mate network were contested, however, both by processors unwilling to engage in capital investment and by those committed to more traditional processing standards. Among the former group, the innovations were viewed as unwarranted capital risks. For the latter, even greater risk was associated with the possibility that newer processing methods and associated alterations in production and harvesting practices would affect the flavour profile of their product. Many holding to more traditional processing methods, such as the following processor, believed their target consumers would not approve of the change:

> 'The most important factor in quality is usually the amount of time since the last harvest. We prefer to buy yerba mate that is at least two years old–up to three or four years since the last harvest. The time between harvests is what gives the yerba mate that we sell its distinctive flavour. We maintain these quality standards to meet with the demands of our customers–the import agents' (Canoinhas – November 1998).

The contested nature of yerba mate quality and its relation to industrial process acts as one of several factors limiting the strength of efforts to renegotiate conventions in the network. Thus, the apparent advantage of processors in initiating the negotiation was subject to internal division.

Public Justifications of Domestic Production

Simply investing in new processing technology was not considered a sufficient solution to the challenge of Argentine imports. A further means of incorporating the industrial justifications associated with MERCOSUL involved addressing some of the inefficiencies that processors perceived within the network more generally. As such, processors in Santa Catarina actively engaged in the reconstruction of the conventions under which yerba mate becomes a consumed product. In an effort to develop the strength of their public justifications, many committed to collaborative action with otherwise competing colleagues through more active participation in SIMSC (Yerba Mate Processors Syndicate of Santa Catarina).

The underlying basis for the attempted reconstruction of conventions involved a common feature in the development of AFNs, namely the promotion of the higher quality of the domestic (as local), relative to imported, product. Within the public arena, processors in Santa Catarina strategically employed justifications that valorised domestic yerba mate relative to that imported from Argentina. These justifications reflect some of those commonly applied in public forums (Boltanski and Thévenot 1999; 2006 [1991]). In particular, processors variably employed civic, domestic, market and industrial justifications as they sought to coordinate the actions of significant players in the negotiation of conventions related to the yerba mate network. In the case of each of these sets of actors, the relative power of the processors in the network was challenged by the need to compromise with the contrasting regimes of justification from which the network was experienced and understood.

Justifying New Conventions of Production

From the outset, processors recognised the importance of involving small producers – as the primary source of raw material for the industry – in the negotiation process. Processors, coordinating their actions through SIMSC, sought collaboration from producers by encouraging greater efficiency through more intensive plantation systems and the shifting of responsibility for harvest labour to producers. The principal form of public justification directed towards producers was located in a domestic regime, emphasising the value of relationships between domestic producers and (local) processors. In this regard, the processors attempted to demonstrate their sympathy for the situation of the small producers while enrolling them in a crusade for a more competitive industry. The value of this crusade was reinforced by nationalist arguments that stressed the dichotomy between a good Brazilian product and Argentine imports of questionable quality. Thus, the objective of the SIMSC members was to create a greater awareness of and pride in local production.

The primary emphasis of processors' messages to their suppliers involved the active promotion of greater tree density in yerba mate plantations. Such practices were promoted as being more modern and efficient – especially for harvesting – than traditional, wild management practices in which tree structure and density reflected the existing stock of yerba mate trees at the time of forest clearance. In addition, the following processor listed the suite of practices that formed part of a larger management package directed at a crop that the processors viewed as lacking adequate attention:

'It [yerba mate] is an advantageous alternative for them . . . It is especially advantageous for those who practice harvesting every one to one-and-a-half years. [But] the farmers need to replant to augment the density of existing trees, to fertilize the young trees, and to control weeds in order to facilitate yerba mate's competitive ability' (Catanduvas – December 1998).

By framing existing management practices as backward, processors located the achievement of a significant aspect of efficiencies comparable to those enjoyed by the Argentine industry in the hands of small producers. In other words, pre-MERCOSUL conventions of raw material provision did little to encourage more 'rational' production technologies. Realising the adoption of such technologies in the sector depended on the initiative of the small producer.

However, existing forms of raw material supply and procurement did not favour the alteration of conventions. Whereas yerba mate once occupied a position as a primary cash crop, it was now considered a secondary alternative relative to higher value or mechanised crops such as tobacco or soybeans. The deteriorating status of yerba mate reflected both the weak export markets and the promotion of crops with greater export earning potential and capacity for modernisation. In this process of redefining the values of cash crops, yerba mate has become an untended (or minimally tended) crop, often planted at the margins of more intensively managed fields. Few producers currently associated their management of yerba mate with high levels of skill. Furthermore, the crop was the target of minimal investment (with the expectation of relatively low earnings), providing little motivation to change. From a conventions perspective, it appears that the order of worth (among alternative cash crops) established within the personal justifications of producers contradicts the public domestic justifications of the processors. Thus, the extent to which the intent of the processors may be realised depends on their ability to overcome the poor perception held by producers of yerba mate as an element of farm management.

Regulating an Unfamiliar Product

In addition to producers, members of SIMSC sought to engage the Brazilian government in their renegotiation of conventions, a process which necessarily involved public justifications. Hoping to appeal to the neoliberal perspective in Brasilia, they initially employed a market justification and argued that the price of imported yerba mate did not accurately represent its value relative to costs of production. With the encouragement of Dorli DaCroce (a forestry engineer working at Santa Catarina's agricultural research and extension organisation, EPAGRI) and together with similar groups from Paraná and Rio Grande do Sul States, SIMSC petitioned the federal government to enforce anti-dumping regulations against the imports. The groups argued that the offer of processed yerba mate at a price below the cost of raw materials in Brazil was a deliberate programme by Argentine firms to sell surplus production. They asked that protections in the form of punitive tariffs (guaranteed in the MERCOSUL agreement) be enacted to more realistically represent the cost of the imports. This strategy ultimately failed due to the limited concern among politicians in Brasilia for a relatively unknown and minor agricultural product. The lack of recognition of yerba mate as a regionally viable commodity was especially important given that, for the Brazilian state, MERCO-SUL was largely a strategic initiative aimed at increasing the country's presence on the international economic stage.

As an alternative, SIMSC attempted to establish health regulations that would require the documentation of added sugar in yerba mate, ostensibly on the basis of concern for diabetic consumers. As such, they identified the crux of the threat posed by Argentine yerba mate as an issue of domestic, rather than international competition. The group's leadership argued that, whereas the aged yerba mate marketed by Argentine firms did not conform to Brazilian taste preferences, the threat of imported dried yerba mate lay in its use as an alternative to domestically produced raw material by 'less scrupulous' Brazilian processors. The bitter product from Argentina was made more palatable for the domestic market by the addition of sugar and mixing with supplies of domestic raw material. There appeared to be a greater possibility of success for this latter action as it was potentially incorporated within a more general rewriting of food labelling regulations.

While the SIMSC pursued several policy avenues intended to reduce the financial advantage realised though the utilisation of imported raw material, they faced quite stubborn resistance from the Brazilian state. From a CT perspective, the difficulty in achieving the processors' objectives lay in their inability to garner collaboration on the basis of the justifications employed. In other words, whereas government officials and politicians project public justifications which appear comparable to those employed by the SIMSC, their ordering of the worth of requests for action is dominated by a *regime of fame*. Thus, requests for actions on their part are ordered according to assessments of the relative recognition of the objects of the request in Brazilian society. Yerba mate, as a product of the South, is relatively unknown in the large population centres of São Paulo, Rio de Janeiro, Belo Horizonte and further north. Furthermore, it is not a sufficient source of foreign income to warrant trade action – that is anti-dumping legislation. By contrast, there was a better possibility for mandatory labelling of sugar additions because such regulation could be incorporated within broader food labelling laws. Thus, the processors' project to alter the regulation of yerba mate as a food commodity appears dependent on their ability to establish a wider claim to its relevance as a public issue.

Convincing the Consumer

As an alternative strategy to altering official regulation of yerba mate quality standards, processors in Santa Catarina appealed to a broader public audience by attempting to raise consumers' awareness of the quality distinctions between domestic and imported yerba mate. The processors were very intent on developing formalised means of demarcating these differences which favoured the local industry and employed an industrial regime of justification to convince consumers of the greater worth of the domestic product. Thus, the processors associated the faults of imported yerba mate with characteristics that were readily identifiable by consumers. For example, the marketing of yerba mate in supermarkets as well as advertising in popular media sources promoted smooth, non-bitter flavour profiles and a pronounced greener colour. As noted in the following quote, processors recognise that the latter feature is the result of the abbreviated drying and aging processes used by the majority of Brazilian firms.

'Here the taste is moving toward a greener product... It began 15 years ago with the development of new, more efficient drying technologies. Greener yerba mate was better for the processing firms. Now, the yerba mate isn't stored for very long at all' (Xanxeré – January 1999).

In an effort to establish the link between the characteristics of the domestic processing methods and quality, promotional sampling of brands in supermarkets involved a demonstration of the loose tea's colour and consistency in the bag as well as taste. As a growing number of processors utilised modern drying equipment, the lack of smoky nuances was also promoted as a determinant of quality – although this flavour characteristic was contested among processors as noted above.

In order to establish the relative value of Brazilian – as compared with imported – yerba mate, the processors relied on the collaboration of those who purchase their product. However, this effort was hampered by the fact that few consumers had clearly defined understandings of 'good' yerba mate. Past marketing of the product had focused on local markets as few firms had sufficient output to justify competition in distant markets. As a result, consumer preferences exhibited strong brand name associations and identification of quality in the characteristics of locally accessible products.[3]. Thus, processor efforts to employ an industrial regime of justification were limited by existing references to habituated orders of worth that reflected domestic justifications. These existing orders of worth are the potential basis for consumer-driven negotiation of conventions in the yerba mate network.

Conclusions: Alternative Food Geographies as Emergent Geographies

The analysis of the response of Santa Catarina's yerba mate processors to the new economic environment associated with MERCOSUL demonstrates the potential insights of a *process-oriented* application of CT to agri-food networks. First, the CT perspective recognises the implementation of MERCOSUL as a critical moment, the result of which is an increase in the uncertainty faced by actors within the food network. This uncertainty challenges the stability of existing conventions related to the production, processing, marketing, regulation and consumption of yerba mate. As the actors most cognisant of the impact of increasing competition associated with imported raw material from Argentina, the processors engaged in a range of strategic actions intended to raise the competitive position of domestically produced and processed yerba mate. Despite being an attempt to structure a new yerba mate network, this is not a situation that is typical of AFN research as it does not strictly challenge a 'conventional' food network. It does, however, involve an attempt to retain elements of an existing network that is being challenged by globalising pressures. As such, the processors' attempts provide a corollary instance in which the process of food geography development (and that of associated networks) can be examined.

---

[3]This aspect of yerba mate markets substantiates, to some extent, the claims that addition of sugar and a stronger flavour were equally traditional features of the industry.

Applying concepts and analytical insights from CT, the interplay and negotiation among actors in the yerba mate network are viewed as the strategic application of orders of worth, both personal and public. Under these conditions, the objective of the processors was the continued viability of the domestic industry under conditions that favoured their own position and role in the network. Thus, the processors felt impelled to initiate a strategic response to the perceived threat, including the alteration of individual actions and the negotiation of new conventions of interaction with the remaining actors in the network. In the preceding analysis, the former actions were shown to reflect personal orders of worth and the differential capacity of individuals to find compromises between these and the predominant market and industrial regimes of justification associated with MERCOSUL. The latter actions were revealed to be subject to the collaboration and compromise of other actors, who often pursued objectives that contradicted those of the processors. Thus, the ability of the processors to realise their preferred ends was limited by the equally justifiable orders of worth employed by other actors.

In the preceding representation of Santa Catarina's yerba mate network, the reference to an order of worth based on geographical designation [the value of local (or, at least, domestic) sourcing of raw material in a less overtly alternative project] is of particular interest to the literature on alternative food geographies. Processors, due in part to their more advantageous position in the yerba mate network and their greater access to the sources of knowledge and information relative to the MERCOSUL economic environment, have seized the initiative in renegotiating the quality standards for yerba mate. As such, they appear much better situated to define the parameters of emerging conventions in the commodity network. However, the analysis also demonstrates the extent to which the objectives of the processors are subject to the *collaboration* of other network actors. First, to the extent that yerba mate is a product subject to discernable differentiation, the processors are dependent on a cooperative object. Producers, in their capacity to select among a variety of cash crops, will likely ignore processors' recommendations that require additional commitment of resources to a less esteemed element of farm production. Similarly, it has been difficult to motivate the Brazilian state to alter the regulation of yerba mate as a commodity of secondary importance to the country's society and economy. Finally, due to the lack of a strong tradition of quality differentiation for yerba mate in Brazil's consumer culture, price differences are potentially more likely to influence consumer purchase than characteristics that are of interest to the processors.

As demonstrated in the foregoing discussion, CT emphasises the contested nature of the creation of food networks, whether or not they fit an idealised definition of 'alternative'. Far from dismissing the value of existing research on such networks, the application of a CT perspective enables a broader assessment of the dynamics underlying the processes in the development of food networks. By expanding the object of analysis that is deemed relevant to 'alternative' and more equitable and just forms of food provision, it is possible to go beyond hand-wringing reflections on the ephemeral nature of AFNs. The analysis in this chapter demonstrates that, by acknowledging the micropolitics of the negotiation of conventions within food networks, CT exposes the possibilities for strategic action by a range of actors – an act that potentially opens space for greater 'alternativeness' despite apparent power differentials among network actors. In other words, reflecting on the yerba mate network in Santa Catarina, is it possible to identify possibilities that enable producers, consumers and states to promote a more equitable and sustainable (socially and environmentally) 'alternative' food geography through a strategic engagement with the justifications initially employed by the processors?

References

Barham, E. (2002). Towards a theory of values-based labelling. *Agriculture and Human Values* 19, 349–360.

Barham, E. (2003). Translating terroir: the global challenge of French AOC labelling. *Journal of Rural Studies* 19 (1), 127–138.

Boltanski, L. and Thévenot, L. (1999). The sociology of critical capacity. *European Journal of Social Theory* 2 (3), 359–377.

Boltanski, L. and Thévenot, L. (2006) [1991]. *On Justification: Economies of Worth*. Translated by C. Porter. Princeton, NJ: Princeton University Press.

Busch, L. (2000). The moral economy of grades and standards. *Journal of Rural Studies* 16, 273–283.

Cidell, J.L. and Alberts, H.C. (2006). Constructing quality: the multinational histories of chocolate. *Geoforum* 37, 999–1007.

Dosse, F. (1999). *Empire of Meaning: The Humanization of the Social Sciences*. Minneapolis, MN: University of Minnesota Press.

Favereau, O., Biencourt, O. and Eymard-Duvernay, F. (2002). Where do markets come from? From (quality) conventions! In O. Favereau and E. Lazega. (Eds), *Conventions and Structures in Economic Organization: Markets, Networks and Hierarchies* (pp. 213–252). Cheltenham, UK: Edward Elgar.

Guthman, J. (2004). *Agrarian Dreams: The Paradox of Organic Farming in California*. Berkeley: University of California Press.

Hughes, A. (2005). Geographies of exchange and circulation: alternative trading spaces. *Progress in Human Geography* 29, 496–504.

Kirwan, J. (2006). The interpersonal world of direct marketing: examining conventions of quality at UK farmers' markets. *Journal of Rural Studies* 22, 301–312.

Latour, B. (1998). To modernise or ecologise? That is the question. In B. Braun and N. Castree (Eds), *Remaking Reality: Nature at the Millennium* (pp. 221–242). London: Routledge.

Levy, T. (2002). The theory of conventions and a new theory of the firm. In E. Fullbrook (Ed.), *Intersubjectivity in Economics* (pp. 254–272). London: Routledge.

Mansfield, B. (2003). Spatializing globalization: a "geography of quality" in the seafood industry. *Economic Geography* 79 (1), 1–16.

Marsden, T., Banks, J. and Bristow, G. (2000). Food supply chain approaches: exploring their role in rural development. *Sociologia Ruralis* 40 (4), 424–438.

Morgan, K., Marsden, T. and Murdoch, J. (2006). *Worlds of Food: Place, Power, and Provenance in the Food Chain*. Oxford: Oxford University Press.

Murdoch, J., Marsden, T. and Banks, J. (2000). Quality, nature, and embeddedness: some theoretical considerations in the context of the food sector. *Economic Geography* 76 (2), 107–125.

Murdoch, J. and Miele, M. (1999). 'Back to nature': changing 'worlds of production' in the food sector. *Sociologia Ruralis* 39 (4), 465–483.

Murdoch, J. and Miele, M. (2004). Culinary networks and cultural connections: a conventions perspective. In A. Hughes and S. Reimer (Eds), *Geographies of Commodity Chains* (pp. 102–119). London: Routledge.

Ponte, S. and Gibbon, P. (2005). Quality standards, conventions and the governance of global value chains. *Economy and Society* 34 (1), 1–31.

Raynolds, L.T. (2002). Consumer/producer links in fair trade coffee networks. *Sociologia Ruralis* 42 (4), 404–424.

Raynolds, L.T. (2004). The globalization of organic agro-food networks. *World Development* 32 (5), 725–743.
Renard, M.-C. (2003). Fair trade: quality, market and conventions. *Journal of Rural Studies* 19 (1), 87–96.
Rücker, N.G. de A. (1996). MERCOMATE: *Cooperação na Competitividade*. Curitiba: Secretaria de Estado da Agricultura e do Abastecimento.
Storper, M. (2000). Conventions and institutions: rethinking problems of state reform, governance and policy. In L. Burlamaqui, A.C. Castro and H.-J. Chang (Eds), *Institutions and the Role of the State* (pp. 73–92). Cheltenham, UK: Edward Elgar.
Storper, M. and Salais, R. (1997). *Worlds of Production: The Action Frameworks of the Economy*. Cambridge, MA: Harvard University Press.
Stræte, E.P. (2004). Innovation and changing 'worlds of production': case-studies of Norwegian dairies. *European Urban and Regional Studies* 11 (3), 227–241.
Stræte, E.P. and Marsden, T. (2006). Exploring dimensions of qualities in food. In T. Marsden and J. Murdoch (Eds), *Between the Local and the Global: Confronting Complexity in the Contemporary Agri-Food Sector* (pp. 269–298). Amsterdam: Elsevier.
Thévenot, L. (2001). Organized complexity: conventions of coordination and the composition of economic arrangements. *European Journal of Social Theory* 4 (4), 405–425.
Thévenot, L. (2002). Conventions of co-ordination and the framing of uncertainty. In E. Fullbrook (Ed.), *Intersubjectivity in Economics: Agents and Structures* (pp. 181–197). London: Routledge.
Wagner, P. (1999). After justification: repertoires of evaluation and the sociology of modernity. *European Journal of Social Theory* 2 (3), 341–357.
Watts, D.C.H., Ilbery, B. and Maye, D. (2005). Making reconnections in agro-food geography: alternative systems of food provision. *Progress in Human Geography* 29 (1), 22–40.
Whatmore, S. and Thorne, L. (1997). Nourishing networks: alternative geographies of food. In D. Goodman and M. Watts (Eds), *Globalising Food: Agrarian Questions and Global Restructuring* (pp. 287–304). London: Routledge.
Wilkinson, J. (1997a). A new paradigm for economic analysis? *Economy and Society* 26 (3), 305–339.
Wilkinson, J. (1997b). Regional integration and the family farm in the MERCOSUL countries: new theoretical approaches as supports for alternative strategies. In D. Goodman and M. Watts (Eds), *Globalising Food: Agrarian Questions and Global Restructuring* (pp. 35–55). London: Routledge.
Winter, M. (2004). Geographies of food: agro-food geographies - farming, food and politics. *Progress in Human Geography* 28 (5), 664–670.

Chapter 8

Is Meat the New Militancy? Locating Vegetarianism within the Alternative Food Economy

Carol Morris[1] and James Kirwan[2]
[1] School of Geography, The University of Nottingham, UK
[2] Countryside and Community Research Unit, University of Gloucestershire, Cheltenham, UK

Introduction

Is vegetarianism a mode of eating that aligns it with, or situates it within, the alternative food economy (AFE)? If so, vegetarianism may be acknowledged, even promoted perhaps, as a 'sustainable diet' (Gussow and Clancy 1986, quoted in Wilkins 2005) since the AFE is largely concerned with creating a more sustainable food system. As this chapter will go on to elaborate, the efforts of vegetarianism to create more ethical relationships within the food system embody much of the contemporary interest in the AFE and its associated food geographies. Indeed, the publication in the 1960s of vegetarian advocacy books such as *Diet for a Small Planet* has been highlighted as being a part of the broader movement for social justice and environmental regulation that stimulated the emergence of alternative food initiatives as we know them today (Allen *et al.* 2003). The oppositional and alternative nature of vegetarianism has been a recurrent theme throughout the history of this ancient consumption practice, as its association with various alternative or counter-cultural movements exemplifies, and it continues to occupy a position in the popular imagination as an 'alternative' practice. Through its rejection of the consumption of meat – the dietary norm – vegetarianism challenges the dominant food ideology of Western culture. Nevertheless, there is also an apparent dissonance between vegetarianism and some of the tenets underlying the AFE. This is worthy of investigation, if the former's place is to be justified within the latter's frame of reference, something which this chapter aims to explore.

Although vegetarianism has been the focus of previous research (e.g. Twigg 1983; Adams 1990; Beardsworth and Keil 1992; Spencer 1993; Eder 1996; Maurer 2002;

Smart 2004), this has not been undertaken, on the whole, by geographers interested in the operation of the agro-food system and the reconfiguration of this system along more sustainable lines through the development of alternative production–consumption relationships (notable exceptions include the work of McManus 1999 and Miele 2001). There are various reasons for this neglect. First, vegetarianism is an ancient diet,[1] albeit one that has expanded in popularity in recent years. In contrast, most of the phenomena currently conceptualised as 'alternative' are relatively new to the agro-food scene, at least in terms of their public, policy and academic interest. Second, social science scholarship on vegetarianism has been undertaken by sociologists, historians and anthropologists of food, rather than those concerned with agriculture and the food system, and it is the latter who have been significant in theorising the emergence of, and investigating empirically, agro-food system alternatives. Third, vegetarianism is a predominantly consumer-oriented movement, albeit one that is built upon a set of concerns about production and production practices. It may therefore have been of less immediate relevance to those agro-food scholars who, at least until recently,[2] have been more interested in questions of production than consumption.

In the rest of the chapter, the key features of the AFE will be outlined before discussion takes place of the complementarities between vegetarianism and the AFE and the tensions and discordances in this relationship. In the penultimate section of the chapter, we explore some possible ways in which theoretical sense might be made of these findings and finally make suggestions for further research into the relationship between diet and the AFE.

The AFE and Its Associated Food Geographies

Over the last decade, there has been a growing interest within rural geography and sociology, and related disciplines, in the emergence of 'alternative' food systems and their potential to offer sustainable solutions to some of the environmental, social and economic problems associated with the globalised or mainstream agro-food system (e.g. Whatmore and Thorne 1997; Hinrichs 2000; Murdoch *et al.* 2000; Morris 2002; Allen *et al.* 2003; Renting *et al.* 2003; Whatmore *et al.* 2003; Winter 2003; Goodman 2004; Kirwan 2004).

---

[1] The word 'diet' comes from the Greek *diaita*, meaning mode of living, although it is now more usually used with specific reference to food (Chambers Cambridge 1988). Individual dietary choices may be the result of religious prescription, levels of income, nutrition, health, perceptions of risk and so on. The adoption of a particular diet by consumers will inevitably impact upon production. Lockie *et al.* (2002), for example, argue that consumer motivations for buying organic produce will be crucial in determining future production practices.

[2] A small body of recently published work argues that production and consumption practices are inextricably linked, necessitating their mutual interrogation (e.g. Lockie and Collie 1999; Lockie and Kitto 2000; Lockie *et al.* 2002; Goodman 2002; Bryant and Goodman 2004), a move that represents an attempt to overcome the disciplinary divide within agro-food studies between the work of rural sociologists, who have tended to focus on the organisation of agricultural practices (and production in particular) and the sociologists of food, whose focus has been more on diet, culture and consumption (Tovey 1997).

Examples include organic and other forms of ecological agriculture, direct marketing such as farm shops, farmers' markets and box schemes, fairly traded goods and produce which comes from locally unique and distinctive places of production. While there are certainly differences in the operation of these alternatives, and each has its own geographies, what they have in common is an intention to make more explicit the connections between the production of food and its eventual consumption, or as Whatmore and Thorne (1997) describe it, in re-orientating the 'mode of ordering of connectivity' to more directly associate the impacts of particular consumption decisions on the geographies of production. For example, within organic agriculture, 'the fetishised abstraction of food is intentionally unveiled' (Goodman 1999: 32), and, notwithstanding the ongoing debate about the conventionalisation of this sector, organic farming systems are regarded as embodying values that are 'contradictory' to 'the dominant values of capitalist society' (Tovey 1997: 33). Similarly, within the context of Fair trade produce, the intention is to incorporate the notion of 'fairness' within the agro-food system, as 'Northern' consumers are encouraged to engage in a more equitable relationship with 'Southern' producers (Raynolds 2000; Renard 1999, 2003). In addition, particularly within a Western European context, there is a recognition that through embedding the production and consumption of food within specific places and relationships, rural regions can both add and retain value at a local or regional level (Marsden *et al.* 1999, 2002; Goodman 2003, 2004).

Collectively, these emerging non-conformist approaches to reordering the agro-food system can be understood as components of an AFE that is variously based on valorising places of production, (re)localisation of food networks, (re)connecting the processes of production and consumption (and concomitantly producers and consumers) and (re)integrating nature into food supply chains. Using a variety of secondary sources, including scholarly research on vegetarianism and 'nutrition ecology'[3] and promotional and other information from vegetarian organisations, the following section explores the extent to which vegetarianism both reflects, but also contradicts, these key characteristics of the AFE.

Vegetarianism and the AFE: The Complementarities and Discordances

Vegetarianism involves the exclusion of certain food products, most notably flesh foods, but also, for other vegetarians, dairy products and eggs, or the by-products of slaughtering. Concomitantly, a vegetarian diet entails the relatively greater use (cf. a 'conventional' omnivorous diet) of other food products such as seeds, fruits, pulses, nuts and grains. In the UK, according to a series of recent surveys (Mintel 2000), at least 7 per cent of the UK population is vegetarian, a doubling of the number from the decade earlier. Another source suggests that vegetarianism is currently the fastest growing dietary trend in the UK (Ashley *et al.* 2004). By examining the typical range of

---

[3] Nutrition ecology is a scientific approach to the study of the relationship between diet, nutrition and ecology, and argues that 'diet matters' in the creation of the ecologically sustainable production systems (Goodland 1997).

arguments that are used for the adoption of vegetarianism by individual vegetarians and organisations that promote vegetarianism (Beardsworth and Keil, 1992; Maurer, 2002), the complementarities between vegetarianism and the AFE will firstly be discussed.

According to its adherents and proponents, vegetarianism is both ethically superior and healthier than conventional omnivorous diets. At the core of the vegetarian ideology is a set of moral beliefs and values relating to the use of animals by humans. A number of philosophical and spiritual arguments are marshalled which oppose the killing of animals for food as both cruel and unnecessary in nutritional terms (e.g. Singer 1983; Regan 1984; Shafer-Landau 1994; Alward 2000; Benatar 2001; Zuzworksky 2001). Furthermore, some vegetarian organisations oppose particular means of rearing and keeping animals, such as factory farming. In the vegan case, it is argued that animals should not be kept for food, or other human uses, at all. Alongside animal rights and welfare, a relatively familiar dimension of vegetarianism, the ethical high ground for this dietary practice is also claimed on the basis of human rights and welfare. Vegetarianism asserts that it is morally superior to meat-eating because feeding grain to animals is less efficient than feeding it directly to humans.[4] Grain-fattened animals take more energy and protein from their feed than they return in the form of food for humans. Therefore, the adoption of a vegetarian diet is seen as a means of alleviating some of the problems of world food shortage and starvation that are likely to be exacerbated by the predicted livestock revolution in the developing world, in which meat consumption is projected to rise by 3 per cent per annum until 2020 [CIWF (Compassion in World Farming) 2004].

Ecological arguments are also used to claim that the adoption of a plant-based diet is ethically superior to its meat-based counterpart (Goodland 1997). These encompass a number of environmental concerns about livestock farming, including the production of greenhouse gases,[5] losses of tropical rainforest to cattle ranching, the collapse of global fisheries and water pollution from intensive livestock holdings. A related argument is that a vegetarian diet is a more efficient user of both land and water. For example, research in the US has shown that 0.5 ha of land is required for a meat-based diet, compared with 0.4 ha for a vegetarian-based diet (Pimentel and Pimentel 2003).[6] In terms of water usage, there is a growing concern that water scarcity will become at least as important a constraint on future food production as a lack of available land. By moving down the food chain (i.e. from a meat-based to a vegetarian diet), we could get twice as much nutritional benefit out of each litre of water consumed in food production. Producing 1 kg of beef requires 100 000 l of water, for example, compared with 900 litres for the equivalent amount of wheat.

---

[4]We note that George Monbiot has recently endorsed this position. In an article published in the Guardian newspaper (18 October 2005: 27) on the ethics of eating British beef as opposed to Argentinean beef, he argues that 'we shouldn't be eating beef at all. Because the conversion efficiency of feed to meat is so low in cattle, there is no more wasteful kind of food production'.
[5]According to CIWF (2004), who refer to a number of sources, a little under one-quarter of all methane emissions (an important global warming gas) globally come from livestock. Livestock farming is also a major contributor of other atmospheric pollutants such as ammonia, nitrous oxide and carbon dioxide (which contribute to soil acidification and global warming). It is estimated that 10 per cent of total global greenhouse gases are derived from animal manure.
[6]For an older articulation of this argument, see Keith Mellanby's book *Can Britain Feed Itself?* (1973).

The health argument for vegetarianism centres upon the notion that 'meat is bad for you', although the benefits of vegetable staples are also emphasised. Indeed, various recent scientific studies have suggested that a vegetarian diet is associated with, although not necessarily causally related to, a reduced risk of illnesses such as heart disease and certain cancers (CIWF 2004). For example, CIWF (2004) cite a 12-year study of 6000 vegetarians and 5000 meat-eaters, published in 1994, which showed that vegetarians have 30 per cent less heart disease, and a reduced risk of various types of cancer by up to 40 per cent.

Taken together, these claims suggest a very close alignment between vegetarianism and the AFE as vegetarianism represents an attempt to counteract some of the environmental, social and economic problems that have come to be associated with the mainstream or conventional agro-food system. Furthermore, it offers a challenge to conventional relations between agro-food production and consumption since it draws attention to the implications of dietary choice for the sites and geographies of production in terms of farmed animals and the farming environment (although perhaps less so for the producers themselves), and also for the health and welfare of society. Indeed, we are not the first to observe that the adherents to a vegetarian diet appear to represent a body of 'mobilised' consumers who have made a deliberate decision to modify their dietary practice in response to various concerns about food production and consumption. Cook *et al.* (1998: 164), for example, identify vegetarians, along with ethical shoppers and health 'freaks' (sic), as 'super-knowledgeable' about the origins of the foods they eat. Nevertheless, closer examination of the practice of vegetarianism suggests that its location within the AFE is more problematic than is implied by these initial observations. Rather than the 'moral elegance' attributed to the practice by some (Porritt 2004), vegetarianism can be interpreted as occupying an altogether more ambiguous position in relation to the current concerns of the AFE.

Central to the environmental arguments marshalled by the proponents of a vegetarian diet is that plant-based agriculture is more efficient in terms of energy, land and water use than a diet based on meat. However, according to some recent commentaries, extensive systems of livestock production may ultimately be more sustainable than purely plant-based agricultures (e.g. Schiere *et al.* 2002).[7] While the evidence presented above suggests that a reduction in farm animals would produce environmental benefits, such an approach, in particular localised contexts, is also likely to generate environmental disbenefits, for example where animal husbandry and pasturing practices are key to the maintenance, enhancement or recreation of valuable habitats and landscape features such as biodiverse grasslands. Take, for example, UK environmentally sensitive areas which invariably seek to maintain extensive livestock grazing and the conversion of arable to grassland. Similar links are made clear in work on high natural value farming systems (Hellegers and Godeschalk 1998), upland farming (Bignal and McCracken 1993) and low-intensity farming systems (Beaufroy *et al.* 1994). In short, 'when livestock are raised according to the tenets of good husbandry . . . they hugely increase the overall economy of farming. Agriculture that includes the appropriate number of animals judiciously deployed is *more* efficient, not less, than all-plant agriculture' (Tudge, quoted in Porritt 2004: 5).

Linked to this, the life cycle impact assessments reviewed by Reijnders and Soret (2003: 667s) for vegetarian and meat products reveal how 'long distance air transport,

---

[7] Such a scenario would, if widely adopted, necessarily entail dietary shifts in the form of significantly reduced meat consumption because of the lower output of these extensive systems.

deep freezing and some horticultural practices for producing fresh vegetables may lead to environmental burdens for vegetarian foods exceeding those of locally produced organic meat'. Indeed, while proponents of agro-food system alternatives assert a return to the local or a 'relocalisation' of food production and consumption, research suggests that a contemporary vegetarian diet is highly dependent upon a *global* agro-food system that allows it to draw on a wide variety of food items (Beardsworth and Keil 1992). This proposition is supported by Lockie and Collie (1999), in their assertion that the decline in red meat-eating in Australia is attributable to increased access to a range of alternatives (both meat and non-meat based), supplied by the *globalised* agro-food system, rather than increased unease about red meat and its violent origins. Likewise, analysing the rise of veganism in the UK, Leneman (1999) points to the increased availability of 'ethnic' (sic) foods,[8] in which dairy products have never been a key feature, *from around the world* as offering an important means of moving to an all-plant-based diet.

In addition to these environmental questions surrounding the contemporary practice of vegetarianism, tensions can be identified in the other ethical dimensions of this dietary practice. The AFE emphasises the ethical aspects of food production, and as outlined above this is reflected in the moral arguments for vegetarianism. However, it is clear that in its most popular lacto-ovo and lacto forms,[9] vegetarianism remains highly dependent on animal husbandry. In other words, it requires the keeping of animals for dairy and egg production, not to mention their slaughter at the end of their productive lives, and in the case of dairying the selling off (often into veal production) or slaughter of male calves (Matheny and Chan 2005). Dairy and egg production systems are also often highly intensive, and there are serious animal welfare and environmental concerns associated with them (e.g. Evans 2000). Moreover, some vegetarians, while refusing to consume meat and other flesh foods, continue to wear animal products in the form of leather and wool and in doing so 'occupy a somewhat precarious moral position' (Beardsworth and Keil 1992: 283).[10] Even a shift to a vegan diet, based on an all-plant-based agriculture, has been shown to present ethical dilemmas. American research (Davis 2003) has demonstrated that a diet based on large herbivores would in fact involve fewer animal deaths overall than an all-plant-based, vegan diet. This is because numerous animals that live in and around agricultural fields, for example voles, rabbits, and various species of amphibians, are killed during the multiple field activities that are required in the production of most crops.

There are also tensions involving the human health argument for adopting a vegetarian diet. Growing publicity and concern about the intake of animal fat in our diets, scares about the safety of meat (particularly in a UK context following the BSE scandal) and a desire to consume more natural and less processed foods has created an increased demand for vegetarian and meat-free foods. This has attracted the interest of large-scale processors and retailers within the conventional and industrialised food system (Maurer

---

[8] A term used by this author to include foods such as those influenced by the Indian sub-continent.
[9] This highlights the considerable diversity within the practices and motivations of individual vegetarians. Beardsworth and Keil (1992), for example, have identified six general types of vegetarians ranging from those that still eat some meat to those that only consume products derived from plants.
[10] It was these contradictions in the practice of vegetarianism that lead to the formation of the Vegan Society in 1944.

2002), keen to take advantage of a growing market, which in the UK is now estimated to be worth between £500 and £600 million per annum. One of the key products to emerge during this period of expansion is Quorn (described by the company which produces it as a 'myco-protein' derived from a type of fungus), which is an increasingly popular meat-protein substitute. Significantly, meat substitute products such as Quorn (which is produced in just two factory locations in the UK) are derived from highly industrialised processing systems. While these may be promoted as healthy meat-free alternatives, they are very much the products of the mainstream or conventional food economy and anything but the locally based and natural alternatives that are the focus of those promoting the AFE. Indeed, this observation is supported by Smart's (2004) study of the UK Vegetarian Society, which he now suggests is 'adrift in the mainstream' having endorsed a range of large-scale processor and retailer products as suitable for vegetarians. There are some parallels here with the appropriation and 'conventionalisation' of organic food and agriculture by the mainstream (Guthman 2002, 2004).

The final area of tension between vegetarianism and the AFE concerns the impact of a recently profiled group of 'lifestyle vegetarians' (Tester 1999), whose reasons for adopting the diet appear to be tied to the accumulation of cultural capital and health benefits rather than to the ethical motivations that traditionally have explained vegetarianism. The consequences for the rest of the food system of a lifestyle approach may be the same as the other motivations for becoming vegetarian, that is benefits to animal welfare, the environment and human welfare, but the impacts may be superficial and short-lived as lifestyle vegetarianism may not represent a permanent shift in dietary practice, it being susceptible, like other lifestyle choices, to the vagaries of fashion (Maurer 2002).

The importance of livestock in sustainable farming systems, the moral ambiguities of vegetarianism (the leather and wool, eggs and cheese problem), the reliance of some vegetarians on a globalised and industrialised diet and the fact that many people adopt it as a lifestyle choice suggest that vegetarianism occupies an uneasy, paradoxical and contradictory position in relation to the emergence of oppositional, alternative food networks intent on valorising places of production, relocalising food networks, reconnecting producers and consumers and reintegrating nature into food supply chains. It also leads us to suggest that it might be meat that represents a new food militancy (when previously it was vegetarianism), and that it is meat-eaters who are concerned about the origins of their food and its related geographies and biographies (after Cook *et al.* 1998), rather than vegetarians, who are more comfortably accommodated within the AFE. Indeed, it is interesting to note that the Food Ethics Council (2001) has recently rejected vegetarianism as the basis of a more ethical and sustainable approach to food production systems.[11] This implies that a situation of reduced meat-eating based on animals raised under more extensive and environmentally sensitive conditions (something that is also

---

[11] A co-evolutionary perspective on the relationship between humans and animals is used to arrive at this position: 'Domesticated animals have undergone marked evolutionary changes which, in many cases, make them totally dependent on human care, having largely lost their adaptation to the wild. Their instincts of dominance and territoriality have become greatly diminished and their physical defence mechanisms atrophied... So it would be a totally perverse act, resulting from a misguided sense of compassion, to attempt to return such domesticated animals to 'the wild', even assuming such territory could be found. They simply could not survive' (Food Ethics Council, 2001: 6).

advocated by CIWF in their recent report on the global benefits of eating less meat) may be more compatible with the development of the AFE than vegetarianism based on industrialised and globalised food supply systems.

Making Theoretical Sense of the Relationship Between Vegetarianism and the AFE

In this penultimate section of the chapter, we discuss some ways in which theoretical sense might be made of some of the contradictions in the relationship between vegetarianism and the AFE. To do this, we turn to some recent writing in agro-food studies that has begun to examine consumption and its relationship with production within the context of the AFE. One emerging theme here is the notion of reflexivity.[12] Consumers, it is argued, or at least a certain proportion of consumers, are becoming increasingly reflexive and discerning in response to a variety of concerns about food production and the operation of food supply chains. Notable among these concerns are those relating to the health and safety of food (chemical and bacteriological contamination and whether it has been genetically modified), the environmental impacts of food production, and the location of its production (whether local or not so local) with all that this implies for the traceability of food products and food miles, the conditions under which animals are reared and the social relations of production. Alternative food initiatives provide an important means of addressing many of these concerns and so their development has been seen as tied up with (perhaps even driven by) the emergence of reflexive consumption.

However, as Miele (2006: 350) argues, there is 'a need to balance the reflexive and routine aspects of consumption' when analysing the operation of alternative agro-food chains, and in doing so draws on research into farmers' markets and organic food. In relation to the first of these contexts, she suggests that 'the success or viability of such ventures does not seem to rely (principally) upon a growing number of consumers acting upon an ethically informed choice. Rather, the markets survive by attracting a growing number of ordinary consumers interested in novelty, freshness, quality and in the opportunity for shopping with friends in a friendly atmosphere' (Miele 2006: 351). Such findings, according to Miele (2006: 351), are 'more consistent with theorization of "ordinary consumption" or increasing "omnivorousness"[13] than with the theory of reflexive consumption'. Likewise, by referring to Halkier's (2001) work on consumers of organic food in Denmark, Miele highlights the complex mixture of routinisation and

---

[12] The notion of reflexivity has been utilised by a range of authors, most notably Beck, Giddens, Lash and Bourdieu. An in-depth discussion of its implications is beyond the scope of this chapter, suffice to say that it involves individuals actively reflecting 'on prevailing social arrangements, norms and expectations' (Adkins 2003: 22) as a response to growing complexity and uncertainty within the modern world: indeed, DuPuis (2000) argues that food and dietary choice is of particular interest in this respect, as trust in the conventional food supply chain has been damaged in recent years.

[13] A term coined by Warde (2000) which encapsulates the broadening of tastes within contemporary diets and distinguished from ideas of hierarchical consumption and 'distinction' on the basis of what you eat.

reflexivity in consumption practices. Alongside the fact that purchasing organic (or other 'ethical foods') can become routinised for some consumers, there is evidence that 'even committed green consumers [in this case organic food consumers] balance their ideas about "environmentally correct consumption" and "healthy foods" with more mundane considerations such as convenience, price, conventional quality and taste' (Halkier 2001, quoted in Miele 2006: 351–352). Routines, constraints and preferences operate alongside reflexivity in consumption practices, but this can be forgotten or under-emphasised in studies of agro-food system alternatives. This leads Halkier to propose the concept of 'creolised practices' (while Miele offers the concept of 'hybrid' consumption practices) in overcoming the divide between routinised (or habitual) and reflexive consumption.

Such theoretical observations help to explain at least some of the apparent ambiguities in the practice of vegetarianism outlined above. In the first instance, eating a vegetarian diet might be interpreted as part of the growth of omnivorousness in food consumption as much as it is a reflection of ethical and moral concerns about food production and the desire, therefore, to practice an alternative form of eating. Consumers are simply more willing to try and accept a range of different tastes and foods, including vegetarian foods, than they once were. More importantly, however, the desire to avoid animal products on ethical grounds (relating to animal rights, ecological and human welfare concerns) may be balanced against the convenience, price and taste of buying industrially derived meat substitutes and/or fresh vegetables and other products sourced from distant locations. Likewise, the motivation to avoid *eating* animal products on health grounds may be balanced against the enjoyment of *wearing* animal products because they are warm, comfortable and fashionable. In these ways, the practice of vegetarianism by any one consumer may be conceptualised as both reflexive *and* ordinary (or, alternative and conventional) – a hybridised or creolised consumption practice. In a similar way, Ilbery and Maye (2005) have argued that seemingly alternative food initiatives (such as specialised regional food products) might be better understood as hybrids of the alternative and conventional food systems, since they rarely operate exclusively as alternative forms. The apparent ambiguities between vegetarianism and the AFE may therefore arise because there are limitations to the concept of 'alternativeness'; rarely, it would seem is anything wholly 'alternative', but more likely a mixture of both alternative and conventional.

Nevertheless, these are, at present, theoretical observations, and further empirical investigation into the contemporary *practices* of vegetarianism would be instructive. The empirical focus of this research may be the purchasing practices of individual consumers: how they negotiate the various tensions and ambiguities within their vegetarianism as identified in this chapter, or whether or not they are concerned about the provenance of their food (in geographical, environmental and social terms), and if so how this is weighed up against more mundane concerns such as price and accessibility. Alternatively, research could examine the governance of vegetarian food, such as whether its standards and labelling address its supply, origins and 'natural' credentials. Related to this, investigation of the vegetarian commodity system would be of interest, and the way in which vegetarian food is being processed, distributed and retailed. Finally, further examination of vegetarianism as a social movement might also be worthwhile (Morris and Kirwan 2006). In organisational terms, there is clearly a tension between increasing the cultural acceptance of vegetarianism and making its produce more widely available, and ensuring that over time these converts will be encouraged to engage with its underlying philosophy. It may be that the tensions highlighted within vegetarianism by the emergence of a wider AFE mean that the movement is due for another schism (the last one occurring in 1944 when the Vegan Society was established). Such a rupture could see the creation of a further offshoot of vegetarianism, one which still values a reduction in the quantity of animal products eaten, but one which also gives a relatively higher priority to locally sourced, relatively unprocessed and natural food products.

Conclusions

Through its focus on vegetarianism, this chapter has begun to identify some questions and issues surrounding the relationship between particular diets and the AFE, something that has been neglected to date by agro-food scholars. We would argue that diet, as the contemporary 'obesity crisis' reveals, is too important to be ignored. The widespread adoption of a particular diet, or dietary choice, inevitably shapes the geographies of food supply networks, and increasingly so within the context of the 20th century agro-food system (Lockie et al. 2002). The immensely popular 'Atkins Diet', with its emphasis on the consumption of protein at the expense of carbohydrates, illustrates this very well since this diet was seen as responsible (in part at least) for a 4 per cent increase in the sale of eggs in 2003 and an associated decline in potato sales. Indeed, in the US, potato sales fell by 10 per cent from 2003 to 2004, with the 'Atkins Diet' being seen as a significant contributory factor (BBC News 2003). Meanwhile, a major shift towards a vegetarian diet could have profound implications for rural areas, and the organisation of agriculture in particular (Dietz et al. 1995; Ashley et al. 2004). As Lockie and Collie (1999: 265) assert in their study of the relationship between gender and meat-eating in Australia, 'it is stating the obvious to suggest that the reasons behind changing consumption practices in relation to red meat are of fundamental importance to further restructuring of both the meat industry and the communities that are dependent on it'.

That diet is an element of agro-food system development, and alternative food systems in particular, which is worthy of further attention and consideration, has been noted in the context of a study of the consumers of local foods: 'The products of localised food systems are still *food* products, which means that practical, nutritional and *diet-related* dimensions play an important part in their role as items for exchange, usage and consumption. The future development of theories on exchange and consumption in alternative food systems would benefit from considering how the outputs of these systems integrate, for example, into the *dietary and culinary repertoires of populations*' (Weatherell et al. 2003: 242; emphasis added). However, these observations appear to have had little impact on subsequent research agendas. It is hoped that the analysis of vegetarianism herein will encourage further investigation of the relationship between dietary preferences and practices and the geographies of production and consumption related to the AFE.

References

Adams, C.J. (1990). *The Sexual Politics of Meat*. Cambridge: Polity Press.
Adkins, L. (2003). Reflexivity: freedom or habit of gender? *Theory, Culture & Society* 20, 21–42.

Allen, P., FitzSimmons, M. Goodman, M. and Warner, K. (2003). Shifting plates in the agrifood landscape: the tectonics of alternative agrifood initiatives in California. *Journal of Rural Studies* 19, 61–75.

Alward, P. (2000). The naïve argument against moral vegetarianism. *Environmental Values* 9, 81–89.

Ashley, B., Hollows, J., Jones, S. and Taylor, B. (2004). *Food and Cultural Studies*. London: Routledge.

BBC News (2003, 07.10.2003). Eggs V Potatoes: The Atkins Diet Stand-Off. Retrieved 13.02.2004, from http://news.bbc.co.uk/1/hi/magazine/3167556.stm

Beardsworth, A. and Keil, T. (1992). The vegetarian option: varieties, conversions, motives and careers. *Sociological Review* 40, 253–293.

Beaufroy, G., Baldock, D. and Clark, J. (1994). *The Nature of Farming: Low Intensity Farming Systems in Nine European Countries*. London: IEEP.

Benatar, D. (2001). Why the naïve argument against moral vegetarianism really is naïve. *Environmental Values* 10, 103–112.

Bignal, E. and McCracken, D. (1993). Nature conservation and pastoral farming in the British uplands. *British Wildlife* 4, 367–376.

Bryant, R.L. and Goodman, M.K. (2004). Consuming narratives: the political ecology of 'alternative' consumption. *Transactions of the Institute of British Geographers* 29, 344–366.

Chambers Cambridge (1988). *Chambers English Dictionary*, 7th ed. Cambridge: W & R Chambers Ltd and Cambridge University Press.

CIWF (Compassion in World Farming Trust) (2004). *The Global Benefits of Eating Less Meat*. CIWF Trust, Petersfield, Hampshire GU32 3EH, UK. Compiled and written by Mark Gold.

Cook, I., Crang, P. and Thorpe, M. (1998). Biographies and geographies: consumer understandings of the origins of foods. *British Food Journal* 100, 162–167.

Davis, S.L. (2003). The least harm principle may require that humans consume a diet containing large herbivores, not a vegan diet. *Journal of Agricultural and Environmental Ethics* 16, 387–394.

Dietz, T., Frisch, A., Kalof, L., Stern, P. and Guagnano, G. (1995). Values and vegetarianism: an exploratory analysis. *Rural Sociology* 60, 533–542.

DuPuis, E. (2000). Not in my body: Rbgh and the rise of organic milk. *Agriculture and Human Values* 17, 285–295.

Eder, K. (1996). *The Social Construction of Nature: A Sociology of Ecological Enlightenment*. London: Sage Publications.

Evans, N. (2000). The impact of BSE in cattle on high nature value conservation sites in England. In H. Millward, K. Beesley, B. Ilbery and L. Harrington (Eds), *Agricultural and Environmental Sustainability in the New Countryside* (pp. 92–110). Truro, N.S.: Nova Scotia Agricultural College.

Food Ethics Council (2001). Farming Animals for Food: Towards a Moral Menu. A Report. London: Food Ethics Council.

Goodland, R. (1997). Environmental sustainability in agriculture: diet matters. *Ecological Economics* 23, 189–200.

Goodman, D. (1999). Agro-food studies in the 'age of ecology': nature, corporeality, bio-politics. *Sociologia Ruralis* 39, 17–38.

Goodman, D. (2002). Rethinking food production–consumption: integrative perspectives. *Sociologia Ruralis* 42, 271–277.

Goodman, D. (2003). The quality 'turn' and alternative food practices: reflections and agenda. *Journal of Rural Studies* 19, 1–7.

Goodman, D. (2004). Rural Europe redux? Reflections on alternative agro-food networks and paradigm change. *Sociologia Ruralis* 44, 3–16.

Gussow, J. and Clancy, K. (1986). Dietary guidelines for sustainability. *Journal of Nutrition Education* 18, 209–213.

Guthman, J. (2002). Commodified meanings, meaningful commodities: re-thinking production–consumption links through the organic system of provision. *Sociologia Ruralis* 42, 295–311.

Guthman, J. (2004). The trouble with 'organic lite' in California: a rejoinder to the 'conventionalisation' debate. *Sociologia Ruralis* 44, 301–316.

Halkier, B. (2001). Routinisation or reflexivity? Consumers and normative claims for environmental consideration. In J. Gronow and A. Warde (Eds), *Ordinary Consumption* (pp. 25–45). London: Routledge.

Hellegers, P. and Godeschalk, F. (1998). *Farming in High Nature Value Regions: the Role of Agricultural Policy in Maintaining HNV Farming Systems in Europe*: Onderzoekverslag 165, Agricultural Economics Research Institute, the Hague.

Hinrichs, C. (2000). Embeddedness and local food systems: notes on two types of direct agricultural market. *Journal of Rural Studies* 16, 295–303.

Ilbery, B. and Maye, D. (2005). Alternative (shorter) food supply chains and specialist livestock products in the Scottish–English Borders. *Environment and Planning A* 37, 823–844.

Kirwan, J. (2004). Alternative strategies in the UK agro-food system: interrogating the alterity of farmers' markets. *Sociologia Ruralis* 44, 395–415.

Leneman, L. (1999). No animal food: the road to veganism in Britain, 1909–1944. *Society and Animals* 7, 219–228.

Lockie, S. and Collie, L. (1999). 'Feed the man meat': gendered food and theories of consumption. In D. Burch, J. Goss and G. Lawrence (Eds), *Restructuring Global and Regional Agricultures: Transformations in Australasian Agri-Food Economies and Space* (pp. 255–273). Aldershot: Ashgate.

Lockie, S. and Kitto, S. (2000). Beyond the farm gate: production–consumption networks and agri-food research. *Sociologia Ruralis* 40, 3–19.

Lockie, S., Lyons, K., Lawrence, G. and Mummery, K. (2002). Eating green: motivations behind organic food consumption in Australia. *Sociologia Ruralis* 42, 23–40.

Marsden, T., Banks, J. and Bristow, G. (2002). The social management of rural nature: understanding agrarian-based rural development. *Environment and Planning A* 34, 809–825.

Marsden, T., Murdoch, J. and Morgan, K. (1999). Sustainable agriculture, food supply chains and regional development: editorial introduction. *International Planning Studies* 4, 295–301.

Matheny, G. and Chan, K.M.A. (2005). Human diets and animal welfare: the illogic of the larder. *Journal of Agricultural and Environmental Ethics* 18, 579–594.

Maurer, D. (2002). *Vegetarianism: Movement or Moment?* Philadelphia: Temple University Press.

McManus, P. (1999). *Geographies of Competing Food Networks: Meat and Vegetarian Sausages*, Paper presented to the RGS-IBG annual conference, Leicester, England, 4–7 January 1999.

Mellanby, K. (1975). *Can Britain Feed Itself?* London: Merlin.

Miele, M. (2001). Changing passions for food in Europe. In H. Buller and K. Hoggart (Eds), *Agricultural Transformation, Food and Environment. Perspectives on European Rural Policy and Planning – Volume 1* (pp. 29–50). Aldershot: Ashgate.

Miele, M. (2006). Consumption culture: the case of food. In P. Cloke, T. Marsden and P. Mooney (Eds), *Handbook of Rural Studies* (pp. 344–354). London: Sage Publications.

Mintel (Mintel Marketing Intelligence). (2000). Vegetarian (the): Mintel Market Intelligence Report, June 2000. Retrieved 17 March 2003, from http://reports.mintel.com/sinatra/mintel/subscriber

Morris, C. (2002). *Exploring Food-Environment Linkages within the Alternative Food Economy: The Case of Food Labelling Initiatives*. Paper presented at the Alternative Food Economy: Myths, Realities, Potential Conference, Institute of British Geographers, London 6 March 2002.

Morris, C. and Kirwan, J. (2006). Vegetarians: uninvited, uncomfortable or special guests at the table of the alternative food economy? *Sociologia Ruralis* 46, 192–213.

Murdoch, J., Marsden, T. and Banks, J. (2000). Quality, nature, and embeddedness: some theoretical considerations in the context of the food sector. *Economic Geography* 76, 107–125.

Pimentel, D. and Pimentel, M. (2003). Sustainability of meat-based and plant-based diets and the environment. *The American Journal of Clinical Nutrition* 78, 660S–663.

Porritt, J. (2004). Foreword. In *The Global Benefits of Eating Less Meat* (pp. 4–7). CIWF Trust, Petersfield, Hampshire GU32 3EH, UK.

Raynolds, L. (2000). Re-embedding global agriculture: the international organic and Fair Trade movements. *Agriculture and Human Values* 17, 297–309.

Regan, T. (1984). *The Case for Animal Rights*. London: Routledge.

Reijnders, L. and Soret, S. (2003). Quantification of the environmental impact of different dietary protein choices. *The American Journal of Clinical Nutrition* 78, 664S–668.

Renard, M.-C. (1999). The interstices of globalisation: the example of Fair Coffee. *Sociologia Ruralis* 39, 484–500.

Renard, M.-C. (2003). Fair Trade: quality, market and conventions. *Journal of Rural Studies* 19, 87–96.

Renting, H., Marsden, T. and Banks, J. (2003). Understanding alternative food networks: exploring the role of short food supply chains in rural development. *Environment and Planning A* 35, 393–411.

Schiere, J.B., Ibrahim, M.N.M. and van Keulen, H. (2002). The role of livestock for sustainability in mixed farming: criteria and scenario studies under varying resource allocation. *Agriculture, Ecosystems & Environment* 90, 139–153.

Shafer-Landau, R. (1994). Vegetarianism, causation and ethical theory. *Public Affairs Quarterly* 8, 85–100.

Singer, P. (1983). *Animal Liberation*. London: Cape.

Smart, A. (2004). Adrift in the mainstream: challenges facing the UK Vegetarian Movement. *British Food Journal* 106, 79–92.

Spencer, C. (1993). *The Heretic's Feast: A History of Vegetarianism*. London: Fourth Estate Limited.

Tester, K. (1999). The moral malaise of McDonaldization: the values of vegetarianism. In B. Smart (Ed.), *Re-Visiting McDonaldization* (pp. 207–221). London: Sage.

Tovey, H. (1997). Food, environmentalism and rural sociology: on the organic farming movement in Ireland. *Sociologia Ruralis* 37, 21–37.

Twigg, S. (1983). Vegetarianism and the meanings of meat. In A. Murcott (Ed.), *The Sociology of Food and Eating*. Aldershot: Gower.

Warde, A. (2000). Eating globally: cultural flows and the spread of ethnic restaurants. In D. Kalb, M. van der Land, R. Staring, B. van Steenbergen and N. Wilterdink (Eds), *The Ends of Globalization: Bringing Society Back In* (pp. 299–316). Boulder, CO: Rowman and Littlefield.

Weatherell, C., Tregear, A. and Allinson, J. (2003). In search of the concerned consumer: UK public perceptions of food, farming and buying local. *Journal of Rural Studies* 19, 233–244.

Whatmore, S., Stassart, P. and Renting, H. (2003). Guest editorial: what's alternative about alternative food networks? *Environment and Planning A* 35, 389–391.

Whatmore, S. and Thorne, L. (1997). Nourishing networks: alternative geographies of food. In D. Goodman and M. Watts (Eds), *Globalising Food: Agrarian Questions and Global Restructuring* (pp. 287–304). London: Routledge.

Wilkins, J. (2005). Eating right here: moving from consumer to food citizen. *Agriculture and Human Values* 22, 269–273.

Winter, M. (2003). Embeddedness, the new food economy and defensive localism. *Journal of Rural Studies* 19, 23–32.

Zuzworksky, R. (2001). From the marketplace to the dinner plate: the economy, theology, and factory farming. *Journal of Business Ethics* 29, 177–188.

Part II

Public Policy and Alternative Food Projects

Chapter 9

Regionalisation, Local Foods and Supply Chain Governance: A Case Study from Northumberland, England

Damian Maye[1] and Brian Ilbery[2]
[1]Department of Geography, Environment and Disaster Management, Coventry University, UK
[2]Countryside and Community Research Unit, University of Gloucestershire, UK

Introduction

Conversations about 'local foods' centre around various topics/concerns, including for example environmental management and animal welfare, new forms of consumer activism and the future directions of agricultural production and policy. Public support for local foods in the UK is well documented, typified most recently through *Farmers Weekly's* 'Local Food is Miles Better' campaign.[1] Targeting consumers, as well as supermarkets and others in the food industry, the campaign supports local foods as a pragmatic tool to reduce 'food miles' and consequently limit environmental impacts (caused by food transportation), reduce food security concerns and improve freshness, seasonality and producer/consumer connectedness. Complex local food biographies, which include, *inter alia*, intriguing arguments about environmental additionality, social welfare and transformative economic potentials, have thus recently captured the imagination of rural geographers and sociologists in developed market economies. More specifically, these conversations have brought into academic focus a growing body of agro-food studies dedicated towards critically exploring notions of quality, provenance, traceability and the reconstruction of food chains – including the reported growth in food sales from 'alternative' retail points (see Ilbery and Maye 2006). Running alongside increasing

---

[1]*Farmers Weekly* is the main publication for the UK agricultural industry. For details on the magazine's Food Miles campaign see http://www.fwi.co.uk/gr/foodmiles/index.html (last accessed 7 September 2006).

Alternative Food Geographies
D. Maye, L. Holloway & M. Kneafsey (Editors)
ISBN: 978-0-08-045018-6

Copyright © 2007 by Elsevier Ltd.
All rights of reproduction in any form reserved.

public recognition of local foods, therefore, is growing academic realisation that the shape and composition of local food systems are highly amorphous and in need of critical scrutiny. Contributing further to these debates, this chapter explores the nature and structure of institutional arrangements for local (specialist) food products in Northumberland, a rural county in the north of England.

To date, attempts to 'unpack' local food economies have not examined institutional arrangements (but see Marsden and Sonnino 2006), but have focused instead upon particular nodes in the supply chain (e.g. producer and retailer) or particular 'alternative' spaces (e.g. farmers' market, box scheme) (see, for instance, Dürrschmidt 1999; Morris and Buller 2003; Kirwan 2004). The chapter thus makes an instructive empirical contribution, surveying institutional structures that surround local food systems. More broadly, it also attempts to link work in agricultural geography about local foods to work in economic geography about regionalisation. While both literatures overlap in terms of their scale of spatial analysis, attempts to bring them together have only recently begun (see, for example, DuPuis and Goodman 2005; Watts et al. 2005). Why should such a marriage take place? Using a Northumberland case study, the argument presented here is that local foods provide a useful lens to explore and understand wider aspects of regionalisation and rural governance. Focusing on governance structures can also reveal useful things about local food economies.

It is important, at this stage, to position the contribution this chapter makes to the larger body of work on 'alternative food geographies': the focus here is upon the *local dimension of 'specialist' food production* rather than commodity-based products and, more specifically, the nature of institutional arrangements in place to support specific food chain activities in a marginal rural economy. The chapter begins by providing a review of the two parallel literatures on regionalisation and local foods. It next outlines the case study approach and organisational framework adopted to interpret the shape of governance structures for Northumberland's local/specialist food economy. The main empirical findings are then introduced. The chapter concludes with some overall remarks about local foods and outlines future avenues for research from an institutional perspective.

Regional Governance in England

Many commentators use the term 'new regionalism' to describe the process of increasing decentralisation of government responsibilities taking place across the UK (although similar trends are notable in other parts of Europe). Evidence for this is not hard to find, with the number and influence of institutions of regional governance in England increasing rapidly since the late-1980s (see MacLeod and Jones 2001; Burch and Gomez 2002; Tomaney 2002; Counsell and Haughton 2003; Gough 2003; Jones and MacLeod 2004; Ward and Lowe 2004; Hudson 2005; Pearce et al. 2005). Burch and Gomez (2002) argue that this process has been driven by three main factors: first, the need to create regional governance institutions in order to gain access to, and administer, EU structural funds; second, the growth of 'new regionalism' among groups of *elite* regional 'actors'; and third, the creation, in 1994, of government offices in the regions (GOs) which 'brought the concepts of integrated policy and regionalism together in Whitehall thinking' (Burch and Gomez 2002: 769).

As the powers of GOs increased, two new regional institutions joined them: regional development agencies (RDAs) and regional chambers (or assemblies). The former were appointed to 'improve regional economic performance' (Pearce *et al.* 2005: 198), including administering EU structural funds and implementing the Government's cluster policy[2] (Gough 2003). The latter were created primarily to monitor the activities of the RDAs. The strongest statement defending the Government's rationale for decentralisation was provided in the 2002 White Paper, *Your Region, Your Choice: Revitalising the English Regions*. Here, decentralised structures are asserted to have brought benefits to the regions, especially in terms of joint working and strategic thinking. The paper also introduced proposals for greater accountability through elected regional government. An incremental 'two track' approach was proposed: 'Track 1' would be imposed in all regions, building on the current system of administrative decentralisation, while 'Track 2' provided elected regional assemblies with 'a range of executive and influencing powers, but limited financial resources' (Pearce *et al.* 2005: 198). The latter would be established only where evidence of public support was expressed. The first referendum took place in the North East region of England in November 2004. The strong negative public vote (78 per cent voted 'No') meant that the government had to resort to the 'Track 1' objective, giving the GOs, RDAs and unelected Chambers additional responsibilities for regional decision making, with only further, modest administrative decentralisation.

Not surprisingly, therefore, there is considerable debate about the extent to which such developments represent a real devolution of power to the English regions. Outside London, regional chambers are not directly elected; some members are drawn from local councils, providing what Jeffery and Mawson (2002: 715) describe as a 'thin veneer of regional accountability'. Evidence also suggests that the central government retains considerable influence over regional economic policy and planning (see Burch and Gomez 2002; Counsell and Haughton 2003; Ward and Lowe 2004). Indeed, some authors argue that England's new regionalism is being shaped according to national economic priorities (Tomaney and Ward 2000; MacLeod and Jones 2001). Concerns have also been expressed about the dominance of urban interests within the new regionalism (see, for instance, recent attempts to establish 'City-Regions' across the UK, including Newcastle, Liverpool and Leeds) and the need for better integration of regional and local dimensions into rural policy. On this last point, Ward and Lowe (2004) identify a paradox in the implementation of the rural development regulation (RDR) in England. While on the one hand, this 'second pillar' of the Common Agricultural Policy has strengthened a process of 'Europeanisation' in European rural policy development, on the other hand 'it currently risks being at the cost of distorting sub-national priorities to spread rural development support beyond the farm gate' (Ward and Lowe 2004: 121) and thus remains mostly an instrument of agricultural restructuring.

This short review of regional economic policy usefully informs an institutional survey of local foods, especially in terms of implying how sub-national governance structures might be implicated in supporting local food strategies across the English regions. As noted earlier, a case study of local food chain governance can equally inform debates

---

[2]Cluster development work was initiated across the UK regions by the Department of Trade and Industry (DTI). Adapted from Porter (1990), the aim is to develop recognised sector-based clusters (e.g. food and drink clusters) to improve regional competitiveness and industry growth (see DTI 2001).

about 'new regionalism' in England. So far, there have been few attempts to incorporate local foods into regional studies, and similarly a lack of consideration of the relationship between regional and local institutions, both public and non-public. These silences are of no surprise, with contributions focused around the broad activities of GOs and RDAs and generic regional development concerns. However, given the pace of growth within local food economies, this omission seems especially significant.

Food Relocalisation and Rural Policy in England

Interest in foods of local and regional provenance gathered pace in the wake of the foot-and-mouth outbreak in 2001 and the publication of the *Report on the Future of Farming and Food*, also known as the 'Curry Report' (2002) (by the Policy Commission). Responding to the crisis, this report advocated a need to 'reconnect' UK farming with the rest of the food chain, including the final consumer. Here local foods are championed 'as one of the greatest opportunities for farmers to add value and retain a bigger slice of retail value' (Curry Report 2002: 119). The report further argues that potential countryside benefits were hampered by limited availability, a lack of supply chain integration and poor distribution. To overcome these barriers, Curry declared support for *re-scaling processes* in food governance, recommending that RDAs incorporate regional foods into their economic strategies and put in place the necessary support to enable local food initiatives to nurture and develop (Curry Report 2002: 45–46). The regional policy agenda thus became more visible within rural policy circles after the foot-and-mouth crisis (Winter 2003a; see also Barling and Lang 2003 for a review of UK food policy more generally).

Recommendations to regionalise governance structures were further supported by Lord Haskins' (2003) review of rural policy. The Department of Environment, Food and Rural Affairs' (Defra) *Rural Strategy* also embraced the principles of decentralisation for the implementation of economic and social policy for rural areas, including efforts to strengthen working arrangements between regional bodies (Defra 2004).[3] The role and influence of regional governance structures upon English rural development, including food production, thus appear more pronounced. In tandem, rural policy also publicly embraced the 'food relocalisation' agenda to be delivered, crucially, through a sub-national governance framework. This latter process is of central interest here. Pearce *et al.*'s (2005) attempt to bring rural development into the regionalisation debate is useful, as is Clark's (2006) suggestion that 'regional systems of innovation' can inform, in this case, the way multifunctional agricultural strategies are promoted in England. In both, however, no explicit reference is given to local foods. Marsden and Sonnino (2006), on the other hand, do provide a useful analysis of emerging governance frameworks for local foods, with case studies from Wales and the South West of England. They reveal notable differences in governance structures and institutional contexts between the two regions, even though common policy positions are forged. The analysis appears to give

---

[3]For a full copy of the report see http://www.defra.gov.uk/rural/pdfs/strategy/rural_strategy_2004.pdf (accessed 28 July 2006).

weight to decentralisation arguments and the authors call for further empirical accounts from other UK contexts.

To establish such an account, three points need to be drawn from the agro-food literature. The first point concerns the continued importance of the *food chain* in rural policy contexts. Defra's annual report on agriculture illustrates this well. Each year, since the 1980s, an assessment of the farming economy has included an outline of the 'UK food chain' – this provides an economic measure of gross value added. The recent crisis in farming and food, however, has sparked interest not in the food chain per se, but the *type* of food chain. Two things follow from this. First, Jackson et al. (2006) argue that the 'commodity chain' is now mobilised to foreground specific aspects akin to individual political and institutional contexts. They show, for instance, how Defra delineates food chains in economic competitiveness and technical proficiency terms, in contrast to SUSTAIN which mobilises the term to promote social inclusion and sustainability. Second, 'short food supply chains' are currently being promoted because of the rural development benefits they offer Small and Medium-sized Enterprises (SMEs) (see Renting et al. 2003; Maye and Ilbery 2006). These chains, in theory at least, enable small producers to shift production out of 'industrial modes' to capture a better proportion of value added.

The second issue relates to the characteristics of the *products* passing along these supply chains. As noted in Chapter 1, policy initiatives to promote foods with territorial associations became a key focus of interest for agro-food researchers from the late-1990s. Most prominent among these are the PDO (Protected Designation of Origin) and PGI (Protected Geographical Indication) quality status awarded to dedicated regional foods and, more generally, efforts to encourage economic growth and build competitive advantage through the production of niche market foods (see Ilbery and Kneafsey 2000; Parrott et al. 2002). Attempts to link 'product and place' in this way became known as the 'turn to quality' and widely recognised as a key feature of relocalisation. However, others contend that the 'localness' of the product is more important than a turn to quality based on, for example, organic or ecological principles (Winter 2003b). The key thing that emerges from this debate is the distinction drawn between *'local'* and *'locality'* foods. So, while the former refer to products produced and consumed within a certain distance (e.g. 30 miles), the latter refer to products from further afield, but with an identifiable geographical provenance, e.g. the sale of Newcastle Brown ale in a supermarket in Coventry. Unsurprisingly, locality (or 'regional speciality') foods, which emphasise value added, appear to dominate the economic interests of RDAs.

DuPuis and Goodman (2005: 359) thus claim that localism in Europe 'has emerged in the context of new forms of devolved rural governance in parallel with the slow process of reform of the EU's Common Agricultural Policy (CAP)'. This leads on to the third, and probably most contested, issue which considers the relationship between food relocalisation and 'alternativeness'. In dealing with this, Watts et al. (2005) distinguish between 'weaker' and 'stronger' alternative systems of food provision. The former place emphasis on quality and the labelling features of locality *food* networks (i.e. the product is key), whereas the latter focus on the revalorised and embedded characteristics of local food *networks* (i.e. the supply chain/network and nature of relations are key). Watts et al. (2005: 34) conclude 'alternative food networks can be classified as weaker or stronger on the basis of their engagement with, and potential for subordination by, conventional food supply chains operating in a global, neoliberal polity'. In essence, it is the supply chain and *not* the product that delineates possible geographies of alternativeness (see also Maye and Ilbery 2006).

From this review, it should be clear that policy makers and regional agencies do recognise the economic development potential of local food systems, albeit with a noted bias towards value-added markets. The review thus identifies possible tensions in terms of the products supported and the characteristics defining the institutional landscape for local food governance. Within the context of these debates (on both the new regionalism in England and the local food economy), the rest of the chapter aims to assess the strategies, measures and structures of national, regional and local institutions involved in developing Northumberland's local (specialist) food sector.

Case Study Approach and Institutional Survey

In a recent contribution to the new regionalism debate, Jones and MacLeod (2004) draw a useful distinction between *regional spaces* and *spaces of regionalism*. The former relate to the emergence of new regional spaces in which growth, competitiveness, business support and 'institutional thickness' have created 'sunbelt' districts and clusters of economic activity. In contrast, the latter have political rather than economic overtones and involve 'the (re-) assertion of national and regional claims to citizenship, insurgent forms of political mobilisation and cultural expression and the formation of new contours of territorial government' (Jones and MacLeod, 2004: 435). Thus social capital, cooperation and the involvement of local stakeholders become more important than competition and economic growth. Their conceptualisation is based on an account of institutionalisation in the south west of England, with notable tensions over economic priorities, political representation, territorial shape and cultural vernacular. Although the description of the two spaces is a little fuzzy, the distinction has potential value in local foods research, given the range of activities, practices and bodies currently represented in this part of the food economy. Building on this work, this chapter focuses mainly on 'regional spaces' to identify possible descriptive characteristics of local food governance in Northumberland and also to highlight the potential value of engagements with the regionalisation literature.

The institutional survey had two main objectives: first, to map relevant institutional structures and policies for local foods in Northumberland; and second, to uncover the nature of relations between institutions and any potential tensions. The case study region was selected as part of a larger EU-funded project[4] on food chains in 'lagging rural regions', a description which takes its lead from EU cohesion policy. The emphasis is, therefore, on a geographically peripheral region that is more sparsely populated and less agriculturally productive than more prosperous regions in the EU. The case study region in fact comprised two counties: Northumberland and The Scottish Borders (see Figure 9.1). In this chapter, the focus is on Northumberland, which is part of the North East region of England. No attempt is made to compare differences either side of the international border. The county was badly affected by the 2001 foot-and-mouth outbreak. Indeed, the cause of the disease was traced to a pig farm in Northumberland,

---

[4] SUPPLIERS (Supply chains linking food SMEs in Europe's lagging rural regions).

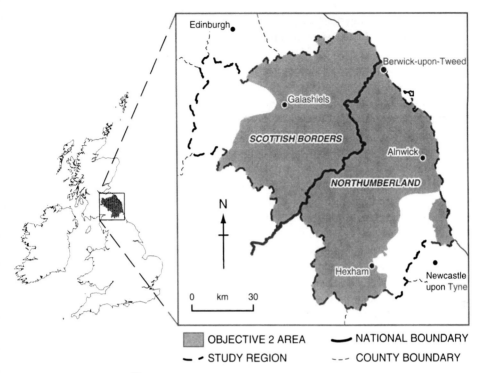

Figure 9.1: The case study region: Northumberland, England.

and the disease devastated numerous farms in the region. After the outbreak, interest in local foods increased thanks in part to producer events organised to promote and support local (mostly specialist) food producers in Northumberland and other parts of the North East region.

Twenty-five institutions were originally surveyed in Northumberland, The Borders and nationally. This chapter is concerned with the six national and 10 regional institutions associated with Northumberland. Most institutions were known to the research team because they were identified during earlier surveys with food producers, intermediaries and retailers in the study region (see Ilbery and Maye 2005, 2006; Maye and Ilbery 2006). Institutions were selected according to their institutional type (e.g. public and voluntary), policy remit (e.g. economic development and rural affairs), geographical coverage and relevance to SMEs/local food production in the region. The link between product type and institution was also important because the project focused on the food supply chains for three dedicated sectors: livestock; bakery, confectionery and preserves; and fish. Advice from a Consultation Panel, consisting of local and regional institutional actors, also helped to finalise the sample of surveyed institutions.

Interviews were conducted on a face-to-face basis with key individuals within the selected institutions. Respondents tended to be senior managers or local development officers who were familiar with the institutions' strategies and measures, and their role in the development and management of food supply chains. In a couple of cases, more than one person was interviewed per institution due to the wide-ranging nature of the food-based agenda. In terms of mapping institutional structures for local foods, the analysis is similar to Pearce *et al.*'s (2005) methodology in the West Midlands region, which mapped institutional structures for rural policy. A key component of the interview thus involved asking respondents to draw their own *'institutional map'*. This process helped to identify the main links that individual institutions had with other institutions in the region or nationally (if applicable) and the nature of those links in relation to local food chains.

The individual diagrams were used to create a composite map to represent the institutional landscape for local food governance in Northumberland. Institutions have been arranged to reflect different *scales* of food governance. Surveyed institutions in the region seem to operate on three different scales: first, absorbing international and national *policy* agendas for regional/rural/food governance; second, transforming those policies into regional/local *strategies*; and third, establishing relevant *projects* at the regional/local level. This organisational framework has been employed as part of an evaluative methodology to 'map' the institutional landscape surrounding local food systems. The three scales are not mutually exclusive, but represent primary roles for an institution to identify characteristics akin to this particular regional economic food space. Scaling food chain governance in this way obviously runs the risk of presenting an overly stylised and linear synopsis, when relations are usually complex and overlapping. Nevertheless, it does begin to reflect an overall pattern of policy implementation, especially for state level institutions. As a final stage in the methodology, respondents were invited to a workshop in Newcastle where survey findings, including the composite institutional map, were presented and finalised.

Scaling Local Food Governance in Northumberland

The rest of this chapter now turns to outline some key findings from the institutional survey. It is divided into three parts: first, the relevant national and regional institutions are introduced; second, the institutional map drawn for Northumberland is presented, according to the three scales of local food governance; and third, a critical assessment is provided of the institutional structures in the county in terms of governance and local food supply. With limited space, it is not possible to explore all of the institutional arrangements or initiatives in detail; instead, the chapter identifies key trends and overall themes that emerged from the interviews.

Surveyed Institutions

Table 9.1 provides a description of each surveyed institution (including acronym where relevant), the year it was established, area of principal activity and geographical scope. The first six national institutions include three voluntary and three industry bodies that are, in different ways, associated with local foods. Different institutional contexts influence attitudes towards local food systems. When interviewed, voluntary organisations thus supported a relocalisation agenda on the basis of perceived environmental, social and economic benefits. This is contrasted with the more economically centred objectives of surveyed industry bodies, which here included the Institute of Grocery Distribution

Table 9.1: Institutional profiles for surveyed institutions.

Institution	Start	Principal activity	Scope
f3 – Foundation for local food initiatives	1999	Promote and support the growth of healthy local food economies. Provides consultancy, training and research	National
Sustain: the alliance for better farming and food	1997	Support and encourage sustainable food and agricultural policies and practices. Includes core policy work and a series of reports and projects	National/ International
Local Food Works (Soil Association)	2002	Foster sustainable systems through local food networks. Provides information and network development services	National
Institute of Grocery Distribution (IGD)	1909	Provide research and education services for the food industry	National
Meat and Livestock Commission (MLC) (catering)	1967	Food service: represent meat producers and satisfy customer/retail demands. Public sector: promote red meat	National
English Beef and Lamb Executive (EBLEX)	2003	Looks after the interests of cattle and sheep producers in England	National
Government Office for the North East (GONE)	1994	Co-ordinates the delivery of Government policy in the region	Regional
One NorthEast (ONE)	1999	Responsible for developing and implementing a regional economic strategy, including farming and food	Regional
Department of Environment, Food and Rural Affairs (DEFRA)	2001	The Rural Development Service delivers the England Rural Development Plan (ERDP)	Regional/National
Northumbria Larder (NL)	2001	Regional food group for speciality producers in the region	Regional
The Countryside Agency (CA)	1999	Statutory advisor on countryside recreation and conservation	Regional/National
North East Land Links (CA sponsored)	2001	Project to develop countryside spaces around towns and foster community linkages, especially in deprived areas	Regional
Business Link Northumberland (BLN)	1992	Promotes economic and business development. Assists businesses with grant applications and strategic plans	Local
Northumberland Farmers' Market Association (NFMA)	2000	Helps to promote, publicise and make permanent farmers' markets in Northumberland	Local

(*Continued*)

Table 9.1: Continued

Institution	Start	Principal activity	Scope
Northumberland County Council (catering)	–	Oversees food catering services/contracts provided for schools and other public bodies	Local
Hadrian's Wall Tourism Partnership (HWTP)	1995	Promotes sustainable tourism and economic development along Hadrian's Wall, including a local food brand	Sub-regional

(IGD), the Meat and Livestock Commission (MLC) and the English Beef and Lamb Executive (EBLEX).

Most of the 10 regional institutions are public bodies and relatively new in terms of when they were established (mostly the 1990s). The remit and objectives are not unique in that similar structures exist across all English regions (Ward and Lowe 2004). The Government Office (GONE) and the RDA (ONE) are thus part of the national decentralisation process. Defra and the Countryside Agency (CA) are the agencies' regional offices, relevant here because of their strategic work in the North East region, including Northumberland. They also provide a useful lens to explore relations between central and regional government structures. Northumbria Larder, meanwhile, was formed after the foot-and-mouth crisis, while the Northumberland Farmers' Market Association (NFMA) came about in response to the increased number of farmers' markets in the area and the need for some form of local co-ordination. Other surveyed institutions such as North East Land Links and Hadrian's Wall Tourism Partnership are individual projects with a more territorially specific food chain component. The next section tries to reflect the relative position of these different institutions in the context of supporting local foods in Northumberland.

Institutional Structures for Local Foods

Surveyed institutions have been arranged in Figure 9.2 to show the institutional structures and mechanisms in place for Northumberland's local food economy. It attempts to show how this process is part of a wider regional and national rural development agenda. The 'map' also includes one or two institutions not spoken to directly, but deemed significant as an outcome of the interviews. Arrows have not been drawn to connect the different institutions. As one workshop discussant put it: '[I]f you tried to put arrows on that diagram, you would find it a complete mishmash because there is such an overlap' (Taped discussion, Newcastle Workshop, June 2003). The results are presented according to the three 'scales' outlined above: policy, strategy and projects. As noted earlier, this framework is intended to help decipher a highly complex set of arrangements and used only to highlight key 'modes of governance' in terms of institutional relations, linkages, etc.

Scale 1: Policy

There appears to be a fairly *stratified* division of labour in terms of national policies, regional strategies and local projects in Northumberland and the broader North East

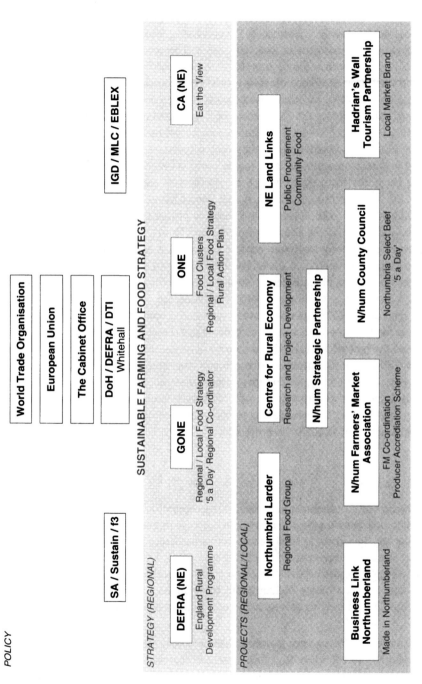

Figure 9.2: Institutional structures for local food economies in Northumberland.

region. The policy part of Figure 9.2 lies 'outside' the region and in this case concerns six surveyed national institutions. It is divided from left to right in terms of voluntary, public and industry bodies. For surveyed institutions in Northumberland and the North East, the crux is the 'public domain', with the WTO and EU at international, and Cabinet Office and Government departments at national, scales. The recognised overarching national public policy framework is the *Sustainable Farming and Food Strategy*. Respondents at the project workshop explained that this provided the 'linchpin' between national policy (from the Cabinet Office to Defra) and regional strategy and local food-related projects.

Figure 9.2 also includes the Department of Trade and Industry (DTI) and the Department of Health (DoH). These two institutions each have their own food policy directives, concerned with food clusters and healthy eating respectively, which again 'feed down' to dedicated regional strategies. Industry bodies are also involved in regional and local food policy in important ways (see Figure 9.2). As the key national food industry research provider, the IGD and its sister company, the Food Chain Centre (formed in May 2002), are significant here.[5] The Food Chain Centre is primarily focused on '... efficiency along the food supply chain and information flow' (Ins4). An IGD representative noted that research and policy activities relating specifically to local foods are piecemeal. The most notable example is a *Local Sourcing Guide* (IGD 2002) developed to showcase examples of local procurement. National voluntary bodies meanwhile work largely as 'policy facilitators'. Made up of 100 or so voluntary organisations, SUSTAIN, for instance, runs six projects that cover issues such as food poverty and urban farming. The sustainable food chain project is most significant in the context of local foods and tackles three broad themes: (i) local food economies, (ii) food miles, and (iii) public procurement. As noted from the interview, the primary objective is to develop localised food systems that support and provide more sustainable social and environmental footprints. In summary, there is clearly a range of stakeholders and interested parties concerned with local food policy. This broadening political scope introduces potential complexities, especially in terms of strategic co-ordination.

Scale 2: Strategy

The 'strategic scale' attempts to represent the incorporation of national food and farming policies into dedicated regional strategies, in this case in the North East of England. Four institutions are significant to food chains in this particular region. Of these, GONE and ONE play the most important roles. In response to the Curry Report (2002), ONE has been tasked (along with GONE) to devise a regional/local food strategy (see Figure 9.2). The contents of this strategy include input from others in the regional steering group (e.g. Defra, CA). Discussing this strategy, one workshop participant noted that:

> 'A big vehicle now is to put together a development plan that contains aspects of regional food, of local food [. . .] the content of that will help us to overcome a lot of the fragmentation and get some clarity. Curry is an enabling framework that

---

[5] For details see the IGD (www.igd.com) and Food Chain Centre (http://www.foodchaincentre.com) websites (last accessed 28 July 2006).

each region has to pick up and deliver... it's very important; it's the key driver' (Taped discussion, Newcastle Workshop, June 2003).

The significant finding which emerged from the interviews is that each regional institution was also keen to outline its own regional/local food agenda. GONE, for example, has a local food remit at two policy and strategic levels: rural affairs and health (Figure 9.2). The rural team promote and implement Defra-related policies (e.g. the Curry Report 2002 and the Defra *Rural Strategy* 2004). Work here has focused primarily on local specialist food production, including close links with the North East Development Service which had the responsibility for delivering the English Rural Development Programme (ERDP). Meanwhile, the health team has a '5 a Day' regional co-ordinator to promote this and other health food initiatives (e.g. The National School Fruit Scheme) to overcome inequalities in health and improve access to nutritional foods, particularly in urban-based 'food deserts' such as deprived parts of Newcastle (see Wrigley 2002). A similar pattern, although more economically focused, emerges in ONE, as it does with other stakeholders involved in supporting local food production in Northumberland.

Scale 3: Projects

The third scale signifies the implementation of regional strategies into 'tangible' local food projects. Figure 9.2 details work developed by institutions in the region to help assist producers and others in the food chain. Projects are divided into 'regional' (North East) and 'local' (Northumberland) food initiatives. At the regional scale, two projects are significant. The first is Northumbria Larder (NL), the region's speciality food and drink group.[6] Formed after the 2001 outbreak of foot-and-mouth disease, the food group's chairman regards it as a 'practical tool' to assist business growth in the North East region. The Larder has received significant institutional backing in terms of both finance and political support. One obvious link is ONE, but the interviewee also identified important links with the Centre for Rural Economy based at Newcastle University which has worked with the food group to improve and develop local supply and distribution networks. North East Land Links is the second regional food project. This project is dedicated to two main strands of local food work: (i) public procurement and (ii) community food. The main concern for the latter especially is to improve access to and availability of fresh/healthy produce. In the project officer's words: 'The priority of local community food initiatives is not provenance; it's affordable fresh produce, it's convenience...' (Ins12). In terms of institutional links, a number of institutions are significant here. The community foods agenda, which supports food cooperatives, community cafes, food delivery schemes and local buying groups, for example, involves local agencies from across the region, including community organisations, local authorities and primary care trusts.

In Northumberland, the main institution offering direct assistance to small (specialist) food companies is Business Link Northumberland (BLN). The main role is to help local producers prepare grant applications and project funding from national organisations

---

[6]For full details, including a list of members, see http://www.northumbria-larder.co.uk (last accessed 28 July 2006).

such as Defra. In terms of food projects, the 'Made in Northumberland' scheme[7] is significant, promoting and marketing speciality products, as well as organising local trade fairs. Various other local food projects were surveyed in the county, including Northumbria Select Beef and Hadrian's Wall's Tourism Partnership. The former is a public procurement initiative run through the County Council's Catering Service to source beef from Northumbrian farms to supply local schools in the county, while the latter was developed to promote local speciality products produced along the wall, including developing a Hadrian's Wall Local Market brand. Both projects have complex institutional links that stretch within and beyond the study region boundaries.

Assessing Institutional Structures for Local Foods

Having now set out the broad nature of institutional arrangements, it is possible to provide a more critical assessment of institutional support for local food systems in Northumberland. A key argument in this chapter is that an institutional survey of local foods can help inform debates about regionalisation. Both themes (regionalisation and local foods) are discussed below to help 'unpack' this particular regional economic space.

From an institutional perspective, two things are significant. The first issue relates to debates about *decentralisation*. From the surveys, it is evident that a number of structures have been put in place to support more regionalised modes of governance. A good example of this is the implementation of the ERDP. Links can also be 'scaled back' to the local level, with Business Link advisors assisting producers with grant applications and business plans. However, over-riding this 'regional-verve' is evidence that regional strategies for local food systems and rural development more generally remain heavily driven from central government departments. Returning to the ERDP example, it was thus noted that:

> 'Whitehall still maintains a stranglehold on policy frameworks and implementation [...] we would like them to give us the policy framework and we would sort out how to implement it. At the moment they are producing the policy framework and telling us how we should implement it' (Ins 9).

This is perhaps no surprise. Ward and Lowe (2004) note that less than 15 per cent of the ERDP's resources are currently progressed at the regional level, and less than 3 per cent are available for non-agricultural developments beyond the farm gate. This trend is explained in terms of the highly centralised nature of Defra and the weak organisation of rural interest groups at the regional scale. Reflecting on this centralisation theme more broadly, a workshop delegate confirmed that 'Any scheme that you look at in the region will have a route to somewhere. People are only doing what they are told... Nothing happens with us lot unless there is a policy directive' (Taped discussion, Newcastle Workshop, June 2003).

The second issue relates to the process of *'institutional thickness'*, defined broadly as the nature of institutional presence in the region (in terms of form, diversity, etc), levels of interaction and partnership working and a general sense of collective purpose

---

[7] See http://made-in-northumberland.co.uk (last accessed 26 July 2006).

in relation to supporting local food economies. The interviews and institutional maps identified clear examples of partnership working and interaction between institutions at different scales. Although not specific to speciality/local foods, a very good example of how institutional thickness is bolstered is the establishment of Regional Rural Affairs Forums and strategic developments like the Rural Action Plan, with representatives from government agencies and rural stakeholder groups. The latter, led by the RDA, was developed in 2001 in response to the North East region's rural economic crisis and identified a key role for food SMEs, including attempts to develop regional food and drink clusters to maximise added value potential, and further develop the speciality/organic food sector. The preparation of the regional food strategy also involved collective input across the agencies. Similar synergistic links also take place between marketing schemes like 'Made in Northumberland' and the regional food group, with the former helping local producers towards developing a 'regional' profile. On these grounds, modes of governance to support local foods thus support a broader decentralisation thesis.

The process of 'mapping' institutions in this way nevertheless raises questions about blurred accountability and duplicatability between institutions. The first point to note in this context is that there appear to be too many strategies surrounding regional economic development, including (specialist) local foods. This point was made clear in an interview with two RDA representatives. After a lengthy discussion, four strategies were eventually identified as relevant to them. The first was the overarching regional economic strategy, *Unlocking Our Potential* (2003), which identifies standard RDA targets, as set down by the DTI, for example increase GDP, reduce unemployment and improve added value. Below this, the second strategy is the North East Food and Drink cluster, the third is involvement in the regional food strategy and the fourth is the Rural Action Plan, noted above (Figure 9.2). One interviewee summarised the position well:

> 'You have all these marvellous documents, all with action plans that need implementing and people are kind of confused. We are stuck with it. DTI is pushing clusters...The RAP has budget lines to implement... the implementation plan for Curry again will have specific areas which need actions and delivery and funding' (Ins8a).

There is not one distinct regional food agenda, but a range of institutional actors and policies. This governance process thus creates a confusing and sometimes overlapping institutional landscape. A final related point here is that there is also duplication between local and regional food projects (when comparing, for e.g., Northumbria Larder, Made in Northumberland and Hadrian's Wall Local Market brand), as well as potential contestation as different regional and local institutions interpret the local foods concept in specific ways (cf. Jackson *et al.* 2006).

This leads to the second key theme underpinning this chapter – local (specialist) foods. It has already been suggested that the institutional context is an important influencing factor. Two further things are significant when characterising local food economies from the lens of surveyed institutions. The first issue concerns the way local foods are perceived as offering *economic potential*. This is neither a surprise nor necessarily problematic, assuming the right practices are supported and promoted. The IGD's (2002) *Local Sourcing Guide*, for example, identifies marketing opportunities for businesses of various sizes, including small producers selling via direct marketing channels, SMEs selling direct to multiples and McDonald's restaurants sourcing lettuce from local suppliers in West Sussex. The crucial message is that'... local foods is not just the preserve of small companies' (IGD 2000, p. 5). The institutional analysis also shows that regional economic agencies drive the local food agenda, promoting regional branding and value-added strategies as tools to help improve competitive advantage. For some respondents, these strategic concerns potentially sideline environmental or social aspects, cornerstones of a more sustainable food economy. Some of these comments come from likely sources (e.g. SUSTAIN), although similar comments are echoed from less predictable sources. A spokesperson from GONE, for example, noted that more needed to be done to establish a 'health thrust' within the then current regional food strategy (Ins7b).

The second issue follows on from this and concerns the *tension between locality foods and local foods*. The institutional survey concerned itself specifically with local/speciality foods. However, the interviews identified a range of interesting local food schemes that stretched beyond this classification, including, for example, the Northumbria Select Beef scheme. It is no surprise that RDA work focuses largely on specialist products, including also an emphasis on improving 'mainstream' commodity markets. However, as the RDA recognised, local food initiatives need further support. Most significant in this debate is the North East Land Links project, which attempts to move away from an emphasis on locality foods and focuses instead on the type of local supply chain, including community food projects. These latter food schemes do not necessarily involve local products, but source fresh produce such as fruit and vegetables (usually via local wholesalers) for local, more economically deprived urban and rural communities.

Conclusions

This chapter provides one of the first attempts to bring local foods into the regionalisation debate, using Northumberland as an explorative case study. It thus identifies a need to move beyond the boundaries of food systems research to inform wider geographical debates that may in turn help re-interpret local, and more broadly, alternative food practices. One of the key findings to emerge from the institutional survey was the complicated and overlapping nature of governance structures. A great deal of institutional activity is clearly undertaken to encourage the growth and development of the study region's specialist/local food sector. This activity sits alongside a wider political-economy agenda which, post-Curry (2002), has encouraged a re-scaled governance structure and a relocalised food chain agenda. To help convey the nature and range of support mechanisms, activities were tentatively organised according to three scales – policy, strategy and projects. Overall, the results indicated a fairly stratified layering according to the 'three-scaled' organisational framework. Arranging institutions using this methodology is obviously not as straightforward as it appears, although it does provide a useful mechanism to delineate particular institutional roles in relation to supporting local food systems and rural development more generally.

The interviews identified various issues relevant to both local foods and regionalisation literatures. Key among these was recognition of partnership working and interaction, despite the possibility of duplication of effort and an overabundance of regional economic strategies, combined with competing institutional views. In terms of local foods, the interviews indicated a significant degree of support and activity across the county, especially in terms of supporting specialist producers. In fact, most work in Northumberland and the wider North East region has concentrated on locality products. Some interviewees argued that such a narrow economic focus potentially ignores the diverse nature and potential of local food economies, characterised by different types of food provisioning that include 'quality' food production but also involve local community food projects, public procurement initiatives and so on. These latter initiatives thus emerged from this research as potentially significant, both in the context of alternative food economy debates and more practically as tools that offer economic *and* social development potentials.

The institutional structure mapped out in this chapter is clearly not static. Since the interviews, the reforms of the CAP, most notably the introduction of the Single Farm Payment, have brought significant changes both institutionally and at a farm/producer level. The ERDP has also reportedly adopted a more 'environmental economics approach', in keeping with the general push towards ecological modernisation and more environmentally sustainable modes of food provisioning. As an outcome of Curry (2002) and Haskins (2003), which both identified an overabundance of institutions and related strategies, mergers and policy directives have also been introduced to streamline strategic delivery (e.g. the recent formation of Natural England merging English Nature with parts of the CA and Rural Development Service). Clearly, more intensive research is required, across different regions, to determine the impact of these changes on both the local food economy and farming and food restructuring more generally. The chapter thus ends by outlining potential avenues for future research in relation to regionalisation and local foods.

Following Jones and MacLeod (2004), this chapter focused mostly upon the economic aspects of regionalisation, neglecting the political dimension included in their initial categorisation of institutionalism. More research is needed to explore this latter aspect in the context of local foods. Bearing in mind DuPuis and Goodman's (2005) call for a more socially centric local food politics, how, for example, do these regional economic spaces influence or perhaps even threaten grassroots-based initiatives? Related to this, greater theoretical and methodological engagement is required with food initiatives that espouse an avowedly community or political based agenda. Such developments have links with economic geographers currently exploring community economies, political activism, social entrepreneurship and so forth. As Hughes (2005) recently argued, more direct engagement and closer dialogue is required between economic geography research on alternative economic spaces and agro-food studies of alternative food networks to help further critique these particular spaces of exchange and circulation.

One final area of research relates to cross-border studies. While this chapter has focused on Northumberland, the larger research project included institutional surveys in The Borders as part of a Scottish/English borders comparison (see Figure 9.1). Preliminary results reveal a distinct border effect between the two counties in relation to local food governance, with a much less stratified institutional architecture in The Borders. The influence of political boundaries on agriculture and food is a currently under-researched, but potentially valuable part of a broader research agenda. A good illustration of this potential is Reitsma's (1986) seminal study of land use differences on the US/Canadian sides of the 49th parallel. The paper showed how farming on the Canadian side was intensive and diversified (including fruits, grapes and vegetables) and how this type of production was virtually absent on the US side. The differences were explained in terms of socio-cultural and political factors, most notably government. For instance, fruit and vegetable growers in Canada were encouraged and protected by tariffs, import quotas and seasonal duties on imports from the US. A re-engagement with border geographies – still dominant in political studies – may thus provide another lens to scrutinise different political economy forces on agricultural sectors and dedicated local/alternative systems of food supply.

Acknowledgements

This chapter derives from the EU-funded project: 'Supply chains linking food SMEs in Europe's lagging rural regions' (SUPPLIERS, QLK5-CT-2000-00841). Collaborating laboratories are SAC, Aberdeen, UK (Co-ordinator); Coventry University, UK; University of Wales, Aberystwyth, UK; Teagasc, Dublin, Ireland; ENITA Clermont-Ferrand, France; University of Patras, Greece; SIRRT, University of Helsinki, Finland; and the Agricultural University of Krakow, Poland. The authors would especially like to thank those who gave up their time to be interviewed and/or to take part in the workshop. The constructive comments made by Lewis Holloway, Moya Kneafsey and one anonymous reviewer on an earlier version of this chapter are also acknowledged with thanks.

References

Barling, D. and Lang, T. (2003). A reluctant food policy? The first five years of food policy under Labour. *The Political Quarterly* 74, 8–18.
Burch, M. and Gomez, R. (2002). The English regions and the European Union. *Regional Studies* 36, 767–778.
Clark, J. (2006). The institutional limits to multifunctional agriculture: subnational governance and regional systems of innovation. *Environment and Planning C* 24, 331–349.
Counsell, D. and Haughton, G. (2003). Regional planning tensions: planning for economic growth and sustainable development in two contrasting English regions. *Environment and Planning C* 21, 225–239.
Curry Report (2002). *Farming and Food: A Sustainable Future*. London: Policy Commission on the Future of Farming and Food, Cabinet Office.
Defra (2004). *Rural Strategy*. London: Defra.
DTI (2001). *Business Clusters in the UK – a First Assessment*. London: DTI.
DuPuis, M. and Goodman, D. (2005). Should we go "home" to eat?: toward a reflexive politics of localism. *Journal of Rural Studies* 21, 259–371.
Dürrschmidt, J. (1999). The 'local' versus the 'global'? 'Individualised milieux' in a complex 'risk society'. The case of organic food box schemes in the South West. *Explorations in Sociology* 55, 131–152.
Gough, J. (2003). The genesis and tensions of the English Regional Development Agencies. *European Urban and Regional Studies* 10, 23–38.
Haskins, C. (2003). *Rural Delivery Review. A Report on the Delivery of Government Policies in Rural England*. London: Defra [http://www.defra.gov.uk/rural/pdfs/ ruraldelivery/haskins_full_report.pdf (accessed: January 2005)]
Hudson, R. (2005). Region and place: devolved regional government and regional economic success? *Progress in Human Geography* 29, 618–625.

Hughes, A. (2005). Geographies of exchange and circulation: alternative trading spaces. *Progress in Human Geography* 29, 496–504.
IGD (2002). *Local Sourcing: Growing Rural Business*. Watford: IGD.
Ilbery, B. and Kneafsey, M. (2000). Producer constructions of quality in regional speciality food production: a case study from south west England. *Journal of Rural Studies* 16, 217–230.
Ilbery, B. and Maye, D. (2005). Food supply chains and sustainability: evidence from specialist food producers in the Scottish/English borders. *Land Use Policy* 22, 331–344.
Ilbery, B. and Maye, D. (2006). Retailing local food in the Scottish-English borders: a supply chain perspective. *Geoforum* 37, 352–367.
Jackson, P., Ward, N. and Russell, P. (2006). Mobilising the commodity chain concept in the politics of food and farming. *Journal of Rural Studies* 22, 129–141.
Jeffery, C. and Mawson, J. (2002). Introduction: beyond the White Paper on the English regions. *Regional Studies* 36, 715–720.
Jones, M. and MacLeod, G. (2004). Regional spaces, spaces of regionalism: territory, insurgent politics and the English question. *Transactions of the Institute of British Geographers* 29, 433–452.
Kirwan, J. (2004). Alternative strategies in the UK agro-food system: interrogating the alterity of farmers' markets. *Sociologia Ruralis* 44, 395–415.
MacLeod, G. and Jones, M. (2001). Renewing the geography of regions. *Environment and Planning D: Society and Space* 19, 669–695.
Marsden, T. and Sonnino, R. (2006). Rural development and agri-food governance in Europe: tracing the development of alternatives. In V. Higgins and G. Lawrence (Eds), *Agricultural Governance: Globalisation and the New Politics of Regulation* (pp. 50–68). London: Routledge.
Maye, D. and Ilbery, B. (2006). Regional economies of local food production: tracing food chain links between 'specialist' producers and intermediaries in the Scottish–English borders. *European Urban and Regional Studies* 13, 337–354.
Morris, C. and Buller, H. (2003). The local food sector. A preliminary assessment of its form and impact in Gloucestershire. *British Food Journal* 105, 559–566.
Parrott, N., Wilson, N. and Murdoch, J. (2002). Spatializing quality: regional protection and the alternative geography of food. *European Urban and Regional Studies* 9, 241–261.
Pearce, G., Ayres, S. and Tricker, M. (2005). Decentralisation and devolution to the English regions: assessing the implications for rural policy and delivery. *Journal of Rural Studies* 21, 197–212.
Porter, M. (1990). *The Competitive Advantage of Nations*. London: Macmillan.
Reitsma, H.A. (1986). Agricultural transboundary differences in the Okanagan region. *Journal of Rural Studies* 2, 53–62.
Renting, H., Marsden, T.K. and Banks, J. (2003). Understanding alternative food networks: exploring the role of short food supply chains in rural development. *Environment and Planning A* 35, 393–411.
Tomaney, J. (2002). Democratically elected Regional Government in England: the work of the North East Constitutional Convention. *Regional Studies* 34, 383–399.
Tomaney, J. and Ward, N. (2000). England and the New Regionalism. *Regional Studies* 34, 471–478.
Ward, N. and Lowe, P. (2004). Europeanizing rural development? Implementing the CAP's second pillar in England. *International Planning Studies* 9, 23–32.
Watts, D.C.H., Ilbery, B. and Maye, D. (2005). Making re-connections in agro-food geography: alternative systems of food provision. *Progress in Human Geography* 29, 22–40.
Winter, M. (2003a). The policy impact of the foot and mouth epidemic. *The Political Quarterly* 74, 47–56.
Winter, M. (2003b). Embeddedness, the new food economy and defensive localism. *Journal of Rural Studies* 19, 23–32.
Wrigley, N. (2002). 'Food deserts' in British cities: policy context and research priorities. *Urban Studies* 39, 2029–2040.

Chapter 10

Governing the Speciality Food Sector: Integrating Supply Chains, Sectors and Scales in West Wales

Catherine Walkley
Institute of Geography and Earth Sciences, University of Wales, Aberystwyth, UK

Introduction

This chapter is concerned with the function of governance in the development and sustainability of speciality food networks. It investigates the provision, priorities and processes of support to speciality food networks, focusing on products that are branded with a local or regional provenance. In particular, the role of institutions in advocating and contributing to integrated supply chains and rural development is explored. Consideration is given to the effectiveness of regulation and aid in overcoming tensions between scales and sectors, and an analysis is undertaken of the balance of support available to actors with different functions within supply chains. The introduction contains an overview of the key challenges associated with governing the speciality food sector. These are subsequently examined in more detail by drawing on empirical evidence relating to three products from rural West Wales.

Alternative food networks are defined as 'a broad embracing term to cover newly emerging networks of producers, consumers and other actors that embody alternatives to the more standardised industrial mode of food supply' (Renting *et al.* 2003: 394). The sector encompasses foods that have distinct product characteristics and network arrangements (Holloway and Kneafsey 2004). In terms of product characteristics, these are commonly differentiated on the basis of branding that defines food as speciality, high quality, geographically distinctive, fairly traded and/or sensitive to environmental impacts and animal welfare. With regard to supply networks, such products are often involved with short food chains that are characterised by close relationships between actors and, in many cases, have a limited number of links within the supply chain. A key feature of this 'shortness' is that the transmission of information enables consumers to make informed purchases on the basis of products' 'biographies'. However, it is important to bear in mind that there is no clear distinction between alternative and conventional foods. So-called 'alternative' products are commonly involved in conventional supply channels and are impacted by mainstream strategies, initiatives and institutions that are not exclusive to this niche sector (Marsden *et al.* 2000). Within this study, the focus is on food and drink businesses that are seen as 'alternative' to those that are dependent on conventional commodity-based markets. More specifically, the focus is on 'speciality' products that are differentiated on the basis of their taste, methods of production and/or their strong local place-based association.

Support to businesses involved with speciality food is highly diverse; it transects scales and sectors and incorporates a wide range of organisations whose form and objectives vary significantly. Food businesses are therefore affected by many lateral processes and are influenced by – and have the potential to contribute to – a number of policy domains, including agriculture, health, rural development, sustainable development, tourism, the environment, community development, consumer rights, planning, food safety and education (see Barling and Lang 2003; Defra 2003). Although this reflects the wide-ranging benefits a thriving specialist food sector can potentially make to economic, social and environmental agendas, integrating these interests to achieve 'multi-sectoral governance' presents specific challenges.

In addition to transecting a wide range of policy agendas, institutional support for speciality food businesses is also driven by actors operating at varying scales. As with mainstream food and agriculture, which is dependent on distanced chain relations and a multinational institutional presence, national and international contexts play a central role in shaping politics and policy for this niche market sector (Marsden 1998; Goodman 2003). However, speciality food networks are also supported by a large number of organisations at local and regional scales, many of which act as a delivery agent for higher level programmes managed by government and its sponsored bodies. The consequence is multilevel governance, which involves an interplay of actors at various scales of decision making. Or, as Marsden (1998: 109) puts it more generally: '... we have to conceive rural spaces as ensembles of local and non-local connections, of combinations of local actions and actions "at a distance", situated in regional economies and different institutional contexts'. Successful governance thus requires a marrying of scales so that higher level policy and strategy are supported by projects and actions 'on the ground'.

The data collection for this research focused on three product groups (bottled water, quality dairy products and Pembrokeshire early potatoes), all located within the rural area of West Wales (see Figure 10.1). These products share certain criteria, such as being distinctive food and drinks of the region, a tendency to be marketed with attention to their place of origin and the dominance of small producers across all three sectors. However, they differ in their supply chain arrangements, institutional support and the perceived benefits they deliver to the local economy. Evidence has been sourced from a series of semi-structured interviews with actors involved with the three products in the study region. Over 100 interviews were undertaken with businesses involved in the production, supply and sale of these products, including producers, processors, wholesalers, distributors, retailers and caterers. A further 15 organisations that provide support and assistance to the food businesses were also interviewed. Institutional surveys comprised local authorities, the regional government, business organisations, trade unions,

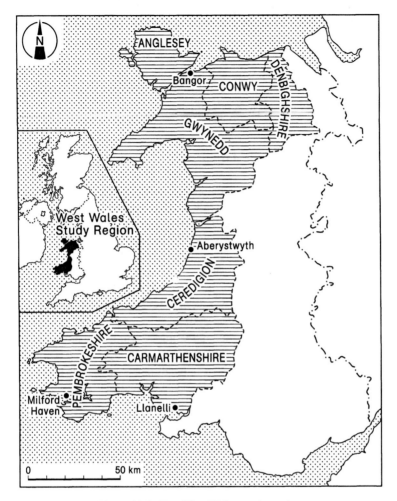

Figure 10.1: The West Wales study region.

regulatory bodies and local development associations, such as LEADER+groups. In addition, two workshops were held with a range of actors involved with the three specialist food networks in order to enhance understanding of the common challenges faced by the sector and to propose potential solutions to improve integration amongst businesses involved with locally branded and speciality products. Collectively, these data provide an insight into the institutional framework for these three products in West Wales.

The methodology adopted a 'whole-chain' approach, which, in broad terms, acknowledges the presence of multidirectional flows of information, knowledge and transactions between supply chain actors. As a result, the findings present a variety of perspectives on the governance of the sector which advances the understanding of support to speciality food businesses within the region. This informs the extent to which economic decisions of businesses involved with speciality food are embedded within an institutional context, and the extent to which the actions of organisations are embedded within broader agendas for the sector and the region (Bristow 2000; Wood and Valler 2001). The volume and effectiveness of support is examined using the concept of 'institutional thickness' which seeks to identify the institutional environment for a given sector or space by examining the density of organisations and the degree to which governance is collective and effective. The data reveal that support available for this sector varies according to geography, product type and supply chain function. This in turn raises questions concerning the extent to which institutions currently adopt an approach that adequately supports integrated rural development.

Institutions and Institutional Thickness

Institutional presence and institutional dynamics are perceived to have a strong influence on regional development (Wood and Valler 2001; Whatmore *et al*. 2003). Such thought has emerged from growing recognition that regional success does not only depend upon economic conditions, but is strongly influenced by the effectiveness and interaction of organisations and levels of social capital within the population (Copus *et al*. 2001). Institutions are therefore judged to condition and regulate economic activity (Amin 2001). Such awareness has inspired an institutional 'turn', or an institutionalist approach in human geography, which centralises the role of these actors in influencing the success of sectors and regions (Amin 2001; Jessop 2001).

The contribution of institutions to economic vibrancy within a local economy can be considered through the concept of 'institutional thickness' (Amin and Thrift 1994). A sense of thickness conveys the strength of local conditions that provide a framework for economic activities within a locality. Like embeddedness and social capital, it recognises the strong associations between the strength of social and cultural activities, economic development and the growth of regions (Granovetter 1985; Putnam 2000). Amin and Thrift (1994) identify four components that constitute thickness in a region. The first of these is a *strong institutional presence*, which is manifested through both the density of organisations within a territory and the diversity of their forms and objectives. Second, *levels of interaction* between institutions are considered important. This relates to formal partnerships and more informal networks, including also communication and dissemination. In the language of social capital, such interactions are expressed as 'bonding capital' which exists between actors with similar objectives and 'bridging capital' which relates to associations that link heterogeneous groups (Putnam 2000). Although it is anticipated that the formation and operation of partnerships will lead to shared rules and conventions, in reality, relationships within partnerships do not occur along an even plane of power. The presence of *structures of domination and patterns of coalition* is therefore considered as a third component of institutional thickness. This is particularly evident through asymmetrical relationships amongst actors that possess different levels of power – also defined as 'linking capital' (see Cote and Healy 2001). The final element of institutional thickness is a *mutual awareness of a common enterprise*, which should be continually reinforced through dynamic interaction between institutions. The processes of developing these four factors are considered to be as important as the inert institutional presence within a region (Amin 2004). The consequence of organisations acting together through social capital is therefore integral to determining the deeper value of a strong institutional presence.

According to definitions within political geography, institutions include formal organisations as well as more informal rules, conventions and knowledges that contribute to a local sense of place-centredness (Amin and Thrift 1994; Henry and Pinch 2001). In this sense, institutions constitute organisations (e.g. firms, households or public enterprises), as well as groups of actors such as cooperatives. The breadth in this classification recognises the contribution of informal actors to defining conventions, impacting behaviour and influencing regulation (Giddens 1984 from Jessop 2001). The rest of this chapter focuses on institutions as a form of governance and is predominantly concerned with a subset of 'hard' institutions that guide economic behaviour (Amin 2001). This incorporates formal organisations, including regulatory agencies, local and regional government departments, voluntary and community sector groups, trade unions and trade associations. Storper (1997) considers such formal organisations to be central to the generation of expectations, preferences and rules within the economy and society. It is, however, also important to recognise the role of other actors, particularly businesses and groups of businesses who are involved in providing a sense of social coherence to economic activity within a given territory or sector (see MacLeod 2001).

Set against this context, the next three sections of this chapter present some results from empirical work undertaken in rural West Wales. The first section provides an analysis of the presence, effectiveness and variation of support along the food chain. This draws on evidence from food business and institutional interviews to examine forms of support available to different actors along the food chain for the three study products introduced above. This is followed by a review of institutional support that emerges from within the food chain, with a particular focus on co-regulation and internal regulation. The third section ends with a discussion about the nature of partnerships between organisations and the challenges of achieving multilevel governance to support specialist food networks.

Support Along the Food Chain

Research undertaken by FLAIR (2003) reveals that a third of businesses involved with the local food sector were not satisfied with the support they received. Similar findings are apparent in West Wales. Overall, just over a half of those businesses interviewed concluded that the institutional framework was inappropriate to the needs and objectives of their enterprise. The forthcoming section disaggregates the survey findings a little further to determine the extent to which the presence, function and effectiveness of support to businesses involved with speciality products varies along the food chain. This is followed by a discussion which considers the rationale behind noted inequities and the consequence this has for rural development in West Wales.

Findings from West Wales

Evidence from West Wales identifies that there is greater policy and institutional weighting towards farmers and producers in relation to both levels of satisfaction and the volume of support. Overall, 61 per cent of producers are satisfied with the assistance available to them, compared to 17 per cent of intermediaries and one-third of retail/catering-based customers. In relation to the amount of support, findings from the institutional interviews present a similar picture. All organisations surveyed seek to support producers and processors; 47 per cent extend their activities to wholesalers; only 33 per cent consider that transport is within their remit; two-fifths are concerned with retailers and customers; and just under half include consumer well-being and education within their range of support functions.

Support targeted at producers is variable according to business form and product type. Farm-based producers are relatively well represented by their unions and eligible for a range of projects that target farm diversification. However, a significant proportion of entrepreneurs involved in the production of speciality food products are not farmers or farm based. Within the quality dairy sector, for example, over a third of producers do not classify themselves as farmers and are only eligible for restricted and piecemeal assistance. There is also frustration from producers that actors involved with distribution are often unable to benefit from support. Intermediary functions are deemed essential to enable local producers to thrive in the region and to help retain value-added within the locality. Gaining access to markets is perceived by producers to be the most severe hindrance faced by their business, given that West Wales is geographically marginal and suffers from poor transport infrastructures.

Consumers have received increased attention in policy documentation in recent years. This has been complemented by a growing institutional presence that is geared towards consumers. In Wales, the Food Directorate, located within the Welsh Development Agency, has responsibility for promoting the identity of Welsh food with a strong emphasis on augmenting consumer recognition of, and demand for, Welsh products. At a more local scale, local authorities and regeneration groups have been instrumental in the production of food directories, developing local branding initiatives and supporting food festivals – here again the intention is to stimulate demand for locally produced food and drink. Pembrokeshire, for example, has developed an identity mark for producers and customers involved with local food. According to the area's local food officer, this has had a positive impact on patterns of consumption, which in turn have impacted on supply side practices:

'People are actively looking for more local produce . . . Restaurants, tea rooms, bed and breakfasts are starting to think "my customers want local, so I need to search for local food to put on my menu"' (Local food officer).

Organisational support in rural Wales tends to target individual nodes within the food chain, rather than embracing a more holistic 'whole-chain' approach. Less than a fifth of the organisations surveyed in West Wales support the whole spectrum of actors from production through to consumption. However, several practitioners, rhetorically at least, are aware of the need for their assistance to be extended in its scope, in order to embrace a whole-chain approach. As one respondent put it:

'We link mainly to the primary producer, but there is no reason that our primary business support should not extend. We are always pushing farmers further and further towards the consumer, and if that's the case we should be supporting them' (Manager of a rural development organisation).

Organisations that do provide whole-chain support tend to encourage the development of links between producers and retailers. The Welsh Development Agency's Trade Development Team, for instance, co-ordinates a 'meet the buyer' programme to encourage producers to establish links with major retailers and wholesalers, most of which are based outside of the region. In contrast, another group of institutions seeks to discourage its producers from dealing with large-scale customers and actively dissuades its clients from entering such chains:

> 'I say to the producers "don't touch them with a barge pole—they can pick you up and drop you, there's no security. If the buyer has a bad day or you piss them off they drop you like a hot potato"' (Business support organisation).

Inconsistencies Explained

Within this section, the uneven nature of support along the food chain for speciality and locally branded foods is explored and some interpretations are offered. The greater depth and breadth of support to producer-based supply chain functions may be attributed to historical processes that continue to have a strong impact on current trends in policy and governance (Copus *et al.* 2001; see also Goodman 2002). As Marsden *et al.* (2002) note, farming interests have dominated rural policy because of the high levels of intervention which have traditionally been provided to the sector. A residual 'agricultural fundamentalism' is still apparent in Wales through the Rural Development Regulation (RDR), which is focused almost entirely on agricultural and agriculture-related development measures. Support from the RDR is consequently prejudiced towards farm businesses at the expense of non-farm producers. As the traditional distinctions within agri-food production become blurred, it appears that institutional support has not fully responded to these changes, with the result that assistance is uneven amongst those who are adding value to Welsh food and drink.

Some organisations and commentators defend a producer bias, making reference to the benefits of stimulating agricultural restructuring within the region on economic grounds and the need to provide support to the 'farming family' to preserve a social and cultural way of life in the Welsh countryside (see Marsden *et al.* 2002). This approach thus regards agriculture as both a cause of deprivation in rural areas and identifies it as a key solution for regeneration in the countryside. Others also advocate targeting support towards the 'head' of the supply chain, on the basis that this is where the foundations of the food sector lie; support here will thus best maximise benefits to supply networks, the agri-food sector and wider rural development.

The implications of this approach are obvious and significant, with a marginalisation of downstream actors and the rejection of whole-chain approaches to rural development (Flynn *et al.* 2003). One other part of the supply chain has, as noted above, captured some attention: the consumer. The growing institutional attentiveness towards the consumer can be attributed to more overt recognition that the trend towards quality foods is, to a large extent, driven by concerns from consumers over food safety and ethical concerns (Sage 2003). Traditionally, government's role in supporting consumers has been to ensure collective food security, especially through intervention at the producer level (Flynn *et al.* 2003). Current provision, however, primarily focuses on ensuring food safety. In 2000, the Food Standards Agency was established to protect the public's health and represent consumer interests, whilst the Food Chain Group was established under MAFF in 1999 to increase understanding amongst the players in the food chain and between industry and consumers. The stronger collective representation and agency for consumers is matched by a growing awareness of the role of branding and regulation within such food networks. There remains, however, only partial understanding of the social, economic and psychological influences on purchasing behaviour.

The overall lack of whole food chain support to the speciality food sector thus reflects a tendency amongst organisations to prioritise either spheres of production or those of consumption (accepting that the latter is limited primarily to food safety). The reluctance for institutions to support and encourage intermediaries and retailers may partly be explained by the fact that such services are part of the wider food system (largely thought to be de-regulated) and will rarely be dedicated to the speciality food sector. In addition, State Aid regulations require that public sector funds do not have inverse consequences on competing firms, which is more likely to occur amongst such functions. The empirical findings also reveal contradictory approaches in the types of marketing channels that are advocated. Combinations of marketing strategies within a locale are considered essential to fully develop a region's speciality food economy and to maximise contribution to regional development (Kneafsey *et al.* 2001). Businesses within the West Wales speciality sector are dependent – to varying degrees – on transactions and relationships that extend beyond the region and beyond the domains of the speciality food sector to more conventional actors. This draws our attention to the imprecise boundaries and characteristics of the speciality food sector and the governance structures which impact upon it.

Governance from within the Food Chain

In addition to support from external actors, an institutional presence exists within the food chain through forms of 'co-regulation' and 'internal regulation'. These two types of governance make distinct contributions to institutional thickness, and this chapter argues that they have a considerable impact on the speciality food sector.

Co-regulation refers to industry-led, representative networks and organisations. In most cases, these institutional arrangements are membership based and have generally been established and sustained by economic actors, rather than the public sector. Evidence from the agri-food sector within Wales reveals that most forms of co-regulation have emerged as a result of an unfavourable power differential within supply chains, or within an industry. They comprise institutions that operate at a range of scales, have diverse objectives and varying degrees of professionalism. Such organisations within West Wales include those that are dedicated to certain business types, such as the National Farmers' Retail and Markets Association (FARMA), which supports farm shops and farmers' markets; those involved with specific products, for example the Specialist Cheesemakers' Association; and those relating to certain chain functions, such as the Ceredigion Producers' Group. These regimes create benefits of scale, associated with the pooling of members' resources, and a greater propensity to attain funding from the public sector as a result of their cooperative structure. In addition, members are able to shape and be part of a formalised common agenda, which may enhance their power within individual food chains (Danson and Whittam 1999; Le Heron and Roche 1999).

A strong marketing agenda is a widely reported benefit of institutions that are based on co-regulation. One such example is the West Wales Good Taste Trail, which was conceived by two local entrepreneurs: a honey producer and the owner of a farm shop. Leaflets and a map have been produced and distributed within Wales and England. The accomplishments of the trail are widely attributed to the strong level of ownership from local food businesses. This case usefully illustrates a form of institutional thickness that is stimulated and sustained by entrepreneurs within food supply chains, rather than formal organisations, reflecting the success and potential of self-governance (see Thompson 2005 for details, including a case study of English National Parks). As with any membership group, this form of organisation is perceived by some to be exclusionary. There are some actors who choose not to be involved with cooperative forms of regulation; others may be excluded, or perceive themselves to be excluded. There is, therefore, a danger of assuming that such bodies are truly representative of entire sectors, scales or supply chain functions.

Internal regulation relates (again) to governance that emanates from actors within the food chain. However, unlike co-regulation, it is not generally associated with formal organisations but is a feature of economic transactions. In the agri-food sector, forms of internal regulation tend to be driven by the dominant actor within the food chain, with consequences on systems of production, product form, quality, design, packaging and branding (Dolan and Humphrey 2000). In the UK, supermarkets are widely renowned for generating buyer-driven chains in which they control the procedures and products of downstream actors (Flynn *et al.* 2003). The extent to which speciality food networks are subject to strong internal regulation varies significantly according to the motivations and decisions of producers, their product type and the form of supply chain with which they are involved.

The survey findings from this research revealed that 70 per cent of Pembrokeshire potato growers sell their locally branded potatoes to multiple retailers. In this case, the supermarkets dictate specifications of the crops, timing of the harvest and traceability systems. In addition, farmers note that the influence of the supermarkets extends beyond the chains in which they are directly involved and purchase the highest quality potatoes and in large enough quantities to dictate prices for the entire market. Furthermore, their preference for pre-packed, clean and uniform potatoes is considered to influence consumers' expectations and demands, especially with regard to product appearance and quality within other channel outlets. In contrast, cheese producers do not tend to be subjected to potent forms of internal regulation. Such businesses are more commonly involved with short food chains that are strongly influenced by the characteristics of production and the decisions of producers who, in this instance, supply less formalised wholesale and retail markets.

These two institutional practices that are derived from within the chain clearly enhance the institutional thickness of the specialist food sector. Both co-regulation and internal regulation thus contribute to the range and number of institutions that impact on speciality and spatially branded foods in West Wales. More specifically, forms of co-regulation promote the redistribution of power within the sector and within specific chains, which in turn addresses the development of a sense of common enterprise. Internal regulation results from an uneven power distribution within food chains and therefore signifies evidence of the third component of institutional thickness: the presence of structures of domination and patterns of coalition. These specific forms of governance appear to be more evident in some specialist sectors in West Wales than others (e.g. Pembrokeshire potatoes and internal regulation).

Scales of Support: Partnership or Competition?

Partnership between organisations is well documented as a key ingredient for successful rural governance (see Goodwin 1998 and Edwards *et al*. 2001 more generally). Interaction, commonality and competition between organisations involved with supporting agri-food development in Wales were central themes of the institutional surveys, the findings of which are discussed in more detail below.

Many programmes specify cooperation between actors and agencies as a prerequisite for being granted aid. Such partnership is deemed to benefit both the process of support (through enhanced resources and knowledge) and also the product, with outcomes more likely to be consensual amongst different organisations. In Wales, the Agri-food Partnership was established in 1999 to provide an integrated approach to the development of the agri-food sector. It is led and supported by the Welsh Development Agency but draws together interest groups from within the food chain who act at various scales. This, it is felt, has resulted in a more 'joined-up' approach to the development of the industry (Agri-food Partnership 2005). A key benefit of overarching organisations, such as this one, is their role in establishing a common agenda, which is seen as a vital component of institutional thickness in terms of providing 'a recognised set of codes of conduct, supports and practices which certain individuals can dip into with relative ease' (Amin and Thrift 1994: 15).

Working in partnership to develop, in this case, the specialist food sector thus involves embracing multilevel governance. One of the most conspicuous consequences of multilevel governance within the speciality food sector is the wide range of scales at which products are marketed. In West Wales, a large number of institutions promote place-based branding which associate products with locations which range from an individual valley or peninsular to those defined more generically as 'Welsh products'. The result is multilevel branding which some respondents argued created a confusing network of labels and typified the often fragmented and complex mosaic of institutional support in place in West Wales. Dialogue between local and regional institutional actors that encourage marketing on the basis of place-based provenance thus reflects hegemonic networks of power within the region. The Welsh Development Agency's Food Directorate, for example, asserts that local brands must substantiate the overall Welsh brand. The majority of local actors consider that these two scales of branding are compatible, although many add that the Welsh brand has a duty to support sub-regional strategies. However, a small proportion of food officers regard the Welsh and more local scales as a trade-off, as the following quote from a local food officer implies: 'I want to see Carmarthenshire getting away from saying it is in Wales. I want to sell Carmarthenshire'. This raises questions of acceptance amongst some local actors of their role within a strengthened, devolved region (Amin 2004). In spite of assumptions that the speciality food sector is highly embedded within a local territory, it is strongly influenced and dependent upon non-local businesses, organisations and policies. It also reflects competing agendas and an absence of a sense of common enterprise or 'industrial purpose' amongst certain regional and local institutional players (Amin and Thrift 1994).

Despite the above comments, there is evidence of partnerships between organisations operating at similar scales, with comparable powers and functions. These are particularly important for institutions that have low staff numbers and budgets, as one county food officer explains: 'I am a one-man band as a food officer in the council, so I need to work in partnership or I couldn't do what I do' (Unitary authority food officer). The majority of representatives from local agencies seek to develop a harmonious and supportive relationship with actors operating at a similar scale. As one local economic development office in Ceredigion put it: 'I don't think we are in the game of saying Ceredigion products are better than anyone else's'. Others, however, have a less harmonious strategy: 'I want to promote this as *the* food county of Wales' (Local authority food officer). This latter quote echoes Hinrichs' (2003) point about the rigid territories within which many (local food) organisations operate, resulting in a 'separatist politics' and an exclusive approach which is in the danger of neglecting wider objectives of the region.

Endogenous approaches to development that prioritise local conventions, are more locally embedded and favour local institutions are deemed as particularly appropriate to supporting food production that has a strong local or regional provenance (Amin 1994; Bowler 1999). Such 'bottom-up' activity is prone to embrace partnership and devolved responsibility. Within West Wales, endogenous approaches to support the sector are increasingly associated with forms of leadership that prioritise the role of local actors. Typically this involves the employment of community-based facilitators (or 'community animateurs') to provide internal, ground-level support in order to mobilise activity. Farming Connect, the predominant farm diversification programme from the Welsh Assembly Government, has thus appointed a set of facilitators who are referred to as the 'farmer's friend' – independent advisors who will provide local access to services. Menter a Busnes, who deliver the Cwysi initiative,[1] aimed at supporting farming families through change and restructuring, also recruited co-ordinators who were well known within the community. The terminology used in these programmes disassociates them from formal government support, whilst it was anticipated that employing people who were already aware of local conventions would reduce scepticism, as a project manager for the initiative noted: 'It diminished the idea that you had someone from outside coming in and giving advice, you had someone who was within the farming community'.

Identifying such initiatives as 'bottom-up' is, however, rhetorical rather than real as a result of their overwhelming dependence on higher level actors for funding, organisational direction, support and so on (see also Day 1998; Little 2001). Nevertheless, initiatives such as Farming Connect do reflect efforts to advance associations between businesses and institutions at a local level. Furthermore, they create mechanisms for a more integrated relationship between policy and delivery functions, as called for by Lord Haskins (2003) in his report on the delivery of government policies in rural England. Facilitative forms of leadership, as discussed above, thus provide a means for organisations to deliver practical support to the speciality food sector and also create a channel for supply chain actors to become more involved in processes of programme and policy delivery.

Overall, there is a large volume of organisations which provide support to speciality food businesses in West Wales and a considerable degree of interaction between them. However, this 'thickness' can be prohibitive to regional and sectoral development. These

---

[1] In more detail, the aim of this project was to support Welsh farming families. It was managed by the business development group Menter A Busnes, who employed a network of local co-ordinators throughout Wales. Activities included promoting on- and off-farm diversification, encouraging farmer co-operation and advising farming families of available funding/support systems.

findings reflect the limitation of assuming that dense institutional structures are necessarily functional, as has been previously identified in particular regions elsewhere in the UK (see, e.g., MacLeod 1997; Hudson, 1994). On this point, Storper (1997: 269) argues that formal organisations are only successful if: 'the rules, procedures, incentives and sanctions they establish are integrated in the conventions that guide people's behaviour'. A widespread criticism from entrepreneurs in West Wales is that they are unable or unwilling to navigate the complex institutional network which currently exists for the speciality food sector. A significant minority of businesses consider that institutions impede them from maximising their potential. Specifically, respondents criticise institutions for their excessive bureaucracy, the inappropriateness of support to small businesses, a lack of co-ordination between institutions, difficulty accessing funding and the prohibitive nature of legislation. Just over a third of businesses complain that they are unaware of available support, uncertain of how to attain funding and unsure how to determine which initiatives are most applicable to their needs. Some organisations have sought to overcome such problems by using so-called 'signposters' to support entrepreneurs and guide them through, what has become, a complex system of governance. Although signposters may resolve some of the above problems, the presence of such 'gatekeepers' only treats the effect of a confusing support network rather than addressing the cause which, on this evidence, is inadequate co-ordination of policies and funding streams. These arrangements also have important implications for the local/specialist food sector more generally, as discussed below.

Conclusion

Food businesses involved with speciality and spatially branded products in West Wales are supported by a diverse organisational framework that is dedicated to the development and regulation of the sector. The density, breadth and effectiveness of support are considered here through the concept of institutional thickness. In assessing the thickness of the speciality food sector in West Wales, it is useful to return to the four components outlined earlier in the chapter. These were a strong institutional presence; interaction between organisations; structures of domination and patterns of coalition; and mutual awareness of common enterprise.

There is evidence of considerable institutional presence for speciality food businesses in West Wales. Of this, only minimal support is dedicated to the speciality food sector or is exclusive to the area of West Wales, which complicates any assessment of its thickness. Forms of governance either tend to be narrower in focus (based around particular products, chain functions or specific territories) or address more extensive remits. Upon disaggregating further the form and motives of organisations, it becomes apparent that there are significant discrepancies in the extent of support, according to product type and, most notably, function within the food chain. There is a strong bias from institutions towards the head of the supply chain. This has resulted in a vacuum of support for intermediaries who are crucial in facilitating access to retail markets. Explanation and justification for this is attributed to two key things. First, the dominance of agricultural interests within rural policy; the political agenda positions family farming as fundamental to conserving social, cultural and environmental aspects of the Welsh countryside. Second, spheres of production are targeted on the basis that this represents the foundations of the food chain from where benefits will be imparted downstream. There is criticism amongst a minority of local and speciality food producers that the institutional network has become too dense, inefficient and complex to navigate. The inference here is that a large number of organisations do not necessarily indicate a sign of successful institutional thickness within a region. Furthermore, the ways in which institutions function and interact are crucial in examining the governance of this sector; these functions are subsequently considered through the three other characteristics of institutional thickness.

Co-ordination between organisations is crucial to developing strong and stable governance for the specialist sector. The findings here reveal ample evidence of interaction between institutions. Co-ordination that transects scales is prevalent, particularly through programmes that are managed by the Welsh Assembly and its sponsored bodies, yet delivered by locally based organisations that are more sensitive to local environments. Interaction between organisations operating at comparable levels is more sporadic. Although the majority of institutions seek to work in partnership with those operating at parallel scales, in order to benefit from enhanced knowledge and resources, for some such relationships are more competitive than collaborative, which is reflected in rivalry between organisations operating at neighbouring scales. In addition to co-ordination across spaces, the integration of the diverse range of sectoral interests that have a stake in the success of speciality food networks is crucial to the thickness of the sector. Currently, there appears to be inadequate complementarity between different interest groups, which prohibits the efficiency of institutional support and undermines efforts to achieve more integrated regional development.

Structures of domination that impact on the local/speciality food sector in West Wales are more evident within the food chain itself, rather than through forms of formal governance. Such structures, referred to here as 'internal regulation', are particularly prominent within food chains that are involved with mainstream supply channels, especially those that deal with multiple retailers. Potato farmers, even when dealing with locally branded crops, are heavily subjected to control from the supermarkets who dictate crop specifications, production methods, prices and so on. In contrast, the consensus amongst quality dairy producers is to actively avoid being subject to these forms of domination, which has influenced producers' decisions to favour supply chains that prioritise direct, less formalised mechanisms for product marketing. This finding, in turn, raises important issues about the diverse nature of the specialist food sector, especially in terms of identifying contrasting supply chain relations and its wider relationship with debates about 'alternativeness'. Clearly such conflations must be treated with caution.

Organisations are unified in recognising that the local/specialist food sector has the potential to benefit the wider rural economy and society. Although there is a mutual awareness of common enterprise, the form and objectives of institutions and the initiatives with which they are involved with vary markedly. To a large extent, this depends upon their motives for supporting the sector, which may relate to economic development concerns (often associated with business development, agricultural diversification and tourism), social concerns (especially improving health and diet) and/or potential environmental gains associated with more localised systems of production. Further, the process and approaches which organisations adopt and advocate differ and, in some cases, are contradictory.

A diverse range of institutional forms are considered a positive element of institutional 'thickness', and this chapter reveals that institutions have an important role to play in stimulating development in the speciality food sector in West Wales; this includes responding to and co-ordinating existing activities across the food supply chain. However, most support tends to favour spaces, actors and sectors that are already accustomed to a strong regional presence. This empirical case study reveals in particular a lack of support to underpresented groups, especially wholesalers and distributors. This scenario seems to replicate findings recently reported in other UK regions (see, e.g., the chapter by Maye and Ilbery in this book). This lack of support to downstream members and limited co-ordination between organisations challenges the effectiveness of institutions charged with promoting speciality food production and marketing as part of a broader goal of integrated rural development.

Acknowledgements

This chapter derives from the EU-funded project: 'Supply chains linking food SMEs in Europe's lagging rural regions' (SUPPLIERS, QLK5-CT-2000-00841). Collaborating laboratories are SAC, Aberdeen, UK (Co-ordinator); Coventry University, UK; University of Wales, Aberystwyth, UK; Teagasc, Dublin, Ireland; ENITA Clermont-Ferrand, France; University of Patras, Greece; SIRRT, University of Helsinki, Finland; and the Agricultural University of Krakow, Poland. The constructive comments made by an anonymous reviewer are also acknowledged with thanks.

References

Agri-food Partnership (2005). *Agri-food Partnership Strategy in Action: Towards 2007*. Cardiff: Agri-food Partnership.
Amin, A. (2001). Moving on: institutionalism in economic geography. *Environment and Planning A* 33, 1237–1241.
Amin, A. (2004). Regions unbound: towards a new politics of place. *Geografiska Annaler B* 86, 33–44.
Amin, A. and Thrift, N. (1994). Living in the global. In A. Amin and N. Thrift (Eds), *Globalization, Institutions, and Regional Development in Europe* (pp. 1–19). Oxford: Oxford University Press.
Barling, D. and Lang, T. (2003). A reluctant food policy? The first five years of food policy under Labour. *The Political Quarterly* 74 (1), 8–18.
Bowler, I. (1999). Endogenous agricultural development in Western Europe. *Tijdschrift voor Economische en Sociale Geografie* 99 (3), 260–271.
Bristow, G. (2000). Structure, strategy and space: issues of progressing integrated rural development in Wales. *European Urban and Regional Studies* 7, 19–33.
Copus, A., Kahlia, P., Jansson, B. and Mariussen, A. (2001). *The Role of Regional Milieux in Rural Economic Development*. Aberdeen: Northern Periphery Programme.
Cote, S. and Healy, T. (2001). *The Well-being of Nations. The Role of Human and Social Capital*. Paris: Organisation for Economic Co-operation and Development.
Danson, M. and Whittam, G. (1999). Regional governance, institutions and development. In S. Loveridge (Ed.), *The Web Book of Regional Science*. Regional Research Institute, West Virginia University. Available at: http://www.rri.wvu.edu/regscweb.htm (accessed: 27/9/06)
Day, G. (1998). Working with the grain? Towards sustainable rural and community development. *Journal of Rural Studies* 14 (1), 89–105.

Defra (2003). *Local Food – A Snapshot of the Sector. Report of the Working Group on Local Food*. London: Defra.
Dolan, C. and Humphrey, J. (2000). Governance and trade in fresh vegetables: the impact of UK supermarkets on the African horticulture industry. *Journal of Development Studies* 37 (2), 147–176.
Edwards, B., Goodwin, M., Pemberton, S. and Woods, M. (2001). Partnerships, power, and scale in rural governance. *Environment and Planning C: Government and Policy* 19, 289–310.
FLAIR (2003). *FLAIR Report 2003 – The Development of the Local Food Sector 2000–2003 and its contribution to Sustainable Development*. Bristol: Foundation for Local Food Initiatives.
Flynn, A., Marsden, T. and Smith, E. (2003). Food regulation and retailing in a new institutional context. *The Political Quarterly* 74 (1), 38–46.
Giddens, A. (1984). The Constitution of Society: Outline of a theory of Structuration. Cambridge: Polity Press.
Goodman, D. (2002). Rethinking food production–consumption: Integrative perspectives. *Sociologia Ruralis* 42 (4), 1–7.
Goodman, D. (2003). The quality 'turn' and alternative food practices: reflections and agenda. *Journal of Rural Studies* 19 (1), 1–7.
Goodwin, M. (1998). The governance of rural areas: some emerging research issues and agendas. *Journal of Rural Studies* 14 (1), 5–12.
Granovetter, M. (1985). Economic action and social structure: the problem of embeddedness. *American Journal of Sociology* 91, 481–510.
Haskins, C. (2003). *Rural Delivery Review – a Report on the Delivery of Government Policies in Rural England*. London: Defra.
Henry, N. and Pinch, S. (2001). Neo-Marshallian nodes, institutional thickness, and Britain's 'Motor Sport Valley': thick or thin? *Environment and Planning A* 33, 1160–1183.
Hinrichs, C. (2003). The practice and politics of food system localization. *Journal of Rural Studies* 19 (1), 33–45.
Holloway, L. and Kneafsey, M. (2004). Producing-consuming food: closeness, connectedness and rurality in four 'alternative' food networks. In Holloway, L. and Kneafsey, M. (Eds), *Geographies of Rural Cultures and Societies* (pp. 262–282). Aldershot: Ashgate.
Hudson, R. (1994). Institutional change, cultural transformation, and economic regeneration: myths and realities from Europe's old industrial areas. In A. Amin and N. Thrift (Eds), *Globalization, Institutions, and Regional Development in Europe* (pp. 196–216). Oxford: Oxford University Press.
Jessop, B. (2001). Institutional re(turns) and the strategic-relational approach. *Environment and Planning A* 33, 1213–1235.
Kneafsey, M., Ilbery, B. and Jenkins, T. (2001). Exploring the dimensions of culture economies in rural West Wales. *Sociologia Ruralis* 41, 296–310.
Le Heron, R. and Roche, M. (1999). Rapid re-regulation, agricultural restructuring, and the reimaging of agriculture in New Zealand. *Rural Sociology* 64 (2), 203–218.
Little, J. (2001). New rural governance? *Progress in Human Geography* 25 (1), 97–102.
MacLeod, G. (1997). 'Institutional thickness' and industrial governance in Lowland Scotland. *Area* 29, 299–311.
MacLeod, G. (2001). Beyond soft institutionalism: accumulation, regulation, and their geographical fixes. *Environment and Planning A* 33, 1145–1167.
Marsden, T. (1998). New rural territories: regulating the differentiated rural space. *Journal of Rural Studies* 14, 107–117.
Marsden, T., Banks, J. and Bristow, G. (2000). Food supply chain approaches: exploring their role in rural development. *Sociologia Ruralis* 40 (4), 424–438.
Marsden, T., Banks, J. and Bristow, G. (2002). The social management of rural nature: understanding agrarian-based rural development. *Environment and Planning A* 34, 809–825.

Putnam, R. (2000). *Bowling Alone: The Collapse and Revival of American Community*. New York: Simon and Schuster.

Renting, H., Marsden, T. and Banks, J. (2003). Understanding alternative food networks: exploring the role of short food supply chains in rural development. *Environment and Planning A* 35, 393–411.

Sage, C. (2003). Social embeddedness and relations of regard: alternative 'good food' networks in south-west Ireland. *Journal of Rural Studies* 19 (1), 47–60.

Storper, M. (1997). *The Regional World*. London: Guildford Press.

Thompson, N. (2005). Inter-institutional relations in the governance of England's national parks: a governmentality perspective. *Journal of Rural Studies* 21, 323–334.

Whatmore, S., Stassart, P. and Renting, H. (2003). What's alternative about alternative food networks? *Environment and Planning A* 35, 389–391.

Wood, A. and Valler, D. (2001). Turn again? Rethinking institutions and the governance of local and regional economies. *Environment and Planning A* 33, 1139–1144.

Chapter 11

Public Sector Food Procurement in the United Kingdom: Examining the Creation of an 'Alternative' and Localised Network in Cornwall

James Kirwan and Carolyn Foster
Countryside and Community Research Unit, University of Gloucestershire, Cheltenham, UK

Introduction

The emergence and evolution of various forms of 'alternative' food networks (AFN) over recent years is suggested as being a critical element of new rural development patterns in Europe (Marsden et al. 2000; Renting et al. 2003). Underlying these AFN is an attempt to re-embed the production and consumption of food within specific places and relationships, in order to add value and "create positive 'defences' for rural regions against the prevailing trends of globalisation and further industrialisation of markets" (Marsden et al. 1999: 295). The literature on alternatives to the conventional (or mainstream) food supply chain (FSC) has tended to focus on short food supply chains, and, in particular, on those involving organic farming, quality production and direct selling initiatives (Renting et al. 2003). However, more recently there has been a growing interest in how (often small-scale) initiatives can scale-up and become more significant within the FSC, as well as how larger-scale initiatives can be re-oriented to become more sustainable in rural developmental terms. This has been the central theme of the European Commission funded project SUS-CHAIN,[1] from which the empirical data within this chapter are taken.

Within the UK, the procurement of food by public sector organisations has emerged as an intriguing example of how attempts are being made to refashion an existing FSC through incorporating the notion of sustainability.[2] The public sector in the UK spends approximately £1.8 billion (ca. €2.65 billion) on food and catering services every year (Sustain 2002), and provision of these services has been dominated by large, highly concentrated multinational companies who supply pre-prepared and standardised food on the basis of 'best value', which does not incorporate notions of sustainability. Indeed, Morgan and Morley (2004: 43) argue that best value, and its predecessor Compulsory Competitive Tendering, have "stripped sustainability" out of public sector catering. Cost has become the predominant criterion when awarding contracts, as public procurers:

"are forced to operate as if they were private businesses, even though their wider objectives – to meet basic standards of nutrition and to promote healthy eating for example – sit uncomfortably with these narrow commercial pressures" (Morgan and Morley 2004: 45).

As labour costs have risen in relation to the cost of food, there has been a rapid de-skilling of catering staff who are able to prepare fresh food, and a growth in the supply of processed food which requires minimal preparation. The latter provides considerable scope for economies of scale, giving large-scale firms a cost advantage even though they may have further to transport the food. The net result has been that smaller, local producers and suppliers find it hard to win public sector contracts and food provenance has all but disappeared as an indicator of quality service provision and ultimately sustainability (Morgan and Morley 2004).

---

[1] SUS-CHAIN (QLRT-2001-01349) 'Marketing sustainable agriculture: an analysis of the potential role of new food supply chains in sustainable rural development'. A major component of this research project has been a diverse range of case studies across the partner countries concerned (The Netherlands, Belgium, Germany, Switzerland, Italy, Latvia and the UK), representing alternative approaches that are innovative and dynamic in terms of enhancing the sustainability of European food supply chains. Following a 'state of the art' investigation of sustainability in UK food supply chains, public sector food procurement was selected as one of the two case studies to be investigated by the UK research team.

[2] The Government's 2005 sustainable development strategy, *Securing the Future*, sets out that sustainable development should be pursued through an integrated, innovative and productive economy that delivers high levels of employment and a just society that promotes social inclusion, sustainable communities and personal well-being. Furthermore, that this will be done in ways that protect and enhance the physical and natural environment, and use resources and energy as efficiently as possible (HM Government 2005). More specifically, DEFRA's Food Industry Sustainability Strategy sets out that the food industry beyond the farm gate (including food service providers) should adopt best practice that ensures economic performance does not compromise sustainable development through exploiting people or the environment (DEFRA 2005).

The report of the Policy Commission on the Future of Farming and Food (the Curry Report) is also concerned that "the current interpretation of best value may be too narrow to allow public bodies to take into account wider sustainable development issues when setting supplier requirements" (DEFRA 2002a: 104).[3] However, the sheer scale of public procurement within the food sector has attracted the attention of those intent on improving the overall sustainability of the FSC, particularly in relation to rural development. The Curry Report, for example, notes that public bodies can help to create a 'critical mass' of purchasing within an area. Moreover, this can have the effect of kick-starting a more localised food economy, which in turn may lead to a reduction in the number of kilometres food travels and an improvement in the local economic multiplier (see also Morgan and Morley 2002). Nevertheless, this requires that public procurers are more flexible in their interpretation of best value, incorporating more than simply cost indicators.

Public sector food procurement (PSFP) in the UK covers a range of institutions, including hospitals and healthcare facilities, schools, prisons, the armed forces and various tiers of government. Research to date on utilising PSFP to positively impact the sustainability of FSCs has tended to focus on the practicalities involved (e.g. Local Food Works 2002; Morgan and Morley 2002; Schiopu 2005; WPI 2005). This has included highlighting the benefits of using more locally supplied food but also looking at the ways in which the supply of local food can be encouraged. Crucially, it has also considered how public sector procurement legislation can be interpreted to facilitate the development of an alternative system of public food supply that is more localised and sustainable. Having said that, a growing number of authors have criticised the simplistic equation that 'localness' equals 'alternative' equals 'sustainable'.

Local is often proposed as a counterpoint to global; a geographical re-orientation that can help facilitate the endogenous development of those areas disadvantaged by processes of globalisation (Ray 1998). However, Winter (2003) found that localism may in fact be more about protection than development; something he describes as 'defensive localism'. Likewise, Hinrichs (2003) distinguishes between 'defensive localisation' and 'diversity-receptive localisation', in arguing that the latter incorporates cultural, social and environmental meanings in addition to the spatial. In so doing, local, and the potential benefits of localisation, can be understood within the wider world in which the local exists. Critically, if the benefits of localisation are to contribute to the sustainable development of a particular area, the wealth created must be retained at a local level by "investing in the local environment, creating/strengthening local institutions, and employing people and their resources" (Marsden and Smith 2005: 441). This requires the development of alternative structures (or networks) that can help inspire a "sense of shared ownership" and the facilitation of synergy within the area concerned (Marsden and Smith 2005: 442). Watts et al. (2005: 34) suggest that there are both

---

[3] The idea of 'interpretation' is significant, in that the 1999 Local Government Act, which replaced Compulsory Competitive Tendering with 'Best Value', specifically states that "best value is not just about economy and efficiency, but also about effectiveness and the quality of local services" (RCHT 2002: 35).

'weaker' and 'stronger' alternatives within the FSC, arguing that they "can be classified as weaker or stronger on the basis of their engagement with, and potential for subordination by, conventional FSCs operating in a globalising, neoliberal polity". Specifically in relation to the public procurement of local food, they suggest that it is necessary to understand the functioning of the networks that underpin this process, in order to understand its potential to contribute to the sustainable development of the area(s) involved.

It is not the purpose of this chapter is to re-examine these issues *per se*, but rather to analyse from within the way in which local-level actors build alternative networks of public sector food provision that actively incorporate notions of sustainability and are not hidebound by narrow cost definitions of best value. In particular, it concentrates on hospital food procurement, using the development of the Cornwall Food Programme (CFP) as its case study. The CFP is an initiative to develop local and sustainable food sourcing across the National Health Service (NHS) in the county of Cornwall. This necessitates a conceptual framework that can encompass the relations between a range of individuals, agencies, technical and biophysical factors, any one of which can affect the stability of the network concerned (either positively or negatively).

Political economy approaches allow for an explanation of the processes involved in the globalisation of the agro-food system, but are arguably less well-suited to an exploration of the context-specific development of 'alternative' systems of food provision. As such, a number of authors have advocated a network approach as a means of incorporating the diversity and complexity of actor interaction within the current agro-food system (e.g. Lowe *et al.* 1995; Murdoch 1995, 2000; Busch and Juska 1997; Goodman 1999, 2001; Lockie and Kitto 2000; Murdoch *et al.* 2000; Kneafsey *et al.* 2001). Network analysis may not be novel within the social sciences, having been extensively utilised to understand relations between social actors (social networks) as well as the take up of new technologies (technological networks) (Murdoch 1995). However, Actor Network Theory (ANT), in particular, has emerged since the 1980s as "a hybrid of these two more traditional forms" (Murdoch 1994: 3) which allows network construction to be viewed in action (Law 1992). ANT has been variously adopted within rural studies. For example, Herbert-Cheshire (2003: 455) has used it to examine the "capacity of local people to negotiate, challenge and ultimately transform rural policy" in relation to the power of the state; Comber *et al.* (2003) to investigate how different features are deemed worthy of inclusion within particular environmental mapping programmes; Burgess *et al.* (2000) to explore different understandings of nature between conservationists and farmers; and Morris (2004) to conceptualise the implementation of the Countryside Stewardship Scheme within the wider social and cultural framework in which it developed.

The next section of the chapter outlines the key elements of ANT, including its conceptual relevance to the analysis of an emerging agro-food network (the CFP) that encompasses a wide range of potential actors. The following section of the chapter draws on empirical data gathered during the SUS-CHAIN project in applying ANT to the development of the CFP as the active construction of an actor network. In the concluding section of the chapter, the complex and contingent nature of this 'alternative' and localised food network is discussed, as well as the appropriateness of using an ANT approach to understand its emergence and development.

The Application of Actor Network Theory (ANT)

ANT, or 'the sociology of translation' (Callon 1986), was conceived by its originators (most notably Michel Callon, Bruno Latour and John Law) as a means of understanding the way in which "particular technological and scientific models gain acceptance as 'normal' custom and practice" (Selman and Wragg 1999: 329). However, ANT's adapted pertinence within rural studies is based to a large extent on its ability to understand how particular ideas are "conceived and then developed by their conceiver, with or without resistance" (Woods 1997: 323) into more or less stable networks. This exercise is known within ANT as 'translation', which refers to processes of negotiation and alignment by which actors enrol heterogeneous others within their network. Networks are likely to contain 'macro-actors' (Selman and Wragg 1999), who may have had the original idea for the development of the network concerned. Woods (1997) refers to this type of actor as the 'originating entity', who, in order to successfully achieve their vision, must convince others to enrol in their network.

Networks do not exist or form in isolation; they are the result of actors deliberately bringing together a range of diverse interests and materials to create a stable network (Latour 1986). The process of translation by which this happens can be understood as having four main 'moments', although in reality not every translation will involve all four stages and on occasions 'moments' may occur simultaneously (Woods 1997). First, the *problematisation* of an issue, where the translator of the network identifies a problem and suggests a solution; second, *interessement*, where the translator attempts to convince other actors that it is in their interests to agree with the problematisation of the issue; third, *enrolment*, where those actors who have become interested in the project are convinced by the values of the proposed network and agree to adhere to them; and fourth, *mobilisation*, at which point an effective network that addresses the identified problem becomes established (Callon *et al.* 1986; Woods 1997). Thus, a network is formed by the originating entity enrolling other actors into their actor world, translating the interests of these actors into their own. However, Callon (1986: 199) cautions that networks need to be understood as "uncertain, ambiguous, and disputable", variably open to negotiation and potential failure should "sufficient actors become disillusioned with its performance and commit 'betrayal' " (Selman 2000: 111); as with the St Brieuc Bay scallops of Callon (1986).[4] As a consequence, Callon identifies a possible fifth moment of 'dissidence', at which point the network may disintegrate or be appropriated by other competing networks.

At the core of the ANT approach is a concern to understand how stable networks are constructed and strengthened over time (Whatmore and Thorne 1997). There is no pre-conceived frame of reference, simply an exploration of network formation that is recognised as negotiated and contingent, whereby "if the proponents of a new theory fail to gather a large enough network of allies then, in the long run, it will be unsuccessful" (Comber *et al.* 2003: 303). The principle of agnosticism within ANT requires

---

[4]In this case, scientists attempted to persuade fishermen to alter their harvesting practices and to cultivate scallops on artificial offshore collectors. The case illustrates the construction of an actor-network but also its collapse once the scallops failed to breed, and the subsequent withdrawal of the fishermen's support.

that no *a priori* assumptions are made between humans, non-humans, institutions or individuals. Indeed, Comber *et al.* (2003: 303) argue that "ANT's theoretical richness derives from its refusal to reduce explanations to just natural, social, or discursive categories while recognising the significance of each". Actor networks may also include intermediaries, which can be understood as "anything passing between actors which defines the relationship between them" (Callon 1991: 134). Intermediaries can help induce network stability and facilitate conciliation between actors, and Burgess *et al.* (2000: 123–124) suggest that "building networks depends on actors' capacities to direct the movement of intermediaries such as texts, technologies, materials and money". The difference between an intermediary and an actor is that "an actor is an intermediary that puts other intermediaries into circulation - an actor is an author" (Callon 1991: 141).

However, there are concerns that while ANT may liberate the way in which networks are analysed, it fails to give any theoretical guidance about how to interpret the resulting data, thereby arguably being "methodologically strong but substantively weak" (Marsden 2000: 24). As such, it does not facilitate a critical appraisal of the ethics and philosophy underlying the networks concerned. Indeed, even Bruno Latour argues that it is more of a method than a theory, "a way for social scientists to access sites ... not an interpretation of what actors do" (Latour 1999: 20). There are also concerns that the agnosticism of ANT, and its recognition that power resides in the relations within networks, rather than in the actors or actants (inanimate entities) themselves (Latour 1991), fails to acknowledge that not all actors and actants have the same degree of agency (or power to act). Woods (1997: 335), for example, in his research on hunting conflicts, argues that although the deer are given a more prominent role within an ANT account than more traditional narratives, they "could neither have instigated or problematised the political network, nor demonstrated consent at the identity ascribed to them". The result being that although actants may be crucial to the overall development of a network their actions should not be seen as conscious nor should they be understood as having agency. Nevertheless, proponents of ANT, while recognising its limitations as a theory, value its "conceptual and metaphorical tools that expose and address the erasures [of] ... the modernist ontology" (Goodman 1999: 34), facilitating an examination of the hybrid and contested nature of network construction. "Strictly speaking, ANT does not qualify as a 'critical theory' but we should recognise that it undoubtedly performs a critical task" (Murdoch 1997: 753).

The purpose of this chapter is to examine the construction and evolution of an actor network that aims to improve the sustainability of food procurement within the NHS in Cornwall. Specifically, it looks at the CFP whose originating entity was then the catering manager for the largest hospital trust in Cornwall. Over time, he has sought to construct a network that can fulfil his ambitions by enrolling a wide range of entities. These include both human and non-human actors and intermediaries that comprise institutions, individuals, money, reports, policies, buildings and geography. The ability of ANT to encompass this diverse range of constituents is critical to an examination of how the CFP functions as a network. Underpinning ANT is the notion of translation. However, as indicated above, the successful translation and stabilisation of a network is not certain, dependent on whether or not actors "conform, and continue to conform, to their allotted roles" within the network (Burgess *et al.* 2000: 124): likewise, the extent to which actors are able to direct intermediaries to their best advantage.

The Cornwall Food Programme: Building the Network

The Cornwall Food Programme (CFP) is an initiative aiming to improve the sustainability of food procurement within the NHS in Cornwall, through the re-orientation of its food purchasing policy. The ultimate goal is to increase the amount of locally and organically produced food procured by Cornwall NHS for patient, staff and visitor meals. This section, based on case study research that included 26 semi-structured interviews with a range of stakeholders, charts the development of the CFP from its early beginnings in 1999, up to the latest data collection in 2006.

Within Cornwall, five NHS Trusts [Royal Cornwall Hospitals Trust (RCHT), Cornwall Partnership Trust (CPT) and three Primary Care Trusts] provide health and care services through 23 hospitals. Catering services across the five trusts are coordinated by Cornwall Healthcare Estates and Support Services (CHESS), involving a combination of conventional cook-serve systems, whereby the food is prepared and cooked in on-site kitchens and cook-freeze systems whereby, pre-prepared, pre-cooked, frozen meals are bought in by hospitals for subsequent on-site regeneration. The cook-freeze meals are provided predominantly through a contract with an international, out-of-county food service company that accounts for about 33% of the county's catering provision. However, within the RCHT, the Royal Cornwall Hospital (RCH) operates a cook-serve system from their on-site kitchens, wherein the food is served directly to patients on the wards, or in the staff and visitor canteens. This accounts for a further 61% of hospital meal provision within the county. The remaining 6% is served as cook-freeze to St Michael's and the RCH Maternity Unit, having been cooked in the RCH's on-site kitchen before being frozen in the RCH's own cook-freeze unit.

Although it was not then known as the CFP, the starting point of the network that was eventually to develop into the CFP can be traced back to 1999, at which time the catering manager (CM) of the RCH began to problematise the issue of hospital food provision within Cornwall, based on the mantra of "local food for local people". Inherent within his vision was a belief in the high quality of Cornish food and a concern to retain the potential benefits associated with 'localising' hospital food procurement within the county of Cornwall, in order to help counter its geographic peripherality and economic marginalisation:[5] "living in a rural community, stuck out on a peninsular has made a difference to my thinking" (Pearson 2004). Despite Cornwall's reputation for quality food production, established NHS procurement practices, whereby food is supplied through national contracts (originating from producers and suppliers outside Cornwall), meant that local food was often inaccessible to Cornish hospitals. The CM became increasingly frustrated that much of the food supplied to the NHS was in fact available "on the doorstep" and was travelling "unnecessarily long distances". As CM for the largest hospital trust in Cornwall, he saw an opportunity to increase the amount of locally supplied and produced food to benefit the hospital, local producers, the environment and the wider community alike. He felt that health services should be

---

[5]Objective One is the highest priority designation for European aid and is targeted at areas where prosperity, measured in Gross Domestic Product (GDP) per head of population, is 75% or less of the EU average. The region of Cornwall and the Isles of Scilly was designated an Objective One region in March 1999 because its average GDP was around 67% of the EU average at that time.

an integral part of the community, able to provide support beyond its obvious healthcare role: to be "part of the solution, not part of the problem" (Pearson 2004).

Initially, the CM did not act upon these concerns, until a series of issues forced him into what he described as a "mind change" that prompted him to actively address them and propose a solution. These issues included the relative geographic isolation of Cornwall, wherein national contracts had on occasions failed to adequately supply some of the community hospitals due to the small quantities involved. A second issue concerned the catering capacity at the RCH, where, in the late 1990s, it became clear that there was a need to 'de-stress' the site: partly due to the space taken up by its new cook-freeze unit, and partly due to wider pressures to physically accommodate further hospital growth. The final catalyst occurred in 1999, when in-house sandwich production at the RCH was curtailed in order to accommodate the new cook-freeze facility to serve St Michael's hospital. At that time, the sandwich contract was outsourced to a national food and catering service company which bought its sandwiches from a company in the north of England. A patient questioned the logic of this, which had the effect of bringing the issue of local sourcing to the attention of the hospital managers, as well as the local media. In other words, as Woods (1997) suggests can happen, actors were starting to become interested in the issue, although not as yet in response to an actor-articulated problematisation. Coincidentally, a local bakery supplier had approached the hospital as a possible outlet and started to develop a relationship with the CM. This was to result in the first directly negotiated contract with a local supplier (in 1999), when an agreement was set up with the bakery to supply the RCHT with sandwiches.

The CM's 'mind change' can be understood as the point at which he, as the 'originating entity', actively problematises the issue and starts to create what will become the network of the CFP. The successful negotiation of the local sandwich contract served as a model on which the CM started to build his ideas for realising a more localised pattern of food procurement for the RCHT. However, as a way of taking these ideas forward and assembling support for his original concerns (i.e. strengthening his problematisation), the CM decided to commission a feasibility study (FS) into the future options for catering within the RCHT, with the possibility of extending the resultant options to all five Cornish NHS Trusts.

The FS can be understood as an important intermediary in the overall translation process of the CFP network, in that the CM was clearly intent on drawing other actors into his nascent network through his identification of the need for a FS. These actors included: the Chief Executive of the RCHT, whose approval was needed in order to apply for grant funding; the CPT's Grants Officer, who worked with the CM to find a source of funding once the Chief Executive had given his consent; and the Cornwall Agricultural Council Development Team (CACDT)[6], whose Project Manager, having been approached by the CPT's Grants Officer, facilitated the link with the Government

---

[6]The CACDT is an EU Objective One funded project set up in November 2000, sponsored by Cornwall Enterprise. It is the organisation responsible for the economic development of Cornwall, and its main function is to create a team of dedicated personnel tasked with assisting the agricultural, horticultural, food and land-based industries to access Objective One funds. Operating under the steer of the Agricultural Task Force, the team ensures that the various industries' strategic objectives are met and that the allocation of funds is fully committed.

Office South West (GOSW) who administer the Objective One programme for Cornwall. Following an appraisal by GOSW, a grant of £17 700 (ca. €26 000) was awarded to carry out the FS. Another important actor in this process was CHESS, who represent the interests of Cornwall NHS. It was they, along with the Department for Environment, Food and Rural Affairs, who provided the match funding for the grant and indeed formally commissioned the FS. The grant itself then becomes another important actor (a non-human one) within the evolution of the CFP, in that without it the FS could not have been undertaken: although clearly it did not have agency.

The FS was completed by the end of 2002, recommending that there was a need to develop a Central Food Production Unit (CFPU) on a dedicated site. The CFPU would provide for the food needs of all NHS facilities in Cornwall through a new cook-freeze facility, whereby the food would be delivered to, prepared and cooked in a central, on-site kitchen and distributed throughout the Cornwall NHS (Harrow 2002). In deciding on the CFPU, the study highlighted the potential positive impacts on the local community, environment and economy with this alternative approach, in that it provided the best opportunity to increase the procurement of locally produced and supplied food.[7] Through the intermediary of the FS, a wide range of stakeholders became interested in the CFP, in that it was perceived as having the potential to address a number of issues they faced. The result of this process was that a stable network began to emerge whose actors were fully engaged with the proposed solution. It is also possible at this stage to identify the objectives and themes around which the stakeholders were aligning (see Figure 11.1), and it is this convergence of interests which leads the actors to actively enrol in the CFP network. In other words, the actors involved had become interested in the project, were convinced by the values of the proposed network and had variously mobilised their resources within the network.

The stability of this emerging network is affirmed by the fact that the preferred outcome of the FS can be viewed as mutually, and collectively, beneficial to those actors involved. The project places 'Cornwall', and how to benefit Cornwall, at its centre. It proposes to meet, in part at least, the goals set by the Objective One programme, while at the same time enabling the CACDT to fulfil its role as facilitator. The concerns of hospital managers and the catering department are met by resolving the requirements for hospital growth, and there is the prospect of food being supplied that is of an improved nutritional quality. The suppliers also viewed it as a positive development, in that it increases the demand (both in terms of scale and regularity) for local supplies of food. With its focus on more sustainable patterns of food procurement, supply and distribution,

---

[7]The FS built on Mike Pearson's 'mind change', providing supporting evidence for his vision of a more sustainable procurement system within Cornwall's NHS. It recognised the importance of efficiency in procurement, and the need to get value for money, but at the same time argued that "the encouragement of social, economic and physical health will have a direct positive impact on the community and therefore have a reduction in patient care costs" (RCHT 2002: 2). It also argued that through increasing the proportion of locally purchased produce and goods it had the potential "to provide additional skilled jobs... reduce the amount of produce transported over long distances and... aid... in the creation of a sustainable economy for Cornwall" (RCHT 2002: 23), enabling the valorisation of Cornwall's endogenous assets.

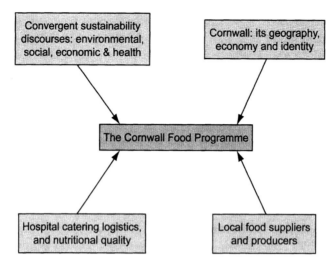

Figure 11.1: Support for the Cornwall Food Programme.

the CFP also addresses the needs of policy makers and environmentally oriented NGOs alike.[8]

Once the FS outcome had been approved, the attention of those involved in the CFP began to focus on how to make it happen. The recommended option to build a CFPU implied a massive and systemic transformation of activities. In order to implement the proposed solution, several critical elements had to be in place for the network to function effectively which involved interesting and enrolling further actors into the network. The immediate task was to secure funding to keep the CFP's Project Manager (PM) – author of the FS – in post. Interim Objective One funding was allocated for this purpose because the CACDT and GOSW felt that the CFP was a 'strategically good fit' with the goals of the Objective One programme, and that the expertise developed by the PM through conducting the FS should not be lost. However, in order to secure his post over the longer term, match funders were required, with the five NHS Trusts the most likely source. The FS enabled a persuasive case to be put to the management of the four remaining Trusts, as well as consolidating support from the RCHT's Chief Executive.

---

[8]The Government's response to the Curry Report (whose remit was to "advise the Government on how we can create a sustainable, competitive and diverse farming and food sector which contributes to a thriving and sustainable rural economy [and] advances environmental, economic, health and animal welfare goals" (DEFRA 2002a: 5), was DEFRA's Strategy for Sustainable Farming and Food (SSFF) (DEFRA 2002b). Government Offices (e.g. GOSW) and Regional Development Agencies (e.g. South West RDA) are responsible for implementing the national SSFF through their regional delivery plans. Public sector procurement features in the South West's plan, and largely as a result of the CFP the South West Public Sector Procurement Strategy Group (SWPSPSG) was formed, on the basis that its constituent members recognised the potential of public procurement to link in with their own agendas on local and regional food. The SWPSPSG includes: GOSW, SWRDA, the Countryside Agency, South West Food and Drink, SW Local Food Partnership, Public Health Observatory, Five-A-Day and National Farmers' Union and Sustain.

CHESS played a key role in facilitating links and negotiating support from the Trusts' management, and from March 2003, the PM's post was extended under Objective One with match funding provided by the Cornwall Health Care Community (through CHESS) and DEFRA. The PM's main remit was to draft an Outline Business Case for the CFPU (including a cost-benefit analysis and risk assessment), which would form the basis of a funding application to build the CFPU.

At this point, the CPT took on the role of CFP project host, and their enrolment into the project was formalised and further stabilised by the creation of a Project Board to oversee the progress of the CFPU. Composed of representatives from the executive levels of all five NHS Trusts, as well as the Chief Executive of PASA,[9] it is responsible for delivering results for the CFP via CHESS. The creation of the Board also signalled the enrolment of all five Trusts within the Cornwall Health Care Community into the CFP network, rather than solely the RCHT.

Critically, the demand implications for locally produced food, of deciding to build a CFPU, meant that there needed to be a corresponding development of the local supply chain. To this end, a Sustainable Food Development Manager (SFDM) was appointed in April 2004 to identify sources of local and organic produce, and to investigate ways of introducing it into patient meals and staff/visitor restaurants. His post was funded by Organic South West (OSW), a Soil Association managed project, joint funded by DEFRA and Objective One. The objective of OSW is to grow the organic market in Cornwall and the Isles of Scilly, and in particular the food service sector. The CFP was recognised by OSW as having the potential to contribute to this objective, and initially they committed 12 months of funding, although this has now been extended to 2008. Interestingly, however, following this extension, the SFDM's role changed whereby his time is now evenly split: two and a half days a week looking at how to introduce local and organic produce into the NHS (i.e. much the same as for the first 12 months); and two and a half days on getting organic produce into other sectors of the industry (outside the NHS), including local hotelier groups, restaurant groups and other public and private sector organisations. This change in emphasis is because the NHS is recognised as being limited in relation to the aims of OSW, primarily because they do not have the budget to buy more than a relatively limited quantity of organic produce (Heath 2005). This demonstrates how actors will only support a network if it continues to provide what they expect from it. This is not to suggest that OSW are exhibiting dissidence (in Callon's 1986 terms), necessarily, simply that the stability of networks is contingent upon the continued support of their constituent actors.

The final key element that arose from the FS was the need to secure capital funding to build the CFPU. This necessitated interesting and subsequently enrolling the South West Peninsular Strategic Health Authority (SWPSHA) into the CFP, in that they are the body responsible for creating a coherent and strategic NHS in the region and are accountable to the Secretary of State for Health for the performance of services delivered

---

[9]PASA (NHS Purchasing and Supply Agency) is an executive agency of the Department of Health. It was established in April 2000 to advise on policy and the strategic direction of NHS procurement. However, it is also intended to function as an active participant in the ongoing modernisation of purchasing and supply in the NHS (PASA 2005).

(SWPSHA 2005).[10] In this context, the Outline Business Case (OBC) can be envisaged as a key intermediary (utilised by the CFP team) in garnering the support of the SWPSHA Capital Investment Group (CIG), which is composed of the regional SHA plus the Chief Executives of all the NHS Trusts in Devon and Cornwall. The OBC needed to convince the SWPSHACIG that the vision of the CFPU was sound in economic terms and worthy of their support, which it succeeded in doing. At this point, it became necessary to find a suitable funding stream to finance the £3.65 (€5.35) million that the CFPU would cost to build.

Ordinarily, for a capital build of this nature, if agreed, Objective One would contribute 20%, match funded with 20% from DEFRA; with the remaining 60% from the NHS (as owner of the project). However, because the NHS is a public body, it cannot be match funded with further public money from DEFRA. So in this scenario, 20% would be EU money and 80% NHS. Therefore, in order to maximise the funding opportunity afforded by being an Objective One area, an alternative approach needed to be found. This entailed drawing the NHS Local Improvement Finance Trusts (LIFT) into the network, convincing them to enrol within the CFP network because it accords with their aims.[11] The local LIFT company in Cornwall is Community First Cornwall (CFC), whose management structure is 60% private and 40% PCTs (West Cornwall, Central Cornwall, North and East Cornwall). The ensuing solution was that 90% of the £3.65 million will be borrowed by CFC, with 10% being provided up front. About 60% of this 10% will be private money (i.e. 6% of the total cost), and 40% public money (i.e. 4% of the total cost), supplied by the Department of Health from strategic capital via the PCTs. In other words, 96% of the total build cost will effectively be provided through private funding. This means that the CFP is now able to maximise the potential of Objective One funding by obtaining 20% of the 96% from EU money, match funded by DEFRA [making a total of £1.40 million (€2.06 million)]; whereas, in the absence of this arrangement, only 20% of £3.65 million [£0.73 million (€1.07 million)] would have been available (Harrow 2005).

The CFP submitted its planning application to build the CFPU in December 2005. This involved convincing yet another actor that they should enrol within the CFP network, in this case the local planning authority (Kerrier District Council). The latter proved to be very supportive of the application, seeing it as something that accorded with their planning policies and granted planning permission in March 2006. Currently, the Objective One Business Case is being considered by a Strategic Health Authority Board,

---

[10]Capital builds up to £3 (€4.4) million can be authorised as an internal NHS Trust decision, but projects over £3 million require SHA approval. In this case, the initial projected cost of the CFPU was £4.5 (€6.6) million, although this was later revised downwards to £3.65 (€5.35) million; but still over the £3 million threshold (Harrow 2005).

[11]NHS LIFT aim to develop a new market for investment in primary care and community-based facilities and services in order to improve the quality of service delivered within the community they serve. The NHS LIFT approach involves developing a joint venture between local health bodies and a private sector partner, which at a local level will involve a local LIFT company. A local LIFT will be a public–private partnership, set up as a limited company which will own and lease premises (DoH 2005) – in the case of the CFPU, over a 25-year timeframe. The lessee will be the RCHT and, at the end of the 25 years, they will have the option of either buying the building or continuing the lease (Harrow 2005).

which is due to make a decision at the end of June 2006. Subject to a successful outcome from the SHA Board, the building phase is projected to start in the second week of July 2006, with the CFPU operational in April/May 2007 (Harrow 2006).

Discussion and Some Concluding Remarks

The purpose of examining the creation of the CFP in terms of the translation of a network has been to elucidate how this is complex and highly contingent, dependent upon an ongoing spectrum of negotiation and alignment processes. The narrative above is intended to highlight the key elements that have shaped the CFP from the time of its conception, up until the time of writing this chapter.[12] It has demonstrated that within the construction of the overarching network of the CFP, there are a number of concurrent and linked networks, each of which is formed through a similar process of translation. For example, the FS grant can be understood as an actor (non-human) within the CFP network, which resulted from garnering support from other actors. In other words, the grant is in itself a network, even though as a successful grant application the complexity of its formation is hidden.[13] Within ANT this simplification is known as 'punctualisation', and in this case the grant can be envisaged as a 'punctualised node' within the CFP network (Callon 1991; Law 1992). Critically, as with any network, these 'punctualised nodes' are prone to destabilisation, should any of their components (whether human or non-human, actor or intermediary) dissent from the aims of the network, which may then in turn impact on the stability of the network itself.

It has also shown how actors are often at different stages of engagement with the CFP. Some, who are already interested in the network, may enrol once they are required to play a part in addressing the solution and promoting the values that the network represents. Others, however, may be interested in the potential of the CFP, but so far have been peripheral to its development and not yet enrolled. This is indicative of how the so-called 'moments' of network translation can occur concurrently (Woods 1997), dependent perhaps on the impetus of an intermediary. As Burgess *et al.* (2000) suggest, the successful translation of a network may be determined by the ability of its controlling actors to direct intermediaries to their best advantage. This is most apparent with respect to the FS and the OBC. In the case of the FS, the originating entity used it as a means of strengthening his problematisation of the issue, in order to draw further actors into the CFP network. Likewise, the OBC was pivotal in convincing the SWPSHA to enrol their

---

[12]Nevertheless, it is important to recognise that it is by no means a complete description of the network's formation. For instance, every local supplier and/or producer that has been convinced of the benefits of supplying the CFP will have gone through a process of translation, which ends with their enrolment into the network (i.e. supplying it). Likewise, GOSW in coming to a decision that the CFP is complementary to their aims, and something they wish to support and mobilise their resources within; similarly, OSW, PASA, DEFRA and so on.

[13]The development of the grant network starts with a problematisation of why it is needed. A range of actors are then interested in what the FS might achieve and are prepared to enrol in the network.

support for the CFPU, which in turn is crucial for the future stabilisation of the CFP as a fully mobilised network capable of improving the sustainability of all five NHS Trusts in Cornwall.

It is apparent that the CFP offers a genuine alternative to the conventional system of food provision within the NHS in Cornwall. This is not simply because it has focused on increasing the supply of local produce, but more profoundly because it has sought to develop robust, inclusive networks that can form bridges between what were disjointed, and often undeveloped, elements of the FSC in Cornwall. This has involved local suppliers and producers but also a range of bodies who have an interest in the sustainable development of Cornwall. Around 60% of the RCHT's £1.2 million (€1.75 million) annual food budget is now sourced from Cornwall's producers (30–35%) or suppliers (20–30%) (NAO 2006). Even so, there is no room for complacency, and its ongoing development is dependent upon the continued engagement of the actors involved, as well as the maintenance of control at a local level. The successful completion of the CFPU will be an important step in this direction. Even then, the CFPU will have to tender for hospital contracts in order to not fall foul of EU-tendering rules, indicating that it has to demonstrate a genuine advantage within the wider marketplace and not simply rely on its 'localness', or be overly parochial.

Examining the CFP through the lens of ANT has helped unravel how it has developed from an original problematisation. In so doing, it has shown how it is not enough to simply identify a problem and to then suggest an alternative; the problem needs to be understandable, refined and articulated in such a way that the necessary actors become interested enough in the alternative to subsequently enrol in the emerging network. The communication of what the proposed network can achieve is vital, which includes identifying a commonality of interest amongst the actors involved, at whatever stage or moment within the translation process this may be. In addition, as suggested with punctualised nodes, there is always the potential for actors to feel that the network no longer contributes to their aims and they disengage from it. It may be that these actors are no longer critical to the functioning of the network, in which case their dissension may not lead to destabilisation. However, the dissent of actors that are still critical to the network could lead to profound instability in the network, and perhaps even its disintegration if the resultant network no longer fulfils the expectations of other actors.

ANT may not provide theoretical guidance on how to interpret these findings (Marsden 2000), but it has provided a conceptual framework within which to examine the formation of a network that is dependent on a range of actors and intermediaries. It has highlighted the contingency of network formation and the need to have a strong vision that resonates with the agendas of others, and to steadfastly promote this vision through a process of translating the interests of others into those of the developing network. So, although the CFP arose in response to the practical needs of a hospital catering department, the development of the CFP has been informed throughout by a desire "to enhance the economic, environmental and social well-being of Cornwall" (Harrow 2004), as described earlier. Through articulating how it aimed to achieve this (its vision), it drew a range of actors into its network. In addition, this chapter has shown the importance of constantly ensuring that the original problematisation is still relevant and that the developing network still has the capacity to provide the necessary solution to the problem(s) identified: lessons that can usefully be applied to the development of any AFN.

References

Burgess, J., Clark, J. and Harrison, C. M. (2000). Knowledges in action: an actor network analysis of a wetland agri-environment scheme. *Ecological Economics* 35, 119–132.
Busch, L. and Juska, A. (1997). Beyond political economy: actor-networks and the globalisation of agriculture. *Review of International Political Economy* 4 (4), 668–708.
Callon, M. (1986). Some elements of a sociology of translation: domestication of the scallops and the fishermen of St. Brieuc Bay. In J. Law (Ed), *Power, Action, Belief: A New Sociology of Knowledge?* (pp. 196–233). London: Routledge and Kegan Paul.
Callon, M. (1991). Techno-economic networks and irreversibility. In J. Law (Ed), *A Sociology of Monsters: Essays on Power, Technology and Domination* (pp. 132–161). London: Routledge.
Callon, M., Law, J. and Rip, A. (1986). *Mapping the Dynamics of Science and Technology.* London: Macmillan.
Comber, A., Fisher, P. and Wadsworth, R. (2003). Actor–network theory: a suitable framework to understand how land cover mapping projects develop? *Land Use Policy* 20, 299–309.
DEFRA (Department for Environment Food and Rural Affairs) (2002a). Farming and Food: A Sustainable Future. Retrieved 05.02.2002, January, from http://www.cabinet-office.gov.uk/farming.
DEFRA (Department for Environment, Food and Rural Affairs) (2002b). The Strategy for Sustainable Farming and Food: Facing the Future. Retrieved 12.12.2002, December, from http://www.defra.gov.uk/farm/sustain/newstrategy/strategy.pdf.
DEFRA (Department for Environment, Food and Rural Affairs). (2005). Draft Food Industry Sustainability Strategy (FISS). Retrieved 15.05.2006, from http://www.defra.gov.uk/corporate/consult/fiss/index.htm.
DoH (Department of Health) (2005). NHS Local Improvement Finance Trust (LIFT). Retrieved 09.11.2005, from http://www.dh.gov.uk/ProcurementAndProposals/PublicPrivatePartnership/NHSLIFT/fs/en.
Goodman, D. (1999). Agro-food studies in the 'age of ecology': nature, corporeality, bio-politics. *Sociologia Ruralis* 39 (1), 17–38.
Goodman, D. (2001). Ontology matters: the relational materiality of nature and agro-food studies. *Sociologia Ruralis* 41 (2), 182–200.
Harrow, N. (2002). *Cornwall Community Food Manufacturing & Distribution Study.* Truro: Cornwall Healthcare Estates and Support Services.
Harrow, N. (2004). *'Sustainable' Modernisation of NHS Food Supplies and Food Service in Cornwall: De-Sensitised Outline Business Case.* Truro: Cornwall Healthcare Estates and Support Services.
Harrow, N. (2005). *Project Manager, Cornwall Food Programme.* Face-to-face interview, 21st October.
Harrow, N. (2006). *Project Manager, Cornwall Food Programme.* Telephone conversation, 15th May.
Heath, R. (2005). *Sustainable Food Development Manager, Cornwall Food Programme.* Telephone interview, 25th November.
Herbert-Cheshire, L. (2003). Translating power: power and action in Australia's country towns. *Sociologia Ruralis* 43 (4), 454–473.
Hinrichs, C. (2003). The practice and politics of food system localisation. *Journal of Rural Studies* 19, 33–45.
HM Government (2005). Securing the Future: Delivering UK Sustainable Development Strategy. Executive Summary. Retrieved 15.05.2006, from http://www.sustainable-development.gov.uk/publications/uk-strategy/index.htm.
Kneafsey, M., Ilbery, B. and Jenkins, T. (2001). Exploring the dimensions of culture economies in rural West Wales. *Sociologia Ruralis* 41 (3), 296–310.

Latour, B. (1986). The powers of association. In J. Law (Ed.), *Power, Action, Belief: A New Sociology of Knowledge?* (pp. 264–280). London: Routledge and Kegan Paul.
Latour, B. (1991). Technology is society made durable. In J. Law (Ed.), *A Sociology of Monsters: Essays on Power, Technology and Domination*. London: Routledge.
Latour, B. (1999). On recalling actor network theory. In J. Law and J. Hassard (Eds), *Actor Network Theory and After* (pp. 15–25). Oxford: Blackwell.
Law, J. (1992). Notes on the theory of the actor-network: ordering, strategy and heterogeneity. *Systems Practice* 5 (4), 379–393.
Local Food Works (2002). *Creative Approaches to Local Food in Schools and Hospitals*. Briefing Paper No. 13 ed.: Local Food Works, 42–56 Victoria Street, Bristol BS1 6BY.
Lockie, S. and Kitto, S. (2000). Beyond the farm gate: production-consumption networks and agri-food research. *Sociologia Ruralis* 40 (1), 3–19.
Lowe, P., Murdoch, J. and Ward, N. (1995). Networks in rural development: beyond endogenous and exogenous approaches. In J.D. Van der Ploeg and G. Van Dijk (Eds), *Beyond Modernisation: The Impact of Endogenous Rural Development*. Assen, The Netherlands: Van Gorrum.
Marsden, T. (2000). Food matters and the matter of food: towards a new food governance? *Sociologia Ruralis* 40 (1), 20–29.
Marsden, T. and Smith, E. (2005). Ecological entrepreneurship: sustainable development in local communities through quality food production and local branding. *Geoforum* 36 (4), 440–451.
Marsden, T., Banks, J. and Bristow, G. (2000). Food supply chain approaches: exploring their role in rural development. *Sociologia Ruralis* 40 (4), 424–438.
Marsden, T., Murdoch, J. and Morgan, K. (1999). Sustainable agriculture, food supply chains and regional development: editorial introduction. *International Planning Studies* 4 (3), 295–301.
Morgan, K. and Morley, A. (2002). *Relocalising the Food Chain: The Role of Creative Public Procurement*: The Regeneration Institute, Cardiff University: in association with Powys Food Links, The Soil Association and Sustain.
Morgan, K. and Morley, A. (2004). Creating sustainable food chains: the role of positive public procurement. In M. Thomas and M. Rhisiart (Eds), *Sustainable Regions: Making Sustainable Development Work in Regional Economies*. St. Bride's Major, CF32 0TN, Wales: Aureus Publishing.
Morris, C. (2004). Networks of agri-environmental policy implementation: a case study of England's Countryside Stewardship Scheme. *Land Use Policy* 21, 177–191.
Murdoch, J. (1994). *Weaving the Seamless Web: A Consideration of Network Analysis and Its Potential Application to the Study of the Rural Economy*. Working Paper 3, February University of Newcastle upon Tyne: Centre for Rural Economy, Department of Agricultural Economics and Food Marketing.
Murdoch, J. (1995). Actor-networks and the evolution of economic forms - combining description and explanation in theories of regulation, flexible specialization, and networks. *Environment and Planning A* 27 (5), 731–757.
Murdoch, J. (1997). Inhuman/nonhuman/human: actor-network theory and the prospects for a nondualistic and symmetrical perspective on nature and society. *Environment and Planning D* 15 (6), 731–756.
Murdoch, J. (2000). Networks - a new paradigm of rural development? *Journal of Rural Studies* 16 (4), 407–419.
Murdoch, J., Marsden, T. and Banks, J. (2000). Quality, nature, and embeddedness: some theoretical considerations in the context of the food sector. *Economic Geography* 76 (2), 107–125.
NAO (National Audit Office) (2006). Smarter Food Procurement in the Public Sector: A Good Practice Guide. Retrieved 14.05.2006, from http://www.nao.org.uk/publications/index.htm.
PASA (NHS Purchasing and Supply Agency) (2005). The NHS Purchasing and Supply Agency: What We Do. Retrieved 25.11.2005, from http://www.pasa.doh.gov.uk/whatwedo/.

Pearson, M. (2004). *Head of Facilities Services, Royal Cornwall Hospital Trust*. Face-to-face interview, 24th September.
Ray, C. (1998). Culture, intellectual property and territorial rural development. *Sociologia Ruralis* 38 (1), 3–20.
RCHT (Royal Cornwall Hospital Trust) (2002). *Cornwall Community Food Manufacturing and Distribution Study*. Truro, TR1 3LJ: Catering Services Department. November.
Renting, H., Marsden, T. and Banks, J. (2003). Understanding alternative food networks: exploring the role of short food supply chains in rural development. *Environment and Planning A* 35, 393–411.
Schiopu, T. (2005). *Eating Local Food in Thames Valley Schools*. Berkshire: Buckinghamshire and Milton Keynes and Oxfordshire Food Groups.
Selman, P. (2000). Networks of knowledge and influence: connecting the planners and the planned. *Town Planning Review* 71 (1), 109–121.
Selman, P. and Wragg, A. (1999). Local sustainability planning: from interest driven networks to vision-driven super networks. *Planning Practice and Research* 14 (3), 329–340.
Sustain (2002). Sustainable Food Chains - Briefing Paper 2: Public Sector Catering; Opportunities and Issues Relating to Sustainable Food Procurement. Retrieved 27.05.2003, from http://www.sustainweb.org/pdf/breifing2.pdf.
SWPSHA (South West Peninsular Strategic Health Authority) (2005). About Us. Retrieved 24.11.2005, from http://www.swpsha.nhs.uk/aboutUs/aboutUs.shtml.
Watts, D. C. H., Ilbery, B. and Maye, D. (2005). Making reconnections in agro-food geography: alternative systems of food provision. *Progress in Human Geography* 29 (1), 22–40.
Whatmore, S. and Thorne, L. (1997). Nourishing networks: alternative geographies of food. In D. Goodman and M. Watts (Eds), *Globalising Food: Agrarian Questions and Global Restructuring* (pp. 287–304). London: Routledge.
Winter, M. (2003). Embeddedness, the new food economy and defensive localism. *Journal of Rural Studies* 19, 23–32.
Woods, M. (1997). Researching rural conflicts: hunting, local politics and actor networks. *Journal of Rural Studies* 14 (3), 321–340.
WPI (Welsh Procurement Initiative) (2005). Food for Thought: A New Approach to Public Sector Food Procurement - Case Studies. Retrieved 28.07.2005, from http://www.organic.aber.ac.uk/schoolsnet/casestudies/wpi2005.pdf.

Chapter 12

'Bending Science to Match Their Convictions': Hygienist Conceptions of Food Safety as a Challenge to Alternative Food Enterprises in Ireland

Colin Sage
Department of Geography, University College Cork, Republic of Ireland

Introduction

A series of food safety crises in the UK and to a lesser extent elsewhere in Europe during the 1980s and 1990s introduced far-reaching regulatory and institutional reform. The enormous cost of the BSE crisis – fiscal, political and, critically, that sustained by consumer confidence – initiated a process involving the creation of food safety authorities at national and at EU level (Marsden *et al.* 2000; van Zwanenberg and Millstone 2003). These statutory, science-based and ostensibly 'independent' organisations were responsible for ensuring a seamless guarantee of food safety from farm to fork. The reallocation of responsibilities for food governance that had hitherto been in the hands of the old agricultural ministries was designed to restore the faith of consumers in the food supply chain (Barling 2004). Yet the adoption of a rather narrow and conservative approach to protecting the consumer against the risks caused by the way food is produced or supplied has tended to encourage the defence of conventional agriculture and the mainstream food system (Barling and Lang 2003).

A feature of the institutional response to public anxieties associated with the food scares of the 1980s and 1990s has been the emergence and consolidation of a science-based approach to food safety regulation. Frequent reference to 'sound science' and 'scientifically based evidence' has become a prevailing feature of the contemporary food safety discourse.[1] A science-based approach to food safety is principally concerned with risk assessment, which involves the identification of hazards (e.g. microbial infections, pesticide residues, allergens), determining exposure of the population to that hazard and calculating the balance of risk to benefit and cost (Nestle 2003). Clearly, this places emphasis upon probabilities and statistics and places considerable emphasis upon the characteristics of the risk itself: the number of cases of illness and death, the fiscal costs associated with medical treatment, the number of days lost from work and so on. This measurable quality of risk is what enables science-based food safety to represent itself as objective and neutral.

Given the power of numbers in the practice of governance (Enticott 2001), the generation and accumulation of statistics through monitoring and surveillance by science confers considerable legitimacy for a food regulatory regime that determines what constitutes 'safe' food. Thus, it places emphasis on the microbiological dangers of food contaminated with pathogens such as *Listeria* or *Salmonella*, but not on the risks presented by genetically modified organisms or on the use of recombinant growth hormones in livestock. By defining the narrow technical area that it considers appropriate, effectively excluding other issues from public debate, science provides for politicians the authoritative reassurance needed to assuage public anxieties about food enabling them to say: "trust us, it's safe to eat". Yet it is evident that not only is science used as a political resource (van Zwanenberg and Millstone 2003), but also that a reductionist science is unable to assess adequately wider social and moral dimensions of risks associated with new innovations and practices (Macnaghten and Urry 1998).

In recent years there has been a considerable interest in foods of local origin, culinary tradition, regional speciality and gastronomic merit. Across Europe, North America and elsewhere small- to medium-sized enterprises engaged in artisan production of low volume, but high-quality foods, have flourished.[2] The rise of 'alternative' food networks has been heralded as evidence of a 'turn to quality' with consumers expressing an interest in recovering a connection with the provenance of food and a sense of trust in its production. This has led to a demand for gastronomic products that display a higher content of 'raw nature', such as wild foods, or those subject to limited processing before consumption. While it may appear 'irrational' to food science, many people "find the benefits of eating raw fish or raw milk cheese to exceed the small but finite risk of ingesting harmful microbial contaminants; the choice is voluntary and the foods are familiar" (Nestle 2003: 21). Yet for food safety authorities, with a hygienist regulatory mindset and in pursuit of the goal of 'no-germ' food, such products may appear to represent a real threat to public health.

One product that captures this sharp polarisation in perspective is raw milk cheese (RMC), which is frequently attributed by food safety authorities to be the cause of food-borne illness due to the presence of pathogenic bacteria. Yet, as we shall see, the

---

[1] For example, the Food Safety Authority of Ireland (FSAI) describes itself as a "statutory, independent and science-based body" while Flynn *et al.* (2003: 40) note that the UK Food Standards Agency promotes itself as "a science-based body able to determine risks in an objective manner".

[2] Amongst a growing literature on the topic are: Feenstra (1997), Ilbery and Kneafsey (1999), Murdoch *et al.* (2000), van der Ploeg *et al.* (2002), Renting *et al.* (2003), Sage (2003), and Goodman (2004).

microbiology of cheese is a much more complex and contested area of science than is generally known. Aficionados of RMC appreciate the organoleptic qualities of flavour and aroma retained without the denaturing effect of pasteurisation of milk. This market is catered for by a large community of cheesemakers, the majority of whom are producing well-known brands, such as Parmegiano-Reggiano (Italy), Emmental (Switzerland) and Comte and Roquefort (France). Such cheeses have been designated by national bodies, as well as recognised by the EU, as being culturally distinctive products of local origin with clearly defined methods of production. In Northern Europe, however, the landscape is populated by fewer specialist cheesemakers who have not only faced overwhelming market competition from industrial production, but regulatory harassment from the food safety authorities.[3] It is ironic that while the rural development discourse in Europe is increasingly positive about small food businesses providing the basis of economic regeneration in peripheral areas, food safety regulations impose significant obstacles to their economic viability. Consequently, this chapter is concerned with highlighting this disparity, revealing how the regulatory frameworks introduced to protect consumer health in the post-BSE era represent, in some institutional contexts, a serious obstruction to the development of specialist alternative food networks (especially, in this case, small artisan producers) offering greater choice and nutritional benefits to consumers.

Following a review of the food safety issues surrounding raw milk cheese in general, and a discussion of the situation facing farmhouse cheesemakers in Ireland, the chapter provides an account of a recent court case involving a highly regarded raw milk cheesemaker and the Department of Agriculture and Food (DAF). The purpose of the case was to establish whether a batch of cheese was to be considered fit for human consumption. Summarising the main areas of scientific dispute that were revealed in court sheds considerable insight into large areas of ignorance and uncertainty in food science, yet a willingness to assert the scientific superiority of food safety expertise in the face of contrary evidence. Above all, it reveals the extent to which science can be used (or, as shown here, 'bent') to provide endorsement for value-based judgements reached by regulatory authorities with political or other agendas. In the process, the chapter will also help demonstrate the enormous costs under which speciality producers are placed if they are deemed to have fallen foul of the stringent, arguably even prejudicial, food safety regulations.

Science and Food Safety Regulation

Safety is relative; it is not an inherent biological characteristic of a food. A food may be safe for some people but not others, safe at one level of intake but not another, or safe at one point in time but not later. Instead, we can define a safe food as one that does not exceed an *acceptable* level of risk. Decisions about acceptability involve perceptions, opinions, and values, as well as science (Nestle 2003: 16, emphasis in original).

---

[3] Amongst the most well-publicised cases is that of Humphrey Errington, the producer of Lanark Blue Cheese who was accused in 1995 of producing cheese with exceptionally high levels of *Listeria monocytogenes*. Eventually the court found in favour of Mr Errington (see Cunynghame 2001; see also Country Life (2005) for an account of the case).

While food safety is an issue of long standing, it was during the 1990s and the early 2000s that we witnessed particularly intense legislative activity in relation to food quality (Hennessy et al. 2003). Indeed, it might be said that food safety has become the food policy narrative that dwarfs all others (Lang and Rayner 2003) and not only in the UK but across Europe. The formation of European and international regulatory regimes has increased the overall importance of science and risk assessment which play a critical role in establishing regulatory policies and standards (Phillips and Wolfe 2001). Science-based decision-making and risk assessment have become a universal discourse across different levels of food governance. Indeed, science appears to have achieved an unassailable platform of authority and influence in relation to policy-making, with Cameron noting how the Sanitary and Phytosanitary Standards (SPS) Agreement (under the WTO) "exalts the role of science far beyond the point it is appropriate, attempting to eliminate all 'non-science' factors from standard setting" (Cameron, quoted in Buonanno et al. 2001: 7).

Yet science is also beset by significant challenges. The risk assessment process itself is becoming increasingly complex as "disciplinary expertise fragments and specialises, interdisciplinary contributions to food issues increase, and as new products create new hazards and differing levels of exposure" (Khachatourians 2001: 14). There are growing disputes around the different ways of evaluating risks as well as the weight that should be given to public attitudes and preferences (see Ansell and Vogel 2006). Moreover, the experience of the BSE scandal in the UK and the ongoing debate about biotechnology in a number of countries raises questions about the independence of science and its susceptibility to be compromised by commercial pressures and political agendas (Rowell 2003; Skogstad 2006). Other critics argue that risk assessment takes a too-narrow view of food safety, and indeed, returning to Cameron's comment above, fails to address 'non-science' factors.

Nestle (2003) refers to these as value-based approaches, the personal beliefs that depend upon a host of cultural, social and psychological factors. Drawing on research concerned with evaluating people's perception of risk, she further underlines the validity of the 'risk society' thesis (Beck 1992) which suggests that people are less willing to accept risks induced by technology or poorly understood by science. This means that although microbial contaminants might pose the greatest food-borne threat to health, the public rank chemical pesticides and additives, irradiation and genetic engineering high on their list of perceived risks. The reason for this is "largely because exposures to them are invisible, involuntary, imposed, and uncontrollable" (Nestle 2003: 21). Yet it is, as Harvey et al. (2004) argue, unfortunate that so much of the analysis of contemporary food controversies is filtered through the lens of the 'risk society' thesis. While appearing to offer a compelling explanation of the current anxieties presented by modern science, a diverse range of issues around food are reduced to a single overarching preoccupation with the assessment of risk.

Given the ability of consumers to evaluate for themselves the relative merits and risks of consuming a particular product, what is the role of food safety regulations in determining the availability of that product? Working on the basis of being risk averse rather than risk tolerant or risk neutral (Buonanno 2006) and using a high-risk category of consumers as a reference (comprising the elderly, pregnant women, young children and the immunocompromised), food safety regulation works to protect public health with a singular standard apparently based on 'sound science'. Yet, at a time when food markets are differentiating, when the prevailing discourse proclaims the sovereignty of consumer choice, and is reflected by consumers' willingness to pay for different food qualities, a stringent food safety standard has the effect of reducing the variety of foods that might be available. Indeed, the 'heterogeneous consumer' is not only reflected in different food preferences and changing consumption habits, but in different attitudes towards risk as well as differences in their vulnerability towards health risks. In other words, if food safety standards are set at a level to protect this high-risk category identified above, then there is the likelihood that products enjoyed by those in robust good health may be denied the opportunity to consume them. On the other hand, if standards are set too low it is the vulnerable that bear a disproportionate share of the population's food-related health risks (Peltola and Schalin 2003).

While food safety authorities might argue for the sound scientific basis for such standards it is clear that trade-offs between choice and safety involve value judgements about the acceptability of certain foods. Indeed, the focus on science, safety and risk serves to obscure the values that lead the authorities to defend the dominant agro-industrial model over new innovations in the food supply chain in which there may be an attempt to recover a more sustainable, more socially just and more ethical approach to food. One such example of this would be the attitudes demonstrated by the UK Food Standards Agency regarding the growing consumer demand for organic food (see Rowell 2003; Barling 2004 for reviews).

Meanwhile, as Marsden (2003) has observed, the food safety paradigm puts into place, in a highly interventionary and bureaucratic fashion, policies that effectively police the food system in ways that make it seem more hygienic. The effect of the installation of these regulatory safeguards is to 'ratchet up' the costs of compliance for those small producers and food processors that, ironically, were entirely marginal to the causes of health fears in the first place. This is done by creating, as Marsden (2003: 170) puts it:

> a kind of Foucaultian panopticon around the agricultural and agro-food sector such that it is increasingly watched, monitored and regulated at arms length... (and) provides a new raison d'être for the old productivist-driven agricultural ministries. They now espouse consumer-driven needs through a highly bureaucratic and rationalistic approach.

Unsurprisingly, this process is at odds with encouraging regional culinary heritage and gastronomic traditions and places a hygienist conception of safety before concerns for sustainability. Thus it appears to profoundly hamper other policy initiatives designed to encourage the revival of more local and environmentally sensitive efforts to improve food quality. The case study below provides an illustration of such a process.

Raw Milk Cheese

Cheese is the most diverse group of dairy products and is reputedly the most scientifically challenging of all foodstuffs. Unlike many other foods which are relatively stable, cheeses are biologically and biochemically dynamic. Cheese contains complex bacterial ecosystems and changes considerably during maturation. If these changes are balanced they lead to the production of a very large number of flavour compounds and the highly desirable aromas and flavours of different cheeses (Fox and McSweeney 2004).

Approximately 16.5 million tonnes of cheese is produced worldwide annually of which Europe accounts for just over half (Fox and McSweeney 2004); this figure includes 700 000 tonnes made from raw milk (Grappin and Beuvier 1997). Cheese is generally a very safe product for which the infrequency of large, cheese-associated outbreaks of human illness is notable, especially when compared to the record for cooked meat products. Between 1973 and 1992 there were 32 cheese-associated cases reported to the US Centre for Disease Control of which not one involved raw milk cheese aged for a minimum of 60 days (Donnelly 2001).

However, the microbiological safety of raw milk cheese remains a highly contested issue. In the early 1990s, for example, there was debate within the EU and the Codex Alimentarius Commission for the mandatory pasteurisation of all dairy products and there remains a powerful scientific discourse opposed to the continued availability of raw milk cheese (Dixon 2000). Yet an alternative view argues that, far from being the guarantor of food safety, blind faith in pasteurisation can lead to more problems than it solves and serve to limit efforts to improve total hygiene in food production (Wheelock 1997). This is because in the majority of cases of food poisoning outbreaks confounding factors other than the use of raw milk, most commonly involving post-production contamination, contributed to pathogens being present at the time of consumption. As Donnelly (2001) concludes in a report that comprehensively summarises previous reviews of the epidemiological literature related to outbreaks of human illness involving raw milk cheese, there is no compelling data to indicate that mandatory pasteurisation would lead to a safer product. Indeed, some microbiologists argue that the use of pasteurised milk in cheese-making may provide an environment which provides for optimal growth of pathogens in contrast to raw milk where natural inhibitors provide a margin of control over pathogen growth – the principle of 'competitive exclusion' (Donnelly 2001).

Nevertheless, pasteurisation remains the central technique by which nature is outflanked and milk products made safe for human consumption. It is therefore an indispensable weapon of dairy science and the regulatory authorities in their common fight against food-borne disease.[4] In this respect pasteurisation has taken on the characteristics of a 'black box' for many scientists for which it is simply unimaginable that it would be circumvented, especially with regard to the consumption of unpasteurised liquid milk.[5] Yet there remains considerable divergence in the different positions of EU member states with respect to raw milk. While its sale is prohibited in Ireland, Finland, the Netherlands and Scotland, in Austria up to 10% of milk produced is consumed as unpasteurised liquid or dairy products (Dixon 2000). Moreover, many of its customers

---

[4] The most usual form of pasteurisation in the modern dairy industry involves heating the milk to 72 °C for 15 s followed by cooling to 4 °C. All pathogenic microorganisms (such as *Mycobacterium tuberculosis, Brucella, Staphylococcus aureus, E. coli, Listeria, Salmonella, Campylobacter*) are reputedly destroyed.

[5] This is exemplified by Dr William Keene, a food safety expert based in Oregon, who notes that: "There is no mystery about why raw milk is a common vehicle for salmonellosis and other enteric infections; after all, dairy milk is essentially a suspension of fecal and other microorganisms in a nutrient broth" (Nestle 2003: 127).

firmly believe in a 'lay immunology' that associates regular consumption of liquid raw milk with good health (see Enticott 2003a,b; Weston Price Foundation 2005).

For producers and consumers that relish the sensory characteristics of raw milk cheese, pasteurisation is believed to kill the natural liveliness of the product. It is widely acknowledged that pasteurisation influences the biochemistry of ripening by altering the indigenous microflora of the milk, partially or completely inactivating certain enzymes that contribute to ripening (Rance 1989; Grappin and Beuvier 1997; McSweeney and Sousa 1999). In effect, it is a standardising procedure with which to create uniform milk suitable for industrial-scale manufacturing of dairy products (Fox and McSweeney 2004). Retaining the natural characteristics of the milk that originates in a specific area of production is a particularly desirable feature of traditional and artisan cheeses that reflect the composition and quality of grazing and variations by season (Dixon 2000).

While no single technology can ensure a safe cheese, a high degree of safety can be achieved in a system that combines a number of antimicrobial effects (Schällibaum 1999). These include the synergistic effects arising from the use of starter cultures, the relatively high salt content, the acidification process (with pH falling to 5.5 or lower, depending on the cheese), declining moisture content, storage temperature, the development of enzyme systems and the length of the maturation period, all of which serve to eliminate pathogenic bacteria (Kalantzopoulos 1999; Donnelly 2001).

It is important to emphasise that bacterial contamination can take place at many points throughout the food chain, from the milking animals to the consumer's domestic refrigerator. Moreover, unlike industrial plants where a high degree of automation eliminates much of the potential source of contamination from human labour, farmhouse cheese production involves a large number of manual operations requiring scrupulous attention to hygiene and the operation of risk management procedures such as Hazard Analysis Critical Control Points (HACCP) (Unnevehr and Jensen 1999). Hence the limited number of illnesses that arise from the consumption of raw milk cheese in Europe and North America and the high degree of trust shown in farmhouse cheesemakers by consumers of their products.

Raw Milk Cheese in Ireland

Milk and milk products have always fulfilled a significant role in the Irish diet. The lush pastures grazed by milk cows supplied an abundance of liquid that was turned into various sour milk drinks, curds, soft and hard cheeses and butter (Sexton 1998). Their prevalence was not only due to dietary need, but also to pay taxes to the local aristocracy. However, from the seventeenth century decades of political and agrarian upheaval saw the decline of cheese-making and the development of the provision trade in which Ireland increasingly exported its farm produce in the form of salted beef and salted butter. Then, from the early twentieth century, the larger dairy cooperatives began the production of industrial cheddar, a product that had become synonymous as 'Irish cheese', at least until recently.

From the 1970s and 1980s a group of pioneer cheesemakers emerged, many of them non-nationals and most with little prior knowledge or experience and who learnt through experimentation and by attending the occasional course in microbiology. Having bought small farms in remote rural locations, especially in the South West, these individuals began making cheese on their kitchen stoves as a way of creating a livelihood by which to survive. The cheeses that they created were a diverse range, reflecting the personalities and origins of their producers. Gradually their numbers swelled and an association, Cais,[6] was created to act as a forum and lobbying body. The artisan nature of the process, involving a handmade element with upper limits on quantities of milk used and volume of cheese produced, has always been an important defining characteristic of the sector.

The introduction of European Directive 92/46/EEC[7] arrived at a time when there was considerable pressure for the mandatory pasteurisation of milk and the elimination of raw milk cheese from the market. This was transcribed into Irish law as S.I. No. 9 of 1996, which established the criteria and microbiological standards of commercial dairy products. The introduction of these standards in January 1997, together with the monitoring and compliance procedures of the Department of Agriculture and Food (DAF), had a big impact on the farmhouse cheese sector. Many felt obliged to introduce pasteurisation and others left the business altogether, unable to make the investments necessary to achieve the minimum equipment and infrastructural specifications required. Unfortunately, it is fair to say that an institutional culture of deep suspicion exists towards the farmhouse cheesemakers, especially those using raw milk. The prevailing view within DAF is that small-scale producers are unconventional, maverick and 'alternative'. This seems to justify the tighter domestic legislation, which is considerably stricter than the requirements set out in 92/46/EEC.[8] Currently, there are approximately 90 small farmhouse cheesemakers in Ireland using cow, goat and ewe's milk, though less than 10 continue to use raw cow's milk.

One of the particular challenges faced by this small band of cheesemakers is a zoonotic disease which is proving very difficult to eliminate both in the UK and in Ireland. Bovine tuberculosis (*Mycobacterium bovis*) was once a significant cause of human ill health in Ireland and the UK until the 1940s, with 50 000 new infections and 2500 deaths per year in the UK during the 1930s (MAFF 2000). The introduction of a series of public health measures, including pasteurisation of milk, significantly reduced this incidence such that the numbers of people infected with bovine TB represents only a small fraction (ca. 1%) of the total number of tuberculosis cases in recent years.[9]

Nevertheless, it remains a significant animal health problem in Ireland, with a notable failure of the eradication programme to achieve any significant reduction in herd incidence since 1965 (Griffin *et al.* 1993). Annual tuberculin tests are required of all cattle

---

[6] *Cais* is the Irish Farmhouse Cheesemakers Association. For further details see: www.irishcheese.ie (accessed 16 June 2006).
[7] "The health rules for the production and placing on the market of raw milk, heat-treated milk and milk-based products intended for human consumption".
[8] Amongst many issues raised by farmhouse cheese producers, as indicative of this tighter regulation, is their categorisation as 'milk-processing establishments', which puts them on the same basis as industrial dairies, especially in terms of auditing. Elsewhere in the EU such producers are registered as 'limited capacity establishments' where derogations can be sought from aspects of the Directive.
[9] According to an official report on zoonoses in Ireland, there were 408 notifications of human TB in 2002, of which five were confirmed as *M. bovis*. For 2003 there were 421 cases of human TB of which six were *M. bovis* (FSAI 2005a: 40).

and this involves the injection of antigens derived from *M. bovis* under the skin of the animal and an inspection 3 days later for evidence of a local inflammatory reaction. A positive reaction requires animals to be taken out of the herd and slaughtered. Carcasses are then subjected to post-mortem inspections for signs of lesions. Each year around 10 000 herds (of 130 000 nationally) may be restricted due to diagnosis of tuberculosis in cattle, with 40 000 animals slaughtered each year as a result of positive tests (FSAI 2003). While the tuberculin test is the standard procedure for identifying tuberculosis in cattle it does have some limitations. Three are notable here: first, a failure to detect recent (within 6 weeks) infection; secondly, in presenting 'false-positives' for healthy animals; and thirdly, where post-mortem inspections of cattle that were clear at their most recent test show evidence of tuberculosis (FSAI 2003).

Where herds are restricted due to the diagnosis of tuberculosis, milk must be pasteurised before any subsequent use. The carcass of infected animals, providing there are no generalised lesions, is passed fit for human consumption and enters the food chain. In Ireland the DAF requires dairy herds supplying milk for unpasteurised cheese to be subject to two tuberculin tests each year. If even one animal in a herd shows positive, the herd is restricted and milk must be pasteurised before any subsequent use. Crucially, any dairy products made from unpasteurised milk originating in that herd going back to the preceding clear test is subject to a detention order from the DAF which then directs its destruction. As we shall see below, the timing of the tests and their significance as measures of food safety is subject to considerable contestation.

Contesting Interpretations of Science in a Court of Law[10]

Absence of Proof is not Proof of Absence

WCNC is a small business that makes a hard cheese using the Swiss method. This involves using only the morning milking for the duration of the summer season from herds feeding on fresh pasture, then using a thermophilic starter culture and sub-pasteurisation temperatures during the cheese-making process. Until 2002 total annual production was of the order of 7–8 tonnes, with the cheese highly regarded by connoisseurs and receiving various awards over the last decade. As the quality of the cheese is only as good as the milk from which it is made, WCNC, like many artisan cheesemakers without their own dairy herd, invests considerable time and effort carefully selecting the most suitable farmers within the vicinity from which to source. These farmers must not only employ high standards of hygiene in their milking parlours, but must be vigilant in managing all aspects of their dairy herd and the pastures on which they feed.

As WCNC begins cheese-making in July and finishes in October it requires the cattle to be given a clean bill of health at the beginning and again at the close of the season. Consequently, in June 2002 it sought to have the local veterinary inspectors conduct the tuberculin tests on supplier herds, in accordance with an agreement made with the Department some years earlier. Despite frequent requests the main supplier of milk (CM)

---

[10]This section draws upon transcripts of hearings in the District and Circuit Courts, as well as notes made during my attendance on 3 of the 6 days when the case was put before the courts – including the judgements. The scientific literature referred to during the case has also been consulted.

was not tested until 13 August 2002. The tests showed up a high incidence of bovine TB in the herd – 37 of 57 milking cows were positive reactors. Almost immediately the Department served notices of detention on the cheese made between the 15 July when cheese-making began and the date of the test (approximately 2 tonnes). Had WCNC made cheese year-round, as most farmhouse cheesemakers do, the detention order would have required the removal from the food chain of all cheese made since the last clear test on the herd, in January 2002. Elsewhere in the EU it is usual for cheesemakers to cease production from the date of discovery of TB in the supplier herd. But in Ireland a much stricter approach is taken with no compensation payable to the cheesemaker for loss of stock (IFJ 2005).

While complying with the order, WCNC appealed it to the District Court. The case took place over several days stretching from December 2003 until the judgement was delivered in April 2004. Both sides called several scientific witnesses and the notable lack of agreement between them highlighted the very considerable uncertainties that lie at the heart of many food safety issues. Evidence presented and cross-examination within court was ultimately designed to establish the safety of the cheese for human consumption, although the Counsel for the Department outlined that, in relation to the relevant regulations, it effectively came down to reasonable doubt (that the product was not fit). However, this was no simple black and white case but involved a series of scientific issues on which judgements were to be made.

Predictably, witnesses called by the DAF, whose evidence would establish their case for the detention order, were heavily weighted towards expertise in animal health. It was notable the degree to which the risk of infection in the cheese was extended from the positive reactor status of the animals, with the means of transmission being raw milk. Indeed, it was made clear that the inclusion of unpasteurised milk from even one cow infected with *M. bovis* would be sufficient to condemn the entire cheese as unfit for human consumption. The witnesses called by the Department also set great store by the claim that an absence of proof was not proof of absence. In other words, even if infection was not found in the animals, the milk or the cheese it would not mean that it was not present and therefore could claim continuing 'reasonable doubt' that the cheese was unsafe.[11]

Counsel for WCNC therefore had to do three things: first, challenge the perception that raw milk was a dangerous source of pathogens; secondly, determine the extent of the *M. bovis* disease within the animals and the likelihood of their shedding infection into the milk; and thirdly, establish that the cheese-making process rendered the milk safe and yielded a product that was quite fit for human consumption. In other words, it was for the producer to establish that the cheese was safe, not for DAF to prove that it was otherwise.

---

[11] Cunynghame reports a further example of this 'absence of proof is not proof of absence' approach to food safety. On the basis of a single case of illness, the Department of Health (DOH) in the UK ordered the destruction of £40 000 worth of cheese. This was "[d]espite the fact that (the condemned) cheese came from a different batch to the suspect cheese and despite extensive tests failing to show anything wrong with the cheese, the DOH insisted that, although the tests did not show any pathogens to be present, this did not mean that they were not present!" (Cunynghame 2001: 19).

First, in relation to raw milk, it was noted that a FSAI study (FSAI 2005b) had reported on the high levels of consumption of unpasteurised milk by farm families with their own dairy herds – 84% of families bring a jug or churn from the milking parlour to the kitchen. Yet if *M. bovis* was so widespread (5 000 infected herds nationally) and following the argument that even the milk of one positive reactor animal in a bulk tank would be sufficient to spread the infection, how did this account, asked counsel for WCNC, for such a low level of infection within the human population? The response by witnesses for the DAF simply referred to the possibility of dormancy in the human population.

This raised a second issue concerning the degree of development of the disease in the infected herd. Here, the court learnt about the contraction of the disease in cattle and the signs of its level of development. This usually begins in the lung with the identification of characteristic lesions in the lymph nodes then progresses through other parts of the body. Given the regime of yearly testing it was noted by witnesses from both sides that advanced infection is no longer common. Thus, it is unusual to find infection as far as the supermammary glands associated with the udder and the milk secretion capabilities of the animal.[12]

The 47 positive reactor animals were slaughtered in late September 2002 and their carcasses examined by a veterinary inspector.[13] His testimony in court was that 18 of the 47 animals showed no visible lesions, and although there was evidence of disease in the lymph nodes, there were no signs of generalised TB. Indeed, apart from the head, lung and other minor tissue, all of the carcasses were passed fit for human consumption and entered the food chain. Unusually, instructions were issued to take samples of the supermammary glands and send them for further analysis to the National Veterinary Laboratory. Tests were conducted during November and December 2002. In January 2003 WCNC received notice from DAF that analysis of the supermammary glands had shown some samples to be positive for *M. bovis*. This was used to offer further justification for the Department's position and perhaps to discourage the appeal against the detention order. Yet, just a few weeks before the first court hearing (in December 2003) it was revealed that a labelling error had been made and that the tests had actually all proven negative.[14]

The third major area of debate concerned the microbiology and chemistry of cheese-making and whether this effectively killed off any pathogens present in the milk at the start of the process. This probably represented one of the most contested areas of all, not least because little research had been done on the survival of *M. bovis* in cheese. This justified the position of DAF captured by the FSAI (2003: 13) report that

---

[12] It was noted by the court that milk from the herd was marked by a low total bacterial count, demonstrating good hygiene, and by a low somatic cell count reinforcing claims that there was an absence of mastitis amongst the animals. Thus it was highly unlikely, argued counsel for WCNC and supported by veterinary experts called by that side, that *M. bovis* was present in the mammary glands and therefore in the milk. The chief scientific witness called by DAF responded thus: "you cannot prove absence".

[13] This included 10 non-milking animals. Slaughter followed 5 weeks after testing positive and it was agreed in court that during this time the disease would have developed further.

[14] WCNC contend that this, alongside other issues, was evidence of systematic harassment by the DAF.

notes: "[t]he impact of the cheese production process on the viability of *M. bovis* is not well defined and the cheese production process has not been demonstrated to eliminate viable *M. bovis*". However, during the court hearings a surrogate measure was used – *Mycobacterium avium* subsp. *paratuberculosis* (MAP).[15] This shows more than 90% similarity except that *M. bovis* is less tolerant of heat and acid.

The court was provided with expert testimony, with one distinguished professor of dairy science representing each side and several scientific papers provided as evidence. Adversarial exchanges centred largely upon the degree to which the characteristics of MAP, especially its reduction times in different cheeses, could be extended to *M. bovis*. The professor representing the DAF (brought from the USA) was taken through one of his own scientific papers that had reported on experimental work in which the combined antimicrobial effects of salt, acid, heat and curing time in a soft cheese eliminated the injected level of MAP within 60 days. When then invited to comment on the conditions in the cheese at issue in the case, which would all point to a more rapid reduction in MAP (or *M. bovis*), he would not be drawn: "It is extremely hazardous to extrapolate a different organism, different cheese, different production". When reminded that *M. bovis* was less tolerant than MAP, he responded:

> The chain of evidence is far too weak, all assumptions can be challenged. It is rational to accept the food safety experts rather than a single research paper which can be fallible.

In other words, the evidence was too weak to prove the cheese was safe, but incontrovertible through assertion that it was unsafe. It is also revealing how this senior scientist was reluctant to stand over his own research, as well as that conducted by other scientists that have come to similar conclusions regarding the elimination of MAP,[16] and instead sought sanctuary in the community of 'food safety experts' who operate that Foucaultian panopticon referred to by Marsden (2003).

The Judgement and Its Consequences

In their summing up, counsel for the two sides provided quite contrasting positions. Counsel for DAF took only a few moments to highlight the relevant points of the Regulations to emphasise that the grounds of 'reasonable doubt' were sufficient to condemn the cheese. By contrast, counsel for WCNC took some time to review the

---

[15] *Mycobacterium avium* subsp. *paratuberculosis* (MAP) is a ruminant infection, also known as Johne's disease, which bears a close similarity to Crohn's disease, a chronic inflammatory bowel disease in humans. Because of the similarity in the two diseases it has raised the question of whether milk, among other factors, could be a vector of transmitting MAP from cattle to humans (Spahr and Schafroth 2001). MAP is therefore a contemporary live issue for scientific research whereas *M. bovis* is not.

[16] Spahr and Schafroth (2001) was widely referenced in court. Their findings demonstrate the synergistic effects of the various antimicrobial processes discussed earlier. Indeed, there is strong evidence, conclude Spahr and Schafroth (2001: 4204), that: "the manufacturing and ripening conditions of hard cheese from raw milk are, in effect, equivalent to pasteurisation as regards *M. avium* subsp. paratuberculosis".

lack of scientific evidence that had been presented to substantiate these grounds and underscored the lack of natural justice implicit in the DAF's policy of detention and destruction without concrete evidence of infection.

Finally, in Skibbereen District Court on 14 April 2004, Judge Finn delivered his verdict, beginning by reviewing the evidence and itemising conclusions that could be drawn from it. The first of these conclusions was that the low cell counts in the milk were sufficient for the Court to assume there was no *M. bovis* in the milk. Secondly, that the weight of scientific evidence regarding the combined antimicrobial effects and the lengthy period of maturation of the detained cheese would render it safe for human consumption. However, Judge Finn went further than simply quashing the Detention Order and drew attention to what he considered were two paradoxes within the position of the DAF.

The first concerned the cut-off point when a herd is no longer considered TB-free. The second paradox, however, concerned the striking differences in the policies of the DAF with regard to milk products and to meat. The carcasses of animals infected with TB were inspected and passed fit for human consumption and entered the food chain in a raw state. Yet the milk, despite processing into cheese, was to be withheld from sale to the public and would have been sent for destruction without this court case. While he did not elaborate, it is clear that there is a striking contradiction in a food safety policy that delegates to consumers the responsibility to ensure that meat is safe for eating by cooking it at a sufficiently high temperature for an adequate period of time, yet they are not permitted to buy a product in which the milk from many animals has been transformed through a microbiological and chemical process that has been scientifically proven to render the milk safe for consumption.[17]

Within a couple of days of Judge Finn's ruling, the FSAI issued a Prohibition Order on the same cheese that had just been released from the Detention Order. This was a new development as hitherto the dispute over the safety of the cheese had only involved the DAF. Indeed, WCNC had met Dr Patrick Wall, then head of the FSAI in September 2002, who financed a risk assessment of the detained cheese by an expert team that reported it safe. Moreover, in July 2003 the FSAI published a report, prepared by its scientific committee, which specifically questioned the policy of withdrawal from trade of cheese made from raw milk produced from an 'officially tuberculosis free' herd but in which there was subsequent detection of new positive tuberculin tests in the herd (as in the case study here). It stated that this "is unlikely to make a significant contribution to the protection of public health" (FSAI 2003: 14). This apparent division between the

---

[17] Interestingly, Sir John Krebs, then Chair of the Food Standards Agency in the UK, in his Wooldridge Lecture on food safety and animal disease in September 2003 noted: "Meat from infected carcasses may currently be sold for human consumption, provided that there is no more than one localised lesion. When the FSA Board looked at our policy about 18 months ago, it asked for more research on the levels of *M. bovis* in animals salvaged for human consumption. The preliminary results, which are still unpublished, show *levels were higher than expected in reactor animals with no visible lesions*. ACMSF's [The Advisory Committee] provisional view is that our results do not change the current risk assessment (which is based on human cases rather than affected cattle). However, with the rise in bovine TB incidence, the risk in the future could increase, so the Agency will consider whether or not we need to do more to manage the risk" (Krebs 2003; emphasis added).

FSAI and the DAF in their interpretation of food safety was reinforced in the District Court hearings by comments made by a DAF staff member who, when quizzed about the disparity between the policy of the FSAI and the DAF responded, "The Department is not dictated to by the Food Safety Authority, the Department is more stringent". Yet clearly, the difficulties that the DAF found itself in with regard to this case meant that it had to enrol the FSAI, an overture also made possible by the departure of Dr Patrick Wall and his replacement by a business executive from a major transnational food corporation.[18]

Unsurprisingly, WCNC appealed the Prohibition Order, which would then have returned to the District Court. However, the legal team representing the FSAI sought to have Judge Finn disqualified from the hearing, arguing that he was too closely involved with the other party. Eventually when the case appeared in court the appeal was not contested and the order was cancelled; the FSAI were left to pay €30 000 in costs. When asked by the *Irish Farmers Journal* (2005: J2 5) how it could defend these actions, their reply was thus: "The FSAI believed that it was acting in the best interests of protecting consumer health". Following this, the DAF then appealed the ruling on the cancellation of the Detention Order to a higher (Circuit) court. The two hearings took place in December 2004 before Judge O'Donnabhain and much of the evidence in court largely covered the same issues as during the District Court hearings. However, cross-examination of the Deputy Chief Executive of the FSAI provided a revealing insight into the values-driven science of the regulatory system. When asked about challenge testing, involving laboratory-based experiments to determine the survival of *M. bovis* through the cheese-making process, he replied that it would be the responsibility of the producer in order to demonstrate safety. The FSAI, he said: "would not have the money".

At the close of the second hearing Judge O'Donnabhain provided his judgement and, in doing so, condemned the scientific credibility of the DAF's case:

> I am convinced the Department are moving with a conviction to which all science must bend. There is a considerable gap in their science in what happens with the milk and the cheese-making process. The Department's evidence is defective in relation to this cheese-making process and is based entirely on speculation and the need to get the science to match the conviction that the cheese is contaminated. On the totality of the evidence and the weighing of the evidence I prefer the evidence of (names of two witnesses called by WCNC). I am convinced having considered same that there is no risk to the human population in eating the cheese. I confirm cancellation of the order. I dismiss the appeal and affirm the order of the District Court.[19]

Table 12.1 summarises the contrasting positions of the two sides in the case.

---

[18]This seems to bear out Taylor's (2003) argument that despite the rhetoric and fanfare heralding a new and 'independent' authority responsible for ensuring food safety in Ireland, the FSAI demonstrates rather more in the way of continuity in the way risk is managed and in political access by influential agribusiness interests.

[19]Drawn from the transcript of the hearing at Cork Circuit Court (13 December 2004).

Table 12.1: Contrasting attitudes expressed in court represented as a hygienist approach to food safety and a qualitative approach to product quality.

Hygienist approach (food safety)	Qualitative approach (product quality)
Unsafe unless proven safe (by the producer)	Safe unless proven unsafe (by the authorities)
Number of positive reactors (one sufficient to condemn cheese)	Extent of *M bovis* infection in the cattle (head and lungs not supermammary glands)
Only pasteurisation of milk makes cheese safe	Complex microbial ecology of cheese (pH, salt, heat, maturation time) as effective
TB-free status of herd: cut-off date is last clear test	Status changes on current positive test (stop using milk for RMC)
Evidence of fecal contamination in milking parlour indicates poor hygiene	Low total bacterial counts in milk evidence of a clean operation
Use of MAP as surrogate measure: Cannot extrapolate from one to another; a single research paper can be fallible; trust the food safety experts	Reasonable to extend the behavioural characteristics of one microorganism to another with a high level of similarity
Absence of proof not proof of absence	Lack of presence implies absence

The hygienist approach is based on comments made by the Counsel and witnesses for the Department of Agriculture and Food (DAF). The qualitative approach is based on comments made by the Counsel and witnesses for WCNC (the raw milk cheese business).

Conclusion

As several of the other chapters in this book demonstrate, there has been much enthusiastic interest in the growth during recent years of 'alternative' food networks characterised by a commitment to quality. Yet relatively little attention has been given to the considerable burden imposed upon small artisan producers by food safety regulations. To some degree this may be due to an unconscious acceptance that the catastrophic mishandling of the BSE episode, as well as other food safety crises, led inevitably to the need to create a stronger regulatory regime. However, as the chapter has demonstrated, the contemporary food safety paradigm, with its attendant preoccupation with microbiological standards, has created a highly interventionary and bureaucratic framework by which to police the food system in ways that make it seem more hygienic.

The paradox at the heart of this hygienist food safety paradigm, of course, is the failure to have in place an effective system of traceability of all ingredients used in manufactured foods produced for a mass market such that toxic contaminants, like Sudan 1, would not find their way into several hundred different food products lining our supermarket shelves (Sage 2005). Meanwhile, its 'policing' serves to impose higher regulatory costs on small food producers who have been marginal to the causes of health risks in the first place.

As the chapter established early on, what most effectively underpins the paradigm and legitimates the actions of agencies responsible for food safety is the recurring mantra of 'sound science'. Yet Pielke and Rayner (2004: 355) remind us that, "science is the trump card that we play in disputes about values". Moreover, a distinguished professor of food science notes that:

> The appeal to 'science' will not necessarily resolve disputes. We might assume science speaks a universal language of truth, but it does not. Scientific knowledge is especially contested in such complex domains as human health. Citizens often ask questions to which science can have no answers, which simply highlights that scientific risk assessments often are forced to make implicit value judgements to come to a conclusion (Kachatourians 2001: 21).

The use of scientists and their expertise as a political resource was noted earlier, with reference to van Zwanenburg and Millstone's (2003) analysis of how policy decisions on BSE subsequently sought scientific endorsement. Such expertise can also be used as a means to strengthen cases before courts required to adjudicate on matters of food safety and public health. As Buonanno *et al.* (2001) note, a consequence of this is to privilege those actors more easily able to access technical expertise and with the means to press their case by submitting expensive scientific evidence. Considering this statement in light of the case described earlier demonstrates the considerable obstacles that had to be overcome and the efforts needed in order for WCNC to be successful in securing the cancellation of the detention order.[20]

The chapter has demonstrated that hygienist food safety regulations are significantly at variance with the character of the hazards presented by raw milk cheese produced according to recognised standards and procedures. This was the view also taken by two independent members of the judiciary. Of course, it might be argued that the over-riding commitment of the DAF to protect public health required them to seek detention of the cheese in the absence of scientific proof but with full regard to the precautionary principle. Were this to be the case then the process of risk assessment of the cheese would be conjoined with a re-evaluation of scientific knowledge and policies: the 'doubting' and the 'acting' would proceed concurrently (Dratwa 2002). Evidently, this was not the case, with Judge O'Donnabhain concluding from the evidence that the DAF were bending the science to match their own convictions that the cheese was unsafe.

Clearly, there is a need for a food safety regime to protect public health but one that avoids the implementation of a risk management model preoccupied with achieving 'no-germ' food. While insisting on the highest technical standards appropriate to the scale of production and the implementation of good practice in quality management, a revised food safety paradigm, informed by a precautionary approach, would acknowledge the competence, craftsmanship and vernacular knowledge of farmhouse cheesemakers and others in the supply chain. Of course, this approach would inevitably challenge the dominance of the industrial convention of quality based on price and standardisation and might help introduce other tests of quality involving place, ethics, justice and inspiration.

---

[20] Although the direct legal costs of representation in the court cases were awarded to WCNC, there is no provision under the law for compensation. The significant indirect costs, including reduced cash flow arising from the detention of the cheese, has seriously compromised the commercial prospects of the company in the short to medium term.

Above all, it might even help recover an appreciation for the sensual, nutritional and biological attributes of food.

Acknowledgements

The research on which this paper is based is supported under the Higher Education Authority's Programme for Research in Third Level Institutions, Cycle 3. I would particularly like to thank Bill Hogan and Jeffa Gill for their reactions to an earlier draft, Ursula Kilkelly for legal advice and the constructive comments of an anonymous reviewer. However, I take full responsibility for all matters of fact and interpretation expressed here.

References

Ansell, C. and Vogel, D. (2006). The contested governance of European food safety. In C. Ansell and D. Vogel (Eds), *What's the Beef? The Contested Governance of European Food Safety* (pp. 3–32). Boston, MA: MIT Press.

Barling, D. (2004). Food agencies as an institutional response to policy failure by the UK and the EU. In M. Harvey, A. McMeekin and A. Warde (Eds), *Qualities of Food* (pp. 108–128). Manchester: Manchester University Press.

Barling, D. and Lang, T. (2003). A reluctant food policy? The first five years of food policy under Labour. *The Political Quarterly* 74 (1), 8–18.

Beck, U. (1992). *The Risk Society: Towards a New Modernity*. London: Sage.

Buonanno, L. (2006). The creation of the European Food Safety Authority. In C. Ansell and D. Vogel (Eds), *What's the Beef? The Contested Governance of European Food Safety* (pp. 259–278). Boston, MA: MIT Press.

Buonanno, L., Zablotney, S. and Keefer, R. (2001). Politics versus science in the making of a new regulatory regime for food in Europe. *European Integration Online Papers* (EIoP) 5 (12). Available online at: http://eiop.or.at/eiop/texte/2001-012a.htm (accessed 6/6/2006).

Country Life (2005). Against all odds. Available online at: http://www.countrylife.co.uk/lifecountry/food/lanarkblue.php (accessed 20/12/2005).

Cunynghame, A. (2001). British cheese makers under threat. *The Ecologist Report*, June, 18–20.

Dixon, P. (2000). *European Systems for the Safe Production of Raw Milk Cheese*. Report presented to the Vermont Cheese Council, Vermont, New England.

Donnelly, C. (2001). Factors associated with hygienic control and quality of cheeses prepared from raw milk: a review. *Bulletin of the International Dairy Federation* 369, 16–27.

Dratwa, J. (2002). Taking risks with the precautionary principle: food (and the environment) for thought at the European Commission. *Journal of Environmental Policy and Planning* 4, 197–213.

Enticott, G. (2001). Calculating nature: the case of badgers, bovine tuberculosis and cattle. *Journal of Rural Studies* 17, 149–164.

Enticott, G. (2003a). Lay immunology, local foods and rural identity: defending unpasteurised milk in England. *Sociologia Ruralis* 43 (3), 257–270.

Enticott, G. (2003b). Risking the rural: nature, morality and the consumption of unpasteurised milk. *Journal of Rural Studies* 19 (4), 411–424.

Feenstra, G. (1997). Local food systems and sustainable communities. *American Journal of Alternative Agriculture* 12 (1), 28–36.

Flynn, A., Marsden, T. and Smith, E. (2003). Food regulation and retailing in a new institutional context. *The Political Quarterly* 74 (1), 8–18.

Food Safety Authority of Ireland (FSAI) (2003). *Zoonotic Tuberculosis and Food Safety*. Dublin: FSAI.

Food Safety Authority of Ireland (FSAI) (2005a). *Report on Zoonoses in Ireland: 2002 and 2003*. Dublin: FSAI.

Food Safety Authority of Ireland (FSAI) (2005b). *On Farm Study of Consumption of Unpasteurised Milk*. See: www.fsai.ie/publications/other/unpasteurised_milk.asp (accessed 30/11/2005).

Fox, P. and McSweeney, P. (2004). Cheese: an overview. In P. Fox, P. McSweeney, T. Cogan and T. Guinee (Eds), *Cheese: Chemistry, Physics and Microbiology: General Aspects, Volume 1* (pp. 1–36). London: Elsevier.

Goodman, D. (2004). Rural Europe redux? Reflections on alternative agro-food networks and paradigm change. *Sociologia Ruralis* 44 (1), 3–16.

Grappin, R. and Beuvier, E. (1997). Possible implications of milk pasteurisation on the manufacture and sensory quality of ripened cheese. *International Dairy Journal* 7, 751–761.

Griffin, J., Hahesy, T., Lynch, K., Salman, M., McCarthy, J. and Hurley, T. (1993). The association of cattle husbandry practices, environmental factors and farmer characteristics with the occurrence of chronic bovine tuberculosis in dairy herds in the Republic of Ireland. *Preventive Veterinary Medicine* 17, 145–160.

Harvey, M., McMeekin, A. and Warde, A. (2004). Conclusion: quality and processes of qualification. In M. Harvey, A. McMeekin and A. Warde (Eds), *Qualities of Food* (pp. 192–207). Manchester: Manchester University Press.

Hennessy, D., Roosen, J. and Jensen, H. (2003). Systemic failure in the provision of safe food. *Food Policy* 28, 77–96.

Ilbery, B and Kneafsey, M. (1999). Niche markets and regional speciality food products in Europe: towards a research agenda. *Environment and Planning A* 31, 2207–2222.

Irish Farmers Journal (2005). David and Goliath. *Farmers Journal* 2, 2 April, 4–5.

Kalantzopoulos, G. (1999). Microbial ecology of traditional raw milk cheeses. In *Proceedings of the Symposium on Quality and Microbiology of Traditional and Raw Milk Cheeses* (pp. 39–52), 30 November and 28 December 1998, Dijon, France. COST Action 95: Improvements of the Quality of the Production of Raw Milk Cheeses. Office for Official Publications of the European Communities, Luxembourg.

Khachatourians, G. (2001). How well understood is the 'science' of food safety? In P. Phillips and R. Wolfe (Eds), *Governing Food: Science, Safety and Trade* (pp. 13–23). Montreal: McGill-Queen's University Press.

Krebs, J. (2003). The Wooldridge Lecture on food safety and animal disease. Food Standards Agency, UK. Available online at: http://www.food.gov.uk/news/newsarchive/2003/sep/wooldridgelecturenews (accessed 30/11/2005).

Lang, T. and Rayner, G. (2003). Food and health strategy in the UK: a policy impact analysis. *The Political Quarterly* 74 (1), 66–75.

Macnaghten, P. and Urry, J. (1998). *Contested Natures*. London: Sage.

MAFF (Ministry of Agriculture, Fisheries and Food) (2000). *TB in Cattle: Protecting Human Health. Fact Sheet 2*. London: MAFF.

Marsden, T. (2003). *The Condition of Rural Sustainability*. Assen, The Netherlands: Royal van Gorcum Press.

Marsden, T., Flynn, A. and Harrison, M. (2000). *Consuming Interests: The Social Provision of Foods*. London: UCL Press.

McSweeney, P. and Sousa, M. (1999). Biochemical pathways for the production of flavour compounds in raw and pasteurised milk cheeses during ripening. In *Proceedings of the Symposium on Quality and Microbiology of Traditional and Raw Milk Cheeses* (pp. 73–126), 30 November and 28 December 1998, Dijon, France. COST Action 95: Improvements of the Quality of the

Production of Raw Milk Cheeses. Office for Official Publications of the European Communities, Luxembourg.
Murdoch, J., Marsden, T. and Banks, J. (2000). Quality, nature and embeddedness: some theoretical considerations in the context of the food sector. *Economic Geography* 76 (2), 107–125.
Nestle, M. (2003). *Safe Food: Bacteria, Biotechnology and Bioterrorism*. Berkeley, CA: University of California Press.
Peltola, J. and Schalin, M. (2003). Food safety regulation: freedom of choice vs. safety regulation. In *Proceedings 4th Congress of the European Society for Agricultural and Food Ethics* (pp. 120–123). Toulouse: INRA/EURSAFE.
Phillips, P. and Wolfe, R. (2001). Governing food in the 21st century: the globalization of risk analysis. In P. Phillips and R. Wolfe (Eds), *Governing Food: Science, Safety and Trade* (pp. 1–10). Montreal: McGill-Queen's University Press.
Pielke, R. and Rayner, S. (2004). Editors' introduction. *Environmental Science and Policy* 7, 355–356.
Rance, P. (1989). *The French Cheese Book*. London: Macmillan.
Renting, H., Marsden, T. and Banks, J. (2003). Understanding alternative food networks: exploring the role of short food supply chains in rural development. *Environment and Planning A* 35 (3), 393–411.
Rowell, A. (2003). *Don't Worry its Safe to Eat: The True Story of GM Food, BSE and Foot and Mouth*. London: Earthscan.
Sage, C. (2003). Social embeddedness and relations of regard: alternative 'good food' networks in south-west Ireland. *Journal of Rural Studies* 19, 47–60.
Sage, C. (2005). Food for thought. *Irish Times*, 28 June 2005.
Schällibaum, M. (1999). Hygienic and legal aspects of raw milk cheeses. In *Proceedings of the Symposium on Quality and Microbiology of Traditional and Raw Milk Cheeses* (pp. 31–38), 30 November and 28 December 1998, Dijon, France. COST Action 95: Improvements of the Quality of the Production of Raw Milk Cheeses. Office for Official Publications of the European Communities, Luxembourg.
Sexton, R. (1998). *A Little History of Irish Food*. London: Kyle Cathie Ltd.
Skogstad (2006). Regulating food safety risks in the European Union: a comparative perspective. In C. Ansell and D. Vogel (Eds), *What's the Beef? The Contested Governance of European Food Safety* (pp. 213–236). Boston, MA: MIT Press.
Spahr, U. and Schafroth, K. (2001). Fate of *Mycobacterium avium* subsp. *paratuberculosis* in Swiss hard and semi-hard cheese manufactured from raw milk. *Applied and Environmental Microbiology* 67 (9), 4199–4205.
Taylor, G. (2003). 'From the curry to the slurry': the politics of food regulation and reform in Ireland. *Irish Studies in International Affairs* 14, 149–164.
Unnevehr, L. and Jensen, H. (1999). The economic implications of using HACCP as a food safety regulatory standard. *Food Policy* 24, 625–635.
van der Ploeg, J.D., Long, A. and Banks, J. (2002). *Living Countrysides: Rural Development Processes in Europe: The State of the Art*. Amsterdam: Elsevier.
van Zwanenberg, P. and Millstone, E. (2003). BSE: a paradigm of policy failure. *The Political Quarterly* 74 (1), 27–37.
Weston Price Foundation (2005). *A Campaign for Real Milk*. Available online at: http://www.realmilk.com (accessed 12/12/2005).
Wheelock, V. (1997). *Raw Milk and Cheese Production: A Critical Evaluation of Scientific Research*. Skipton: Verner Wheelock Associates.

Chapter 13

Market-Oriented Initiatives for Agri-Environmental Governance: Environmental Management Systems in Australia

Vaughan Higgins[1], Jacqui Dibden[2] and Chris Cocklin[3]
[1]School of Humanities, Communications and Social Sciences, Monash University, Australia
[2]School of Geography and Environmental Science, Monash University, Australia
[3]Faculty of Science, Engineering and Information Technology, James Cook University, Australia

Introduction

A range of environmental and social problems are concomitants of conventional high-input/high-output approaches to the production of food, including land and water degradation, falling farm incomes and exits from farming (e.g. Buttel *et al.* 1990; Goodman and Watts 1997; Gray and Lawrence 2001; Cocklin 2005). The externalisation of environmental and social impacts has, in many countries, prompted the state to correct this market failure through various forms of support to farmers and the farming sector more broadly. In Europe, this support has been manifested in a range of agri-environmental programmes aimed at promoting multi-functional landscapes (see Ilbery and Bowler 1998; Wilson 2001; Potter and Burney 2002). Australia has taken a somewhat different approach. Consistent with the neoliberalist promotion of entrepreneurial and 'self-reliant' farmers (Higgins 2002), state agencies have sought to facilitate participatory and 'bottom-up' solutions that address the negative environmental consequences associated with productivist agriculture. The National Landcare Program (NLP) and Property Management Planning (PMP) represent the main vehicles of agri-environmental governance (Campbell 1994; Curtis and De Lacy 1996; Lockie and Vanclay 2000).

However, while state-based approaches to agri-environmental management in Europe and Australia are broadly divergent in their philosophy and techniques, both appear to be faced by a similar regulatory contradiction. As Dibden and Cocklin (2005) argue, the problem in Europe is how to sustain a multi-functional agriculture under the liberalised market regimes encouraged through the WTO and reform of the Common Agricultural Policy (Potter and Burney 2002), while Australia is grappling with the opposite problem – how to combine an already liberalised economy with the need to move towards more sustainable land management and viable rural communities.

Under these circumstances, market-oriented initiatives (MOIs) for sustainable food production appear to be an attractive means for governments to arrest problems that existing agri-environmental schemes have limited capacity to address. The purpose of this chapter is to examine how one MOI – Environmental Management Systems (EMS) – is being promoted by Australian governments as *the* way forward for improving natural resource management and farm viability in Australia. While broadly consistent with the 'conventional' productivist approach dominant in Australian agricultural policy, we argue that EMS represents a notable alternative through its attempts to address agri-environmental management as an integral part of farm production. This is not dissimilar from MOIs in Europe where the market is used increasingly "to reconcile agricultural production and environmental protection" (Buller and Morris 2004: 1067) and to promote more sustainable rural spaces. Nevertheless, the extent to which EMS both differs from, and is complementary to, conventional production is a key issue addressed in this chapter. We draw upon our research on a pilot project – the Gippsland EMS – operating in the Gippsland region in the state of Victoria to explore the implementation of EMS at the farm level. A focus on farm-level change is crucial in not only assessing the *quality* of environmental benefits (Wilson and Hart 2001), but also in understanding the often complex ways in which farmers 'translate' programmes into practice (Higgins 2006).

MOIs and Agri-Environmental Governance

MOIs are initiatives in which the "incentive for food producers to manage the environment positively comes directly through the harvesting of market benefits" (Buller and Morris 2004: 1067). These initiatives mark an attempt to move away from conventional modes of food production and their associated environmental and social problems. MOIs are associated with a broader shift from the mass production of standardised commodities towards new *differentiated* markets based around notions of 'quality', 'nature', 'place' and 'sustainability' (Marsden and Arce 1995; Murdoch and Miele 1999; Goodman 2003; Renting et al. 2003; Marsden and Smith 2005; Marsden and Sonnino 2005). In a global environment characterised by increasing consumer concern over food safety and quality, and growing political and social interest in sustainability, MOIs have attracted considerable interest from government, industry and farming organisations. This is the case in Europe particularly, where MOIs have been promoted as key strategies for a safer and more 'natural' alternative to conventional mass-produced food while at the same time providing broader environmental benefits in agri-food production (e.g. Holloway and Kneafsey 2000; Ilbery and Kneafsey 2000; Morris 2000; Renting et al. 2003; Buller and Morris 2004; Marsden and Sonnino 2005).

European scholars argue that MOIs represent a strategy for reconnecting domestic consumers to food production and enhancing a 'multi-functional' countryside (Ilbery and Bowler 1998; Wilson 2001; Potter and Burney 2002; Marsden and Sonnino 2005). In addition, MOIs are viewed as part of a multi-level and multi-interest system of regulation aimed at improving the capacity of agri-food industries not only to pre-empt environmental regulation but also to avert future crises such as food scares (Morris 2000; Morris and Winter 2002; Smith *et al.* 2004). This interest within Europe and elsewhere in improved food safety and quality has repercussions for Australian agriculture, which is highly sensitive to fluctuations in consumer preferences in the major markets for its produce. The promotion of MOIs in the Australian context has focused particularly on the pre-empting of regulation within *overseas* markets rather than domestically; proponents argue that MOIs are likely to assist in improving and assuring continued access to these markets (Dibden and Cocklin 2005). In other words, MOIs are regarded within Australia as a means of improving the efficiency of (productivist) export-driven agriculture.

Both government and industry are increasingly interested in MOIs as a means of: demonstrating the 'clean and green' credentials of Australian agri-food commodities in response to the threat of cheap imports from Asia (e.g. Mech 2004a); pre-empting private regulation, particularly by European retailers (DAFF 2002); and providing a positive framework of incentives for farmers to manage the environment on behalf of the community[1] (e.g. VFF and Victorian Government 2003; Mech 2004b). To ensure that agri-food commodities do indeed meet these credentials, Australian MOIs have been linked to internationally recognised standards and systems of auditing. In this paper, we explore EMS as an exemplary case of how one particular MOI has been singled out by governments and other authorities as *the* way forward in: (1) promoting improved natural resource management; (2) improving competitiveness through better market access; and (3) better meeting community expectations regarding responsible environmental management (NRMMC 2002). We are particularly interested in examining why producers become involved in EMS, whether the use of an EMS challenges productivist farming practices and the limitations of EMS as a tool of agri-environmental governance.

EMS is of considerable interest to the Australian government because it integrates a range of environmental, food quality, trade and farm management issues (Mech and Young 2001; NRMMC 2002; Mech 2004a,b; Carruthers 2005). It is claimed to:

- improve farm productivity while minimising negative environmental externalities;
- address natural resource management at a range of scales;
- improve competitiveness and market access at an industry level through the verification of 'clean and green' credentials; and
- help primary producers to meet market demands for quality and environmental assurance.

---

[1] While various participatory agri-environmental schemes, such as Landcare and Property Management Planning (PMP), have sought to address the adverse effects associated with the 'productivist' approach (e.g. Lockie and Vanclay 2000; Gray and Lawrence 2001), they have been criticised for not going far enough in reversing widespread environmental damage or in providing broader incentives for farmers to undertake environmental work (Higgins and Lockie 2002). Additionally, there has been the expectation that farmers in these schemes provide positive off-site environmental benefits while internalising the associated costs (Lockie and Higgins 2007).

Underpinned by the international standard ISO14001, it is a process designed to:

> ... assist with the recognition of environmental impacts, compliance issues, risks and liabilities; develop an environmental policy that clearly states the aspirations, methods and timeframes to address these impacts; gather and refine the information and data needed to meet the policy aspirations; and, review and evaluate management choices in light of whether the outcomes achieved met those specified within the environmental policy (Carruthers 2005: 1).

Recent promotion by the federal government indicates that EMS is poised to become a key component of natural resource management policy. For instance, a National Framework for Environmental Management Systems in Agriculture was developed in 2002 by ministers for agriculture and natural resource management from across the country. This framework is compatible with ISO14001 to enable external auditing, and thereby verification of farmer compliance.

As part of the encouragement of EMS, the Australian government launched an AUS$8.5 million National Pilot Program in April 2003 "to test and enhance the potential of EMS as a business management tool for primary production" (DAFF 2004). Funding was provided for 16 pilot projects running across Australia, representing a broad range of industries, regions and environmental conditions. While EMS is not regarded as a panacea for the woes of agriculture (see Mech and Young 2001; NRMMC 2002), it is regarded by government as providing incentives for farmers to improve their environmental and business management. EMS thus occupies a 'hybrid' regulatory space (Ilbery and Maye 2005) in working within the dominant productivist agricultural system, while at the same time seeking to 'green' and address the negative externalities created by this system of production. In the following section, we provide an analysis of the farm-level impacts of EMS through an examination of one of the federally-funded EMS pilot programmes – the Gippsland EMS (see also Figure 13.1 below).

Figure 13.1: A Gippsland EMS farm promoting its environmental credentials.

Case Study: The Gippsland EMS

Although the Gippsland EMS received federal government funding in 2003 under the EMS National Pilot Program, its history dates back to the late 1990s when members of Gippsbeef, a producer alliance, were looking for ways of meeting future market demands. Around the same time, Meat and Livestock Australia (MLA), a producer-owned company that provides services to livestock producers, processors, exporters and retailers, was searching for producers to 'test' the application of EMS. Seven of the original alliance members became involved in the MLA pilot project and spent 2 years (the duration of the pilot) developing an EMS relevant to Gippsland beef producers. In order to build the market profile of the Gippsbeef brand, the alliance changed its name to Gippsland Natural (Williams 2004). Subsequently, in 2002, Gippsland Natural applied successfully for funding to participate in the federal government's EMS National Pilot Program. The pilot project was launched in 2004 under the banner 'Gippsland EMS' with the meat marketed using the 'Enviromeat' brand. As of August 2005, there were 80 producers from 59 businesses involved in the project (O'Sullivan and Williams 2005: 3), although most were still engaged in training courses teaching procedures required for EMS certification.

It is important to note that while Enviromeat builds on the attributes marketed through the Gippsland Natural brand, they are completely separate. Whereas the Gippsland Natural brand focuses on prime pasture-fed beef that is quality assured (through Meat Standards Australia) and hormone free, Enviromeat emphasises, in addition, environmental management, protection of biodiversity and water quality and external auditing to ensure the integrity of the EMS. Enviromeat, originally sold on a trial basis through a local farmers' market, is now available through a speciality food retailer in Melbourne and butchers in two seaside towns and there are plans to expand to other gourmet food stores. Although the brand Enviromeat does not incorporate a regional association (Gippsland), authenticity of the product's local origins is communicated to consumers through informal practices such as displaying photographs of the specific farm locality and providing the names and property details of the farmers supplying the meat on sale. Additionally, profiles of all the farmers involved in meat production are displayed on the Enviromeat website (see also Figure 13.2).[2]

In order to explore how the Gippsland EMS was implemented at a farm level we adopted an actor-oriented analytical approach and used an ethnographic methodology incorporating in-depth semi-structured interviews. Interviews were conducted between March and November 2005 with 18 farm management teams throughout the Gippsland region comprising a total of 28 participants. The purpose of the interviews was to gain insights into the often complex ways in which producers act as 'active strategisers' in programmes of regulation (Long 2001; Long and Long 1992).

Participation in the EMS Process

The problems of agri-environmental schemes in enlisting participants and in contributing to long-term and widespread changes in the way that producers think about, and

---

[2] For further information see also http://www.enviromeat.com.au/suppliers.php (accessed 1/6/06).

Supplier Profile:
Bob & Anne Davie,
ENVIROMEAT Bimbadeen Brangus, Phillip Island

Bob and Anne Davie run a Brangus cattle stud on Phillip Island, one and a half hours drive southeast of Melbourne. With a commitment to responsibly managing their land and improving the biodiversity, the property "Bimbadeen" (Aboriginal for place of good view), is an excellent example of combining productivity and conservation.

Bob and Anne's commitment to the environment was recently recognised with the 2005 State Landcare Primary Producer Award.

The Enviromeat in store today is supplied from the Davie's farm.

Looking after the land and wildlife
The Davies have planted over 26,000 indigenous grasses, shrubs and trees as well as protected all existing remnant vegetation on their property. Bimbadeen is an accredited 'Land for Wildlife' property. Birdlife flock to the wetlands they have protected on the historic McHaffie property.

Land Management Challenges
Salinity is a common problem on the island and the Davies have played a leading role in conducting research over many years to try and find the best ways to manage this risk. Solutions so far have included planting native trees and shrubs, trialing salt-tolerant pasture varieties and pumping groundwater to remove saline ground water from plant roots.

Climate change is also a big issue identified by Bob and Anne. They have been exploring how they can reduce greenhouse gas emission by participating in the national Greenhouse gas challenge.

The Davies are always keen to share their knowledge and experience and have regular visits from farmer and Landcare groups and both tertiary and secondary school students and teachers.

Care for cattle
Bob and Anne have been breeding Brangus cattle for 35 years. "Bimbadeen" cattle are renowned Australia wide for their high growth rate performance and carcase quality - well proven in National Steer Trials and Carcase competitions. Their steers have won Domestic, Korean Export and Japanese Export in a combination of growth rate and carcase quality in trials throughout Australia.

Bob and Anne Davie are proud to provide you with the best quality beef so you can enjoy the taste and tenderness while knowing the cattle are well looked after, as is their land and wildlife.

- ✓ Environmentally certified
- ✓ No artificial hormones
- ✓ Free range
- ✓ Guaranteed tender

We are farming for the long term to guarantee you have the best quality meat both now and into the future.

Figure 13.2: Example of a Gippsland EMS producer profile.

therefore manage, natural resources are well documented (e.g. Winter 1997; Lobley and Potter 1998; Lockie 1999; Curtis and De Lacy 2000; Wilson and Hart 2001; Morris and Winter 2002). It is worth considering, therefore, the issues that farmers see as important in attracting them to market-oriented environmental initiatives such as EMS and how farmers who already participate rationalise their involvement.

One of the key reasons producers gave for their involvement in the Gippsland EMS was that it was consistent with their existing 'philosophy' or general approach to farming, a finding that has been observed in other studies (see, for e.g., Wilson and Hart 2001; Morris and Winter 2002). It was clear that all farmers interviewed held attitudes towards environmental issues that played a role in attracting them to the EMS. Little change was needed in order to integrate EMS into existing practices. For instance, farmers stated that EMS was "along the lines of what I believe in" (F03-m),[3] and "the EMS was really just spelling out the philosophy that I already had" (F04-m). When asked why they became involved, others simply stated that they "were doing it anyway" (F08-m). However, farmers did not necessarily identify themselves as environmentalists. Most combined an approach to farm management that was sympathetic to the environment with an economic 'rationality'. The combination of reasons for participation accords closely with the 'active adopters' described in Wilson and Hart's (2001: 263) study of British agri-environmental schemes where "financial reasons or ease of fit with management plans are important but where conservation reasons are equally important in [farmers'] decision to participate". The influence of existing attitudes and practices is further evident in the high participation rates of Gippsland Natural members in EMS, with many farmers stating that the environmental focus of EMS was a logical extension of the 'quality' issues covered in Gippsland Natural, enabling farmers to prove that their meat was not only 'clean' but also 'green'.

Pre-empting regulation provided a second rationale for participation in the Gippsland EMS. With greater direct regulation of environmental activities in Australia (Lockie and Higgins 2007) and private retailers increasingly setting their own standards for environmental 'best practice' (Morris 2000; Busch and Bain 2004; Mutersbaugh 2005; Campbell et al. 2006), many farmers observed that EMS would be a crucial way to pre-empt tough regulations anticipated in the future. As one farmer stated:

> It's a way for us as farmers under this system to develop our rules to some extent, so we're really, you know, establishing the way that we're going to manage our activities ourselves, rather than having those sorts of other people imposing regulations on us (F10-m).

Underpinning these sentiments was the concern that tougher environmental regulations would contribute to a possible loss of market access. Thus, as one farmer pointed out,

---

[3] For the purposes of reporting, the 18 farm management teams whom we interviewed are allocated a code ranging from F01 to F18. The gender of the informants is signified by an 'm' (male) or 'f' (female) following the code (e.g. F01-m). Not all of the farm management teams whom we interviewed are drawn upon in this chapter.

"there's a trend towards quality assurance and environmental responsibility and maybe if we're not heading in that path we'll lose access to certain markets" (F13-m).

EMS were also regarded as a means of deflecting public criticism of farming activities. With 'the rural' assuming prominence as a multi-functional space (e.g. Potter and Burney 2002), pressure by non-governmental organisations on producers and retailers for environmentally sound food production (Freidberg 2004; Lang and Heasman 2004) and community expectations that farmers will provide responsible environmental stewardship of the land (see Marsden 2003; Lockie and Higgins 2007), EMS might be viewed as one way in which to not only act 'responsibly' but to demonstrate this in a formal manner. Thus:

> Sometimes farmers are seen, I think, as the bad guys, the way we manage things, and maybe the EMS is also a way for us to show that, you know, we're not the bad guys and we've got systems in place to do the best we can (F10-m).

Other farmers emphasised the significance of EMS in responding to the increased demand by educated consumers for differentiated products where claims of 'quality' and environmental assurance can be supported. For example:

> ... whenever I see things on the horizon, always, you know, lights go up and bells ring and you think that's the way to go because I do think that ... the consumer is going to become more and more conscious or mindful of the product that they're purchasing (F01-f).

The 'quality turn' is a trend in food production–consumption that is already well established in European research (e.g. Marsden and Arce 1995; Ilbery and Kneafsey 2000; Murdoch *et al.* 2000; Goodman 2003). In the case of an agricultural exporter such as Australia, which has historically relied on the production of undifferentiated bulk commodities, the turn to quality presents implications for existing practices. However, interestingly, the farmers we interviewed see EMS as a strategy for meeting a niche in the domestic and local markets rather than ensuring that they do not lose access to international markets. The question to be addressed in the following section of the article is whether EMS challenges productivist practices.

EMS and Productivist Farming Practices

As an emerging strategy of natural resource management in Australia, EMS is expected to assist farmers in placing environmental issues at the centre rather than remaining on the periphery of farm management (NRMMC 2002; Mech 2004a). This is assumed to occur through market incentives to encourage environmentally sustainable production. On this basis, it is worth reflecting on how EMS might 'green' those productivist farming practices that have contributed to environmental degradation. In other words, if environmental considerations are built into otherwise 'conventional' production, does this provide the basis for fundamental changes in farming practices?

The research found that EMS does indeed contribute to positive changes in how farmers think about, and conduct, environmental management.[4] As a formal process, EMS was regarded as useful in "formulating what you're doing and documenting it" (F04-m) and prioritising "some of the projects that we needed to carry out to manage our business" (F10-m). For one farmer, EMS increased the level of personal accountability, making him "more disciplined in when you're going to do things and if you don't do them, then you have to write down and justify why you're doing it, which is a discipline I don't mind" (F12-m). While record-keeping and calculation may seem like relatively mundane aspects of farm management, Higgins (2005) has recently argued that they are important tools of agri-governance. Drawing upon the Foucauldian-inspired literature on governmentality, Higgins (2005: 129) notes that these 'technologies' of governing can have far-reaching and productive effects in rendering "certain farming practices visible, and, through statistical representations, encouraging farmers to reflect on their management and planning in an advanced liberal way". Similarly, EMS might also be viewed as a technology of governing that, in making certain aspects of farm management visible, provides a mechanism for farmers to reflect on, change and self-regulate their conduct in new ways.

Since farmers involved in the Gippsland EMS were already engaged in a range of environmentally beneficial practices, the 'greening' of a productivist style of farming may not seem significant. However, the main benefit was in terms of questioning existing practices and providing the calculative tools for re-assessing these, a finding documented elsewhere (e.g. Higgins 2005). This contributed to real benefits for some farmers. For instance, one farmer claimed the EMS was useful in helping him to recognise the "natural limits" of production and "basically maintaining a bigger buffer zone between you and the environment as in not stretching everything to the limit" (F14-m). The EMS process thus provided tools for farmers to formalise, reflect on, prioritise and act on resource management issues. In fact, it resulted in concrete changes in decision-making in areas such as land use (e.g. revegetating degraded land), environmental issues (e.g. greater focus on native vegetation management) and stocking rates. Nevertheless, the fact that almost all the farmers we interviewed were already engaged in some form of environmental management prior to undertaking an EMS begs the question of how effective EMS actually is as a mechanism of agri-environmental governance.

The Limitations of EMS as a Tool of Agri-Environmental Governance

Since EMS is promoted as a way forward for natural resource management in Australia and is in its formative stages of adoption by farmers, our research sought to examine its limitations as a tool of agri-environmental governance. This task is of particular significance given the claimed capacity of EMS to integrate the often-competing concerns for farmers of environmental management and financial viability. A number of 'barriers' to adoption of EMS have already been documented by other authors (e.g. Mech and

---

[4]There are close similarities between our findings in this section and Morris and Winter's (2002) research on farm biodiversity action plans in the United Kingdom.

Hugo 2004; Carruthers 2005). The relevance of these barriers to the existing research is considered below.

EMS is relatively new to the agricultural sector in Australia, and therefore the benefits in terms of improving market access or providing a premium to producers are considered elusive (e.g. Mech and Hugo 2004), although some 'success stories' are told of individual farm businesses that have gained marketplace benefits from using an EMS (Carruthers 2003). This lack of demonstrable market benefits was not evident in our research since Enviromeat suppliers were gaining a premium of up to 35% above conventional market prices. However, the more fundamental issue for the farmers whom we interviewed was the small number of suppliers, due to the lack of significant consumer demand – hence only a handful of producers actually received a premium. At the time of the interviews, negotiations for the supply of Enviromeat were occurring with a number of city butchers. As a consequence, there existed little demand for the product. While the lack of existing consumer demand was recognised by the participants in our research as an issue, this was not regarded as a significant barrier to adopting an EMS. In the short term, farmers were prepared to use EMS as one of a range of mechanisms for putting their farming philosophy into practice as well as providing a mechanism for 'quality' meat production and selling to niche domestic markets. In other words, marketplace benefits, while considered important, were not necessarily the core drivers of EMS adoption. As one farmer noted:

> Well, there's been no financial gain, and how do you measure mental gain, or satisfaction? How do you measure that? . . . You do it because of your philosophy. You can't feed, you can't eat your philosophy, but if we were just here for the money, we wouldn't be doing it probably (F07-f).

The reason that a lack of short-term financial benefits did not represent a significant barrier could be that the Gippsland EMS attracted farmers who already had a strong commitment to addressing agri-environmental issues.[5] Gippsland EMS was clearly 'preaching to the converted' to some extent. The capacity of EMS to be adopted by farmers more broadly is likely to depend on how well it 'sells' production, rather than just environmental, benefits (see, for e.g. Lockie 1999).

A second limitation referred to was that many farmers were interested in EMS as a planning tool, but did not necessarily wish to pursue full ISO certification. Although almost all farmers had experienced a (second party) peer audit and many had undergone external (third party) auditing, full ISO14001 certification was not necessarily regarded as the most desirable outcome from the EMS process. One of the reasons for this was a lack of existing marketplace benefits for EMS accreditation, and specifically for Enviromeat – the eco-label which farmers are permitted to use once they gain, and maintain, full ISO certification. In the absence of any financial benefits, the 10 farmers interviewed who were members of both Gippsland EMS and Gippsland Natural, generally found the latter to be a far more reliable way of selling meat. As one farmer stated:

---

[5] Wilson and Hart (2001) make a similar argument.

You don't have to have the EMS. It's definitely much easier, yes, to meet the requirements [of Gippsland Natural]. That's why, if we get the Enviromeat going, it has to be at significantly increased margins to make it worthwhile because, yes, there is definitely increased time, cost involvement (F06-f).

Other farmers who were not members of Gippsland Natural also questioned the necessity of EMS accreditation in obtaining a premium price. For example, if a farmer practising an EMS produces a good quality 'beast':

... you finish up with a good price, but you know, we don't need an EMS programme to tell us that. We get a very sharp reminder of whether it's any good or not when the buyers look it over (F08-m).

From the perspective of many producers, EMS worked better as a process for environmental planning and for business and risk management rather than as a means of marketing meat. This is not necessarily a limitation, and in fact might be viewed as a positive outcome. According to one farmer:

... it opens your eyes to other potential risks that are out there from an environmental point of view and really helps you getting down to a bit of paper and a computer and have a look and do some assessing (F09-m).

Another farmer, who was also a member of Gippsland Natural, stated that:

It's a system, really, that we haven't really got with the Gippsland [Natural] – so the Gippsland Natural bit is ... more a marketing alliance, where this is more like a Whole Farm care program, if you like (F10-m).

A third limitation noted by farmers is a fear of increased paperwork. However, Carruthers (2005) found in her research that farmers believed that record-keeping was useful, particularly in providing 'proof' of good environmental stewardship. Similarly, the farmers interviewed found the formalisation of their beliefs and practices through the EMS to be a worthwhile exercise. The issue is the *increased* amount of paperwork associated with the auditing process. Despite the value of the EMS process, all farmers interviewed found the paperwork both onerous and, in the eyes of many, unnecessarily detailed.

It's very tedious. My husband ... he wouldn't have done it ... because some of it was so often very obvious ... I kept on going but he threw his hands up – it's funny, because it's not that it's hard, it's just tedious (F01-f).

The sheer volume of paperwork has implications for adoption of EMS. Those farmers we interviewed already had record-keeping systems in place and most were used to quality assurance procedures through Gippsland Natural and other schemes. For many farmers who do not have this experience, and who keep records in their head, rather than formally on paper or computer, EMS is likely to represent too much of a shift in practice. As one farmer observed:

... it's not for the Average Joe farmer, if I can use that expression, because unless you're good at bookkeeping and are prepared to put in the effort, it's basically too much hassle (F13-m).

The fourth major barrier to adoption identified in our study is the ongoing cost of auditing and compliance (see also Mech and Hugo 2004; Carruthers 2005). As one of the federal government's pilot projects, the costs of training and auditing for Gippsland EMS participants were covered. However, now that the funding for these pilot projects has ended, producers are expected to cover the costs themselves – presumably through price premiums. Farmers interviewed for our study argued strongly that, since marketplace benefits from EMS are elusive, it was important that the government continued to provide funding to ensure broader participation in, and uptake of, EMS. They considered that "without that funding support the cost, particularly for beef farmers, would be too great a barrier for most" (F04-m), "no-one could undertake... [the EMS] process unless it was financed by the government, because the bottom line is so thin, that education process needs to be paid for" (F03-m) and if the government grant "wasn't there, I don't think a lot of farmers would carry on with it". In the absence of discernible marketplace benefits, it appears that continued government funding will be necessary to support EMS initiatives. Despite farmer informants having a commitment to the EMS process, most would discontinue involvement if they had to pay for the cost of training and auditing – unless they began to receive significant price premiums.

Conclusion

This chapter demonstrates the increasing significance of EMS as a MOI for sustainable food production and agri-governance in Australia. Unlike Europe, where MOIs are oriented towards the creation of a multi-functional countryside, Australian governments have sought to use EMS as a mechanism to extend international market access, at the same time as addressing domestic issues of environmental degradation and farm viability. However, our focus on farm-level responses to EMS has identified issues and limitations that need to be considered if this strategy is to be workable.

First, EMS clearly has the potential to 'green' productivist farming practices. However, as we found in our research, it appears of most interest to farmers who are already engaged in more environmentally sound practices (see also Morris and Winter 2002). As a consequence, EMS may only be attractive to farmers already engaged in conservation and environmental management initiatives, rather than contributing to a more widespread 'greening' of Australian agriculture. Indeed, it seems that EMS has a broader 'image problem' in attracting those producers who would gain most from improved environmental management. An added problem is the dominance of certified organic production in defining for consumers and some producers the meaning of 'alternative' and environmentally friendly production. Secondly, while the EMS participants we interviewed had a strong commitment to environmental improvement on their properties, this did not mean that they were prepared to spend time and money in the long term in order to gain ISO14001 accreditation. Farmers were keen to use EMS as a means of gaining market advantage and, ideally, a price premium, but many believed that this was unlikely to occur in the short term (if at all), and therefore questioned the necessity of full accreditation – particularly as they were more interested in selling their produce domestically, where other means of gaining consumer trust were available (see Kirwan 2004), than on overseas markets, where certification was likely to become a requirement. This indicates a need at government level for continued support in order to not only 'sell' the EMS message, but to underwrite the costs of training and auditing – possibly indefinitely – in order to achieve the national (rather than personal) benefits of improved environmental management and export performance.

In summary, our findings add weight to the claim that MOIs, such as EMS, can usefully be conceptualised as 'hybrid' or 'intermediate' regulatory spaces (see Ilbery and Maye 2005; Lockie and Higgins 2007). This is particularly the case in Australia where EMS is aimed at both supporting the continuation, as well as the 'greening', of 'conventional' export-oriented systems of production. A focus on these hybrid strategies of governing is crucial in examining how the boundaries between 'conventional' and 'alternative' agri-food practices are played out at a range of scales, the role of standards and auditing in this process, and the actual effects on farming practices.

Acknowledgement

The authors acknowledge the assistance of a Monash University Small Grant in funding the research on which this chapter is based. Cocklin and Dibden participated in the research as part of a project on agri-environmental governance and multi-functionality funded by the Australian Research Council.

References

Buller, H. and Morris, C. (2004). Growing goods: the market, the state and sustainable food production. *Environment and Planning A* 36, 1065–1084.

Busch, L. and Bain, C. (2004). New! Improved? The transformation of the global agrifood system. *Rural Sociology* 69(3), 321–346.

Buttel, F., Larson, O. and Gillespie, G. (1990). *The Sociology of Agriculture*. New York: Greenwood Press.

Campbell, A. (1994). *Landcare: Communities Shaping the Land and the Future*. Sydney: Allen and Unwin.

Campbell, H., Lawrence, G. and Smith, K. (2006). Audit cultures and the antipodes: the implications of Eurep-GAP for New Zealand and Australian agri-food industries. In T. Marsden and J. Murdoch (Eds), *Between the Local and the Global: confronting complexity in the contemporary Agri-food sector* (pp. 69–93). Oxford: Elsevier.

Carruthers, G. (2003). *Adoption of Environmental Management Systems in Agriculture. Part 1: Case Studies from Australian and New Zealand Farms*. Canberra: Rural Industries Research and Development Corporation.

Carruthers, G. (2005). *Adoption of Environmental Management Systems in Agriculture. Part 2: An Analysis of 40 Case Studies*. Canberra: Rural Industries Research and Development Corporation.

Cocklin, C. (2005). Natural capital and the sustainability of rural communities. In C. Cocklin and J. Dibden (Eds), *Sustainability and Change in Rural Australia* (pp. 171–191). Sydney: UNSW Press.

Curtis, A. and De Lacy, T. (1996). Landcare in Australia: does it make a difference? *Journal of Environmental Management* 46, 119–137.
Curtis, A, and De Lacy, T. (2000). Examining the assumptions underlying landcare. In S. Lockie and F. Vanclay (Eds), *Critical Landcare* (New edn, pp. 185–199). Wagga Wagga: Centre for Rural Social Research, Charles Sturt University.
Department of Agriculture, Forestry and Fisheries [DAFF] (2002). *National Food Industry Strategy*. Canberra: Australian Government.
Department of Agriculture, Forestry and Fisheries Australia [DAFF] (2004). *Environmental Management Systems – National Pilot Program*. Available online at: http://www.affa.gov.au (accessed 25/11/2004).
Dibden, J. and Cocklin, C. (2005). Agri-environmental governance. In V. Higgins and G. Lawrence (Eds), *Agricultural Governance: Globalization and the New Politics of Regulation* (pp. 135–152). London: Routledge.
Freidberg, S. (2004). The ethical complex of corporate food power. *Environment and Planning D: Society and Space* 22, 513–531.
Goodman, D. (2003). Editorial – the quality 'turn' and alternative food practices: reflections and agenda. *Journal of Rural Studies* 19, 1–7.
Goodman, D. and Watts, M. (Eds) (1997). *Globalising Food: Agrarian Questions and Global Restructuring*. London: Routledge.
Gray, I. and Lawrence, G. (2001). *A Future for Regional Australia: Escaping Global Misfortune*. Cambridge: Cambridge University Press.
Higgins, V. (2002). *Constructing Reform: Economic Expertise and the Governing of Agricultural Change in Australia*. New York: Nova Science.
Higgins, V. (2005). Governing agriculture through the managerial capacities of farmers: the role of calculation. In V. Higgins and G. Lawrence (Eds), *Agricultural Governance: Globalization and the New Politics of Regulation* (pp. 118–132). London: Routledge.
Higgins, V. (2006). Re-figuring the problem of farmer agency in agri-food studies: a translation approach. *Agriculture and Human Values* 26(1), 51–62.
Higgins, V. and Lockie, S. (2002). Re-discovering the social: neo-liberalism and hybrid practices of governing in rural natural resource management. *Journal of Rural Studies* 18, 419–428.
Holloway, L. and Kneafsey, M. (2000). Reading the space of the farmers' market: a preliminary investigation from the UK. *Sociologia Ruralis* 40(3), 285–299.
Ilbery, B. and Bowler, I. (1998). From agricultural productivism to post-productivism. In B. Ilbery (Ed), *The Geography of Rural Change* (pp. 57–84). London: Longman.
Ilbery, B. and Kneafsey, M. (2000). Producer constructions of quality in regional speciality food production: a case study from South West England. *Journal of Rural Studies* 16(2), 217–230.
Ilbery, B. and Maye, D. (2005). Alternative (shorter) food supply chains and specialist livestock products in the Scottish–English Borders. *Environment and Planning A* 37, 823–844.
Kirwan, J. (2004). Alternative strategies in the UK agro-food system: interrogating the alterity of farmers' markets. *Sociologia Ruralis* 44 (4), 395–415.
Lang, T. and Heasman, M. (2004). *Food Wars: The Global Battle for Mouths, Minds and Markets*. London: Earthscan.
Lobley, M. and Potter, C. (1998). Environmental stewardship in UK agriculture: a comparison of the environmentally sensitive area programme and the Countryside Stewardship Scheme in South East England. *Geoforum* 29 (4), 413–432.
Lockie, S. (1999). The state, rural environments, and globalisation: 'action at a distance' via the Australian Landcare Program. *Environment and Planning A* 31, 597–611.
Lockie, S. and Higgins, V. (2007). Roll-out neoliberalism and hybrid practices of regulation in Australian Agri-environmental governance. *Journal of Rural Studies* 23, 1–11.
Lockie, S. and Vanclay, F. (Eds) (2000). *Critical Landcare* (New edn). Wagga Wagga: Centre for Rural Social Research, Charles Sturt University.

Long, N. (2001). *Development Sociology: Actor Perspectives*. London: Routledge.
Long, N. and Long, A. (Eds) (1992). *Battlefields of Knowledge: The Interlocking of Theory and Practice in Social Research and Development*. London: Routledge.
Marsden, T. (2003). *The Condition of Rural Sustainability*. Assen: Van Gorcum.
Marsden, T. and Arce, A. (1995). Constructing quality: emerging food networks in the rural transition. *Environment and Planning A* 27, 1261–1279.
Marsden, T. and Smith, E. (2005). Ecological entrepreneurship: sustainable development in local communities through quality food production and local branding. *Geoforum* 36, 440–451.
Marsden, T. and Sonnino, R. (2005). Rural development and agri-food governance in Europe: tracing the development of alternatives. In V. Higgins and G. Lawrence (Eds), *Agricultural Governance: Globalization and the New Politics of Regulation* (pp. 50–68). London: Routledge.
Mech, T. (2004a). *Linking Environmental Improvement in Agriculture with Marketplace Benefits: The Role of Environmental Standards*. Canberra: Rural Industries Research and Development Corporation.
Mech, T. (2004b). *Environmental Management in Agriculture and the Rural Industries: Voluntary Approaches to Sustainability and Globalisation Imperatives*. Canberra: Rural Industries Research and Development Corporation.
Mech, T. and Hugo, L. (2004). *Environmental Management Systems Implementation in Agriculture: Identifying and Overcoming the Barriers*. Canberra: Rural Industries Research and Development Corporation
Mech, T. and Young, M.D. (2001). *VEMAs: Designing Voluntary Environmental Management Arrangements to Improve Natural Resource Management in Agriculture and Allied Industries*. Canberra: Rural Industries Research and Development Corporation.
Morris, C. (2000). Quality assurance schemes: a new way of delivering environmental benefits in food production? *Journal of Environmental Planning and Management* 43 (3), 433–448.
Morris, C. and Winter, M. (2002). Barn owls, bumble bees and beetles: UK agriculture, biodiversity and biodiversity action planning. *Journal of Environmental Planning and Management* 45 (5), 653–671.
Murdoch, J. and Miele, M. (1999). 'Back to nature': changing 'worlds of production' in the food sector. *Sociologia Ruralis* 39 (4), 465–483.
Murdoch, J., Marsden, T. and Banks, J. (2000). Quality, nature, and embeddedness: some theoretical considerations in the context of the food sector. *Economic Geography* 76 (2), 107–125.
Mutersbaugh, T. (2005). Just-in-space: certified rural products, labour of quality, and regulatory spaces. *Journal of Rural Studies* 21, 389–402.
Natural Resource Management Ministerial Council [NRMMC] (2002). *Australia's National Framework for Environmental Management Systems in Agriculture*. Canberra: Australian Government.
O'Sullivan, J. and Williams, J. (2005). *Gippsland Beef and Lamb EMS Forum 2005: What is the Value of EMS?* Unpublished forum brochure.
Potter, C. and Burney, J. (2002). Agricultural multifunctionality in the WTO – legitimate non-trade concern or disguised protectionism? *Journal of Rural Studies* 18, 35–47.
Renting, H., Marsden, T. and Banks, J. (2003). Understanding alternative food networks: exploring the role of short food supply chains in rural development. *Environment and Planning A* 35, 393–411.
Smith, E., Marsden, T., Flynn, A. and Percival, A. (2004). Regulating food risks: rebuilding confidence in Europe's food? *Environment and Planning C: Government and Policy* 22, 543–567.
Victorian Farmers Federation and the Victorian Government (2003). *The Way Forward: An Action Plan for Adoption of Environmental Management Systems in Victorian Agriculture*. Melbourne: VFF.

Williams, J. (2004). The Gippsland Beef and Lamb EMS Story so far... *Gippsland EMS: An Overview of the Gippsland Beef and Lamb Environmental Management System Pilot Project*, February.

Wilson, G. (2001). From productivism to post-productivism... and back again? exploring the (un)changed natural and mental landscapes of European agriculture. *Transactions of the Institute of British Geographers* 26 (1), 77–102.

Wilson, G. and Hart, K. (2001). Farmer participation in agri-environmental schemes: towards conservation-oriented thinking? *Sociologia Ruralis* 41 (2), 254–274.

Winter, M. (1997). New policies and new skills: agricultural change and technology transfer. *Sociologia Ruralis* 37 (3), 363–381.

Part III

Practising Alternative Food Geographies

Chapter 14

From the Ground Up: California Organics and the Making of 'Yuppie Chow'

Julie Guthman
Department of Community Studies, University of California,
Santa Cruz, USA

> Now if every school had a lunch program that served its students only local products that had been sustainably farmed, imagine what it would mean for agriculture. Today, twenty percent of the population of the United States is in school. If all these students were eating lunch together, consuming local, organic food, agriculture would change overnight to meet the demand. Our domestic food culture would change as well, as people again grew up learning how to cook affordable, wholesome, and delicious food
>
> Alice Waters, executive chef at Chez Panisse and founder of Edible Schoolyard Project, from "A delicious revolution"
>
> Leaders must shift from a perspective of incremental change to a perspective of transformation. The key is to paint a Vivid Picture. The point of the picture is not to accommodate the present; it's to accelerate the arrival of a particular version of the future. We must stand in the future and look back at the way we came
>
> Paul Dolan, biodynamic wine grape grower and leader in Roots of Change project, from "True to Our Roots"

A conceit of many practitioners of alternative food systems is that of a *tabula rasa* – a presumption that it is not only possible, but even desirable to erase the past and start anew – to develop new food systems from the ground up. The latest example of that is a philanthropy-driven initiative (the Roots of Change project) in California that explicitly used a concept called backcasting to develop a 'vivid picture' of a sustainable food and farming system 25 years into the future. Backcasting, according to 'The New Mainstream', which reports the project findings, "starts with a desired condition at some given point in the future, and attempts to imagine what it would take to get there" (Ecotrust 2005: 14). The New Mainstream represents an extreme scaled-up version of this conceit, and Waters and Dolan (cited above), both of whom sit on the council that guides the Roots of Change project, are particularly stylised, albeit highly influential figures in California food politics. Nevertheless, countless other more prosaic examples exist of efforts and rhetorics that are heedless, if not ignorant, of past historical geographies.

The purpose of this chapter is to take issue with blank slate mentalities – to demonstrate that prior histories and geographies shape our possible futures and to suggest that by ignoring those, we re-inscribe the injustices of the past in the development of alternatives. The development of organic food and farming in California provides a particularly instructive example. In regard to how it was constructed as a consumption practice, it put 'good' food out of the economic and cultural reach of non-elites; in regard to how it was constructed as production practice, it failed to bring to scrutiny the labour conditions and land ownership issues under which such food is produced. One unfortunate result is that current movement goals of enhanced incomes for farmers and access to affordable, healthy food for consumers are at odds. Ironically, efforts to address these shortcomings, such as the Vivid Picture project, rely on wishful thinking.

This chapter thus builds on the so-called 'conventionalisation debate' that was spawned by pilot research with my colleagues (Buck *et al.* 1997). While the empirical question at issue is the extent to which organic production has been brought into industrial modes and ideologies of agrarian production (see Lockie and Halpin 2005 for a thorough review), the heart of the debate turns on whether or not organic production will necessarily fall prey to some inevitable logic of capitalism, a position that many of our critics claimed we forwarded (e.g. Coombes and Campbell 1998; DuPuis 2000; Campbell and Liepins 2001; Michelsen 2001). Strangely, though, some have tended to argue this position abstracted from the politics and particularities of place, despite the fact that most of their research reflects geographic particularity (Guthman 2004a). A parallel debate over the meaning of organic food for consumers (James 1993; Morgan and Murdoch 2000; Goodman and DuPuis 2002; Guthman 2002; Lockie *et al.* 2002), while less empirically developed except for Lockie *et al.* in part turns on the politics of what Miele and Murdoch (2002: 312) call the "gastronomic aesthetics of food". Does the new aestheticisation of food hold promise for a more robust transformation in food systems or does it re-inscribe class differences, contributing to a further bifurcation of food ways? Here, again, issues of geographic particularity in cultures of organic consumption are elided. I want to assert that place, history, and politics matter, as a way to argue that social change efforts must take them all into account.

To make this case, I will examine the evolution of organic food in California from what Belasco (1989: 14) called the "counter-cuisine" to what organic growers call "yuppie chow" to suggest that the success of the organic industry was largely wrapped up with gentrification – and the class differentiation that necessarily entailed. To these ends, I will showcase first the provision of a particular commodity (organic salad mix or *mesclun*). In important respects, salad mix gave a jump-start to the California organic sector, which then became what is likely the largest in the world in terms of crop value. Therefore, the production complex around salad mix set a crucial standard in the evolution of the organic sector. Introduced by restaurateurs in the early 1980s, salad mix also helped establish organic food as precious, as a 'niche' product. So successful was organic salad mix as a high-end commodity that it induced major changes in the organic system of provision in the decade that followed. Yet, these changes did not strengthen alternatives; instead, they reinforced some of the worst social injustices of California's farming system. In particular, they contributed to a particular dynamic in land valuation that has tended to lock in historic patterns of labour relations and land ownership. These conclusions are based in part on primary research conducted in 1995 (Buck *et al.* 1997) and then again in 1997 and 1998 (Guthman 2000), involving over 170 semi-structured interviews with organic growers, processors, and industry experts in California. It is supplemented by my participant-observation of the California organic movement in the years since, along with news sources. These findings, then, should not be extrapolated to all places for all times. My purpose is to show how particular cultural and economic histories matter in the construction of alternatives.

Situating Salad Mix

While the organic farming movement has multiple geographic and philosophical origins (see e.g. Peters 1979; Harwood 1990), many of its key institutions and figures were California-based. For example, Alan Chadwick, a British-born Shakespearean actor, set up the first university-run research and extension service devoted solely to organics at the University of California at Santa Cruz in 1967. Having a decidedly counter-cultural milieu, over the years thousands of people have been apprenticed in this programme and many have become full-time farmers. In addition, the first organic certification programme in the United States, California Certified Organic Farmers (CCOF), began in Santa Cruz in 1973, then a rag-tag group of 50 or so self-proclaimed hippie farmers and is now one of the premier certification agencies in the world. The annual 'gathering' of ecological farmers – now a major industry conference – made its home in Asilomar, California. The Capay Valley, a small offshoot of the Sacramento Valley, became an important enclave of subscription farms, where consumers buy in for a weekly box of produce. There are other examples. Most of the organic farmers involved in these formative institutions saw themselves as 'seceding' from the conventional food system (see Kloppenberg *et al.* 1996) to build alternative institutions from the ground up.

Nevertheless, organic agriculture *arrived* in a post-1970s, *post*-counter-cultural climate, in some ways contradicting the sensibilities that some would argue are central to the organic critique. Indeed, this emergence was contingent on bridging the counter-cultural associations of organic food with a new class of eaters, a contingency that was similarly dependent on *where* it occurred: the San Francisco Bay Area. While this region was historically a high-wage economy (for whites), the crucial juncture in terms of contemporary rounds of gentrification was the explosive success of high-tech electronics in Silicon Valley and finance in San Francisco during the 1980s (Walker 1990) and then the 'dot-com' bubble a decade later, with many mini-fortunes made in stock and real estate speculation. Yet, as Walker notes, the Bay Area had long been a centre of personal innovation and indulgence, and cultural nonconformity, as well. It was a local social pundit, Alice Kahn, who coined the word 'yuppie' to connote the emerging group of young urban professionals who as Walker (1990: 22) put it "combin[ed] fierce upward mobility and strong consumerism with some remarkably progressive cultural and political interventions".

From its formative years in the 1850s Gold Rush, San Francisco had also been a restaurant town, an early draw for immigrating French chefs. Unlike most of the rest of the United States, moreover, San Francisco did not shun *haute cuisine* in the era of what Levenstein (1988) calls 'culinary babbitry'. To the contrary, the Bay Area remained a haven of good food sense amid the downward spiral of dietary expectations and food quality that occurred in the middle third of the twentieth century. As early as 1927, a survey of Berkeley diets noted a strong emphasis on fresh fruits and vegetables, especially the leafy and citrus varieties, milk products, and eggs, all in contrast to the average urban diet, which emphasised cheaper cereals and potatoes and spent relatively more for meat (Luck and Woodruff 1931; cited in Levenstein 1993). Proximity to the wine country of Napa and Sonoma counties, as well as a prevalence of truck gardens, contributed to relatively urbane food tastes.

It was a young woman from Berkeley who forged the unlikely connection between this early culinary history, the 1960s counterculture, and the *nouveau riche* of the 1980s. As a young adult, Alice Waters went to France and became enamoured with French rustic cooking. She returned to Berkeley to open a café in 1971 where she served simple meals to her friends. Within a few years of opening, she had pioneered the California version of *nouvelle cuisine*. Feeling that the best food was made from fresh, local, and seasonal ingredients, she bought most of her produce from local farms. In effect, then, she created the first market for salad mix in California.

Making and Remaking Salad Mix

Warren Weber, of Star Route Farms in Bolinas, one of the original self-professed hippie farmers, began to sell cut organic baby greens to Waters in 1981, using the French term *mesclun*. A handful of others soon joined in, some calling it spring mix. In some respects, all were garden-variety organic farmers – relatively small scale, independent, and ideologically motivated. Unlike many other farmers in California, who are multi-generational farmers, many of these early organic farmers were highly educated and had taken up farming on the interstitial spaces of agriculture, the small valleys, hillsides, pastures, and suburban backyards, on land they often had the means to acquire. Here they employed, in Weber's words from an interview, "the time-honoured organic techniques of cover-cropping and composting". So when Waters modified the noun *mesclun* with the word "organic" on the menu in what came to be an upmarket restaurant she started an association that she was only partly conscious of. Not only did Waters inspire a rash of imitation, and quite instrumentally contribute to the diffusion of organic consumption, she also, and in this way, unintentionally, institutionalised a certain set of meanings for organic.

Within a decade after opening, Chez Panisse had become a world-renowned culinary institution. Waters continued to buy local seasonal produce and highlight its organic origins. Many Bay Area chefs trained with Waters and went on to open their own restaurants and become celebrity chefs in their own right. Many also made it a practice to form personal relationships with local farmers to ensure availability of the highest quality ingredients. Following Waters' lead, they wanted organic ingredients, although, crucially, only salad mix was regularly featured as organic. To draw emphasis to the farm–restaurant connection, some featured the name of the farm on the menu, Star Route Farms having received the most notoriety this way.

By the late 1980s, organic salad mix was on the menu of many upmarket restaurants and certainly at those at the cutting edge. Greenleaf, a local Bay Area distributor, and now-defunct Terra Sonoma, a consortium of small growers with personal connections to the restaurant business, made entire businesses out of selling speciality and organic produce directly to restaurants. Because restaurateurs were extraordinarily picky about what they would buy, they enforced a high appearance standard on growers so as not to compromise their own reputations. The need for 'quality' became a major push for technical solutions to organic farming (and processing); at the same time it required an extraordinary amount of care. Growers were pushed to be delicate in their handling of organic produce and to discard (or separate) produce that did not conform to restaurant standards. In turn, organic shed the image of the twisted stunted carrot showing up at the local food-co-op to the splendid display of *mesclun* on a chef's dish.

The specificity of the farm–restaurant connection reinforced another attribute of organic salad mix: that it was necessarily expensive. Restaurants were willing to pay top dollar for the finest, freshest, and eye-pleasing mix – sometimes up to 35 dollars per pound for *mesclun*. The owner of short-lived Kona Kai farms, once situated on a small urban lot in Berkeley, boasted to have made the equivalent of $100 000 per acre in 1 year selling salad mix and herbs to nearby restaurants. While this case was exceptional, such talk nevertheless contributed to the notion that organic salad mix *was* a precious commodity. Upmarket supermarkets picked up on this idea, selling their salad mix as 'custom-made' and pricing it upwards of 12 dollars per pound.

Although a much wider variety of organic produce had long been sold in health food stores, co-operatives, and selected green grocers, the *taste* for organic salad mix was mostly diffused through restaurants, as are many exotic tastes (Warde and Martens 2000). But *sales* of organic salad mix exploded when producers started to infiltrate more mainstream retail establishments. The domestication of salad mix began when two graduates of the University of California at Santa Cruz, Myra and Drew Goodman, who had been selling their own organic berries and lettuce to area restaurants like Chez Panisse, came up with the idea of bagging their lettuce mixes. Adopting the name of Earthbound Farms, from 1986 to 1989 they were the only company selling washed, spun dried, and re-sealable bagged salad mixes to supermarkets. Thereafter, others became involved in retail sales, some imitating the one-meal bags designed by Earthbound and others selling custom mixes in bulk to upmarket supermarkets. The Aldicarb (a fumigant) and Alar (a ripening agent) 'pesticide scares' of 1986 and 1988, respectively, created a surge of growth in the California organic sector at large with certified organic acres quadrupling in 2 years (Schilling 1995). Ultimately, this cause of growth was outlasted by the expansionary activity around salad mix (Klonsky and Tourte 1995), suggesting that food safety was not the only impetus towards organic consumption, at least in this particular period. A leader in one major organic industry organisation was later to quip, "Salad mix has done more to reduce pesticide use in California than all the organising around pesticide reform". In actuality, pesticide use in California increased dramatically in the 1990s (Liebman 1997).

Meanwhile, the equation of organic with high value brought a rash of erstwhile conventional growers into the sector. In the aftermath of the 1980s farm crisis, many growers were looking for higher value cropping or marketing strategies, which occasionally led them to organic production. In California, commercial development pressure on farmland made organic farming especially attractive, a way to reap more crop value per acre in escalating land markets. Many growers simultaneously moved from commodity crops (such as cotton or sugar beets) into fresh vegetables. These new entrants did a major disservice to extant growers, who were eventually faced with unprecedented price competition.

The gradual distancing of salad mix from its earlier movement roots was to have profound implications for the way it was produced. Todd Koons, a former chef at Chez Panisse who started his own brand of mixed greens (TKO), introduced a system of contracting with other growers for the different components of salad mix. Eventually, other salad mix marketers followed suit. Consequently, another set of growers was brought into organic production, this time because they were asked to, as marketers preferred the 'professionalism' and 'reliability' of conventional growers. Koons, along with other key growers, also improved post-harvest processes (washing, spin drying, and bagging), a key value-adding strategy but one that raised the cost of capitalisation and, hence, barriers to entry.

Meanwhile, salad mix production began to stray from agro-ecological principles. Component contracting effectively encouraged mono-cultural production, at the same time it did not preclude suppliers from growing conventional crops on their other fields. Because baby salad greens are picked young, they rarely needed pesticides. Fertility needs, however, were increasingly met with forms of soluble nitrogen such as Chilean nitrate, an allowed but contentious substance within the organic farming community, known to destroy soil microorganisms and contribute to ground water pollution (Conway and Pretty 1991). Because baby greens could be grown quickly, growers could manage several crops per year, contributing to the logic of intensification that has characterised California's salad growing regions. Component production could also move around the state (as well as into Mexico and Arizona), taking advantage of seasonal climatic variation, and allowing salad mix to be produced year round. At the same time, vacuum packing increased storage life and allowed salad mix to be shipped all over the country and into Canada.

And what were working conditions like? Growers in the organic industry continued to rely on the 'time-honoured' exploitation of racialised and marginalised immigrant workers as documented in accounts of the California lettuce industry (Friedland *et al.* 1981; Thomas 1985). Many were hired through labour contractors, a system that keeps wages low through structural over-supply and attempts to remove grower responsibility for ensuring that workers are documented (Martin 2004). To ensure 'care' in weeding, some growers encouraged use of the short-handled hoe, a practice that would have been banned in California were it not for the last minute lobbying of the organic and ornamental flower industries (CCOF 1995). As for the harvest, with hardier components (e.g. radicchio), labour could be partially mechanised, meaning that a conveyor belt was placed in the field, ensuring that each head was cut and packed at a brisk clip; more delicate components were often hand cut with stoop labour. Today, there is little evidence that labour wages and conditions on organic farms is significantly different than on conventional farms (Guthman 2004b; Shreck *et al.* 2006).

TKO itself was to go bankrupt in 1996, attributed to rapid expansion and mismanagement, but the future of salad mix was altered for good. Over the course of 5 years, organic salad mix had gone from a speciality commodity selling for over 12 dollars per pound at retail, to just a commodity at 4 dollars per pound. Extremely low prices squeezed many of the high-end niche growers out of the market, many of whom diversified with other, newly exoticised crops. A later crack down on food safety, after 61 illnesses were linked to bags of salad mix found to be tainted with *E. coli* H157:H7 (Food Chemical News 1998), forced others to get big (for returns to scale on more frequent inspections and more elaborate washing equipment) or get out. As a consequence, salad mix became the province of some of the largest grower-shippers in the state of California. Salinas-based Missionero and Earthbound took up the slack of TKO, buying up its land and taking on the growers it had cultivated, and developed a significant clientele of "white table cloth chains" as well as bagged mix. Major multinationals such as Dole entered the retail salad mix market in force. Meanwhile, Earthbound Farms continued to grow at a rate of at least 50% a year until 1995, when a series of mergers began. Having more capital than organic market potential, Earthbound and its new partners joined forces to create Natural Selection Foods. Thereafter, they became involved in a series of partnerships with major conventional vegetable growers, including Growers' Vegetable Express, and Tanimura and Antle. They continued to grow geographically, with at least 650 ha in production in Baja California where off-season lettuce and tomatoes could be grown; they continued to grow in market share by buying out or contracting with some of their erstwhile competitors. By 2001, they had 2800 ha in organic production, 800 more in transition, and were in contract with dozens of other large acre growers. Natural Selection had become the biggest supplier of speciality lettuces and the largest grower of organic produce in North America (www.ebfarm.com). By 2006, several global corporations – Dole among them – had a solid spot in the salad mix market.

Finally, and perhaps most significantly, salad mix growers in California have not only responded to, but actually contributed to a treadmill of land values that has characterised California agriculture, a reflection of its crop mix, high levels of intensification, and almost-always booming real estate market. Led by various configurations of real estate speculators and large farmers, the first state-developed irrigation projects made possible the shift from extensive grain crops to fruit trees in the late nineteenth century; later projects supported intensive vegetable production in desert areas (Leibman 1983; Pisani 1984; Worster 1985). In the post-war years, coastal vegetable producers adopted practices that speed up crop turnover, including faster-growing varietals, greenhouse, and transplant operations, and heavy use of nitrogen fertilisers allowing them to harvest up to 5 crops per year. In California, agricultural land values generally reflect capitalisation of the highest value crop grown in specific regions and grown with the most state-of-the-art technologies to ensure rapid turnover time. In the last few decades, threats of commercial development added an additional premium to agricultural land (see Guthman 2004c for a sustained discussion of these points). So, of the many consequences of agro-industrialisation, the one that most affects organic farmers – indeed has driven growers to organic farming – is the constant pressure to adopt technologies or cropping systems that create more crop value per hectare. Yet, here again, salad mix – a high-value, high-turnover crop helped lead the way in the latest round of land valuation by setting another benchmark for land valuation.

In short, salad mix was the medium of some dramatic shifts in the politics of organic production. With rampant growth in demand, the production of organic salad mix became increasingly industrialised, with scaled-up growers out-competing some of the earlier movement growers. Many of the practices they incorporated, while in keeping with organic regulations, were not in keeping with organic idioms. But even these 'movement growers' were never really the salt of the earth agrarianists romanticised by the likes of Wendell Berry. For, California had never had a large class of land-holding farmers who had holdings of similar size and nature, where family members performed the necessary labour. Since the Gold Rush, California agriculture was always capitalist agriculture (Walker 2004), with land acquired through appropriation or pre-existing wealth. And, it must be said, a vast majority of land has been owned by whites, disproportionately to not only the population of the state, but certainly those who work in agriculture. Thus, salad mix's contribution to the rarely subsiding treadmill of high land values clearly has not only shaped what can be produced and how, but locked in some of the injustices of the past in terms of access to land and who works it. Meanwhile, the association of organic salad mix with 'yuppie chow' has imparted political ambiguity to organic salad mix in the sphere of consumption, as well.

Eating Salad Mix

In the early days of the organic movement, the shared meanings of organic food suppliers and eaters made for a reasonably coherent movement politics. Salad mix was arguably one of the factors that de-stabilised that coherence, as certain consumers began to see it as a speciality item, rather than a clear alternative to industrialised food. Yet, it is not simply its earlier cost structure that made salad mix seemingly inaccessible to all but the privileged (Allen and Sachs 1993; DeLind 1993), a so-called niche product. Eating organic salad mix was in some sense performative of – indeed helped to animate – the figure of the 'yuppie', the San Francisco Bay version of which was not wholly devoid of social conscience, having grown up in the tumultuous late 1960s and early 1970s, but not shorn of gentrified aspirations either. Thanks to the Alice Waters diaspora and the introduction of ingredient-based menus, this new group of eaters obtained a keener interest in the constituent ingredients of food and how they were put together, in lieu of the haute cuisine pretension of named dishes (Kuh 2001). In that way among others, they helped usher in broader entitlement to luxurious eating. At the same time, they developed their own conceits about taste and brought with them heightened concern with body image that in important respects mapped onto new sensibilities around eating.

Historians of food have shown how the making of taste has been inextricably tied to the conditions and social processes that gave rise to inequitable distributions of food and variations in diet, so that varying levels and practices of food consumption have been shaped by social ranking and identity (Burnett 1966; Mennell 1986; Toussaint-Samat 1994). In that way, taste has come to play a role in defining social ranking and identity (Bourdieu 1984). In particular, taste as an aesthetic has become a sign of privilege, albeit the nature of this aesthetic has changed over time. So, for instance, eighteenth century *nouvelle cuisine* helped usher the aesthetic shift to the visual, in particular "the singularisation of presentation" (Ferguson 1998: 606) that characterises the so-called simplicity of extremely labour-intensive kitchen art (Mennell 1986).

Until the 1960s, dining out in the US (except for at the iconic lunch counter or coffee shop) was largely the purview of the privileged or the middle class enjoying a special occasion (Kuh 2001). Food habits gradually begun to change in the late 1960s, with the expansion of chain restaurants, ethnic restaurants (operated by new migrants), and middle class travel to Europe, creating new interest in fine food (Levenstein 1993). In its frequency, restaurant eating became much more democratised (Mennell 1986). Consequently, as Warde and Martens (2000) show for the UK, *where* to go and *what* to eat became the key indicators of class. And while dining out was never a conscious strategy for social display, the middle class was much more experimental and prone to evaluate the meals they enjoy in aesthetic terms. Following her trip to France, Alice Waters brought *nouvelle cuisine* home and refashioned it as 'California cuisine', launching a trend in food experimentation which evolved into a culinary eclecticism involving "dizzying dives into novel combinations of exotic ingredients" (Levenstein 1993: 24). Northern California's young *nouveau riche* were the primary consumers of this new cuisine, indeed were in some sense defined by it.

Historians of food have also noted that as taste has become a performance of class, gender, and nationality, the body has become a potent symbol of such difference, a way in which one's taste is displayed (Bourdieu 1984: 190). For example, gastronomes – public arbiters of good taste – began to express concern about body weight as an affliction of gourmets in the early nineteenth century, contributing to the trend within *haute cuisine* towards simpler, lighter food and fewer courses (Mennell 1986: 37). Indeed, gastronomy was morally positioned as a model of discipline, control, and moderation, counterpoised to the 'unreflective' and excessive eating of the gourmand (Ferguson 1998: 608–609). During the Victorian era, the bourgeoisie emulated the aristocratic ideal of a graceful and slender body, disdainful of the need to display wealth and power ostentatiously. Women, in particular, were admonished to eat with delicacy, to take in as little as possible, and to display no desire, clearly reflecting extant mores about sexuality and establishing an early link between anorectic self-denial and privilege (Bordo 1993: 191). This is but one example by which good taste became wrapped up in self-surveillance.

Beginning in the 1960s, the links between body norms and taste found a new articulation, when breakthroughs in nutritional science combined with social changes to spur new concern over food intake, particularly in the US. Fresh vegetables came to be routinely available on a mass-market basis, as did chicken, tofu, and other so-called healthy foods. New understandings of heart disease, diabetes, cancer, and so forth, coupled with a round of journalistic muckraking, raised questions regarding the quality of the processed foods that dominated the early post-war era (Levenstein 1988). What Belasco (1989) called the counter-cuisine, which emerged out of the counter-culture movement, emphasised the health-giving properties of relatively unprocessed food. With nutritional ideas increasingly emphasising what should *not* be eaten, exhortations regarding excess weight shifted from the language of aesthetics to that of health (Levenstein 1993). As Levenstein argues, these new ideas about diet fit in well with the moral asceticism of the times, given newly found awareness of international poverty (e.g. Biafra, the 'other' America) and the climate of scarcity that pervaded during the early 1970s energy crisis. Beginning in the late 1970s, body fat came to be relentlessly villainised in the popular media, to the point that 'food replaced sex as a source of guilt' (Levenstein 1993: 212).

Yet, it was more than health concerns (if notions of health can even be disassociated from other cultural constructs) that triggered a shift to near-impossible body ideals in the 1980s. Not only were the success-driven young urbanites helping to shape food tastes, they were helping to define body ideals. Indeed, it is arguable that ideologies of success were directly implicated in the new body ideal of muscular thinness. For example, some of the psychological roots of anorexia nervosa – an extreme form of self-surveillance – are over-achievement, the notion that autonomy, will, and discipline can lead to success, even the idea that tolerance of pain is a sign of strength (Bordo 1993: 178). Price (2000: 92) draws closer the parallels between new body norms and the political economy of the 1980s, juxtaposing the discourse of the tight, thin, sleek body to be made through diet and exercise with that of structural adjustment, e.g. 'tightening their belts', 'cutting the fat', 'shaping up' 'bloated' economies. This discourse was beginning to circulate at the same time that, according to Schlosser (2001), fast, cheap, convenience food was becoming the cornerstone of most working-class American diets and rates of obesity were beginning to soar, particularly among poorer people.

So it was also in this context that *nouvelle cuisine* offered such a 'spectacular challenge' to traditional restaurant cooking, with its emphasis on fresh ingredients, minimum preparation, and an awareness of health considerations (Beardsworth and Keil 1997). When the exhortations of the new cuisine spilled over into North America, it is not coincidental that it was embraced by a new class of over-achievers. For a new generation of well-heeled American eaters, *nouvelle cuisine* was the perfect vehicle to mediate the deeply felt contradictions of food intake and simultaneously enjoy their new class position. It was expensive by nature of its use of the finest ingredients and labour intensiveness, a perfect combination for those whose moral sensibilities increasingly privileged environmental concerns over social ones. Simplicity of ingredients fit well with the asceticism yuppies grew up with, quite different from the stodgy *haute cuisine* of the old rich, at the same time that inventiveness satisfied the craving for difference. And as food came to be presented as art – a sensual *visual* experience – it made it possible for the body-obsessed to enjoy the dining out experience without admitting to the literally visceral sensual experience. In some sense, it made it possible to not be too rich or too thin, the phrase made famous by a New York socialite during the yuppie emergence (Levenstein 1993).

Considered this way, salad mix undoubtedly provided some interesting comfort. As *nouvelle cuisine in extremis*, in its simplicity, perhaps moderated the ambivalence of the new class position. Short of the ability to taste without swallowing (suggesting wine spit jars and aromatherapy lotions as the ultimate pleasures), salad, with its paucity of calories, was a good option for mediating body anxiety. The clincher, though, was organic food's idiomatic associations with health and environmental soundness, perhaps even opposition to fast food. As local food critic and restaurateur, Patricia Unterman (2000) was later to say, "when you choose to buy and eat organic and sustainably raised produce, a little of this karma rubs off on you, which makes everything taste better. A lot of this local, organic stuff does taste better". Eating organic salad mix connoted a political action in its own right, legitimising a practice that few could afford. But the subtle conflation of gastronomic aesthetics and environmental politics added an extra ingredient of desire. It is surely telling that organic farmers themselves began to refer to salad mix as yuppie chow.

One of the ironies of this connotation is that it necessarily limited market size to those who identified themselves in these terms. Consciously attempting to appeal to mass market tastes in order to expand the market, the major producers in the US, including Natural Selection, started marketing non-organic salad mix under several other brand names, especially because prices no longer warranted the cost and hassle of organic certification. Occasionally packaged with a packet of salad dressing, bagged salad mix was increasingly marketed as a convenience food. Pavich Family Farms, another major organic producer introduced organic iceberg lettuce, another way of de-coupling the notion of organic from yuppie. Most recently, Walmart has announced its plans to feature organic produce and will undoubtedly develop a marketing strategy to undermine its yuppie associations. Curiously, only upmarket restaurants continued to consistently modify the menu item of salad greens with the adjective 'organic', suggesting some persistence in the relationships between organics, distinction, and eating out.

Although only one organic commodity among many, salad mix nevertheless had attributes that bore directly on the politics of organic consumption. Diffused through restaurateurs, it was an elite commodity from the onset, playing into yuppie sensibilities, including the desire to control one's body shape. Once produced in relatively more ecologically sensitive ways, it eventually became an output of mass production methods, albeit reaching a broader group of consumers, many who simply want food grown without pesticides. Still, it continued to contribute to organic food's strong position as a niche market, and whether or not organic food is necessarily more expensive to produce – which has not been demonstrated – retail prices for organics remain at significantly higher levels than conventional foods. Even farmers selling in direct markets, such as farmers' markets and community-supported agriculture systems, sell organics at relatively high prices. While they do so to stay in business in competitive markets, salad mix helped set the terms by which they could.

Moving Forward without Looking Backward?

Recently social movement actors, including the likes of Dolan and Waters, have noted the social costs of organic food's positioning as a niche commodity. While yuppies eat this 'good' food, low-income consumers continue to eat lower cost and nutritionally debilitated fast, processed food. Most recently, the so-called epidemic of obesity has further galvanised social movement actors to address the bifurcation of food consumption. Spurred by these concerns, the community food security and sustainable agriculture movements in California have made a fairly explicit strategic alliance. The hope is that by eliminating market intermediaries, both small-scale farmers *and* low-income consumers will have improved outcomes over the conventional food market, although it has yet to be demonstrated that the problem can be solved through the market (Allen 2004; Guthman *et al.* 2006). Noting as she does in the opening epigraph that school children need to learn how to cook affordable, wholesome, and delicious food, Alice Waters has made her Edible Schoolyard a model for other farm-to-school programmes. However, recent research demonstrates that because farm-to-school programmes are pursued on a voluntary basis and thus rely on idiosyncratic initiative and funding (such as Waters'), most have developed in affluent areas and/or university towns, often, that is, where children already have exposure and access to relatively fresh and healthy food (Allen and Guthman 2006). Meanwhile, the Roots of Change project has tried to galvanise this energy, with a vision that promises to scale up these sorts of ideas to become 'the new mainstream'.

Yet, many unresolved questions remain. On the production side, these include issues of access to land, the cost of land, insufficient farmer income, and, most strikingly for California, the persistence of a farm labour system that plays on the vulnerabilities of migrant groups, made all the worse by contemporary border and immigration policy in the United States. On the consumption side, these include insufficient entitlements and/or income for many residents to afford this healthy, organic food, poor access to such food where it is most needed, and a tricky history where white, middle-class social reformists have told others what to eat. Meanwhile, many so-called alternatives depend on the market to solve these problems, when, indeed the market helped create them, as this very place-based story of organic salad mix so clearly shows. It is in this light that efforts such as the Edible Schoolyard and the Vivid Picture seem so problematic. Transforming food systems involves political engagement with states and other institutions of governance that shape the structural conditions (such as subsidies and regulations) under which food is produced, distributed, and consumed in the market. Transforming food systems also involves paying attention to the historical geographies that got us to where we are in order to address the inequities that have endured. Wishful thinking will simply not do.

Acknowledgments

A major portion of this chapter was previously published in 2003 as "Fast food/organic food: reflexive tastes and the making of 'yuppie chow'" in *Social and Cultural Geography* 4, 43–56. The journal can be found at http://www.tandf.co.uk.

References

Allen, P. (2004). *Together at the Table: Sustainability and Sustenance in the American Agrifood System*. State College, PA: Pennsylvania State University Press.
Allen, P. and Guthman, J. (2006). From 'old school' to 'farm-to-school': neoliberalization from the ground up. *Agriculture and Human Values* 23 (4), 401–415.
Allen, P. and Sachs, C. (1993). Sustainable agriculture in the United States: engagements, silences, and possibilities for transformation. In P. Allen (Ed.), *Food for the Future* (pp. 139–167). New York: John Wiley and Sons.
Beardsworth, A. and Keil, T. (1997). *Sociology on the Menu*. London: Routledge.
Belasco, W.J. (1989). *Appetite for Change*. New York: Pantheon.
Bordo, S. (1993). *Unbearable Weight: Feminism, Western Culture, and the Body*. Berkeley, CA: University of California Press.
Bourdieu, P. (1984). *Distinction: A Social Critique of the Judgment of Taste*. Cambridge, MA: Harvard University Press.
Buck, D., Getz, C. and Guthman, J. (1997). From farm to table: the organic vegetable commodity chain of northern California. *Sociologia Ruralis* 37, 3–20.
Burnett, J. (1966). *Plenty and Want*. London: Nelson.
Campbell, H. and Liepins, R. (2001). Naming organics: understanding organic standards in New Zealand as a discursive field. *Sociologia Ruralis* 41, 21–39.
CCOF (1995). CCOF influences defeat of bill to ban hand weeding. *California Certified Organic Farmers Statewide Newsletter* (p. 9).
Conway, G.R. and Pretty, J.N. (1991). *Unwelcome Harvest: Agriculture and Pollution*. London: Earthscan Publications.
Coombes, B. and Campbell, H. (1998). Dependent reproduction of alternative modes of agriculture: organic farming in New Zealand. *Sociologia Ruralis* 38, 127–145.
DeLind, L.B. (1993). Market niches, 'cul de sacs', and social context: alternative systems of food production. *Culture and Agriculture* 97, 7–12.

DuPuis, E.M. (2000). Not in my body: rBGH and the rise of organic milk. *Agriculture and Human Values* 17, 285–295.

Ecotrust (2005). *The Vivid Picture Project*. The new mainstream: a sustainable food agenda for California for review by the Roots of Change Council and the Roots of Change Fund. Portland, OR: Ecotrust. www.vividpicture.net/documents/The_New_Mainstream.pdf (accessed 1/5/2006).

Ferguson, P.P. (1998). A cultural field in the making: gastronomy in 19th-century France. *American Journal of Sociology* 104, 597–641.

Food Chemical News (1998). California Gourmet Salad Processor Charged with Food Safety Violation. *Food Chemical News* January 19.

Friedland, W.H., Barton, A.E. and Thomas, R.J. (1981). *Manufacturing Green Gold*. Cambridge: Cambridge University Press.

Goodman, D. and DuPuis, E.M. (2002). Knowing food and growing food: beyond the production–consumption debate in the sociology of agriculture. *Sociologia Ruralis* 42, 5–22.

Guthman, J. (2000). *Agrarian Dreams? The Paradox of Organic Farming in California*. Unpublished PhD dissertation, University of California.

Guthman, J. (2002). Commodified meanings, meaningful commodities: re-thinking production–consumption links through the organic system of provision. *Sociologia Ruralis* 42, 295–311.

Guthman, J. (2004a). The trouble with 'organic lite' in California: a rejoinder to the 'conventionalisation' debate. *Sociologia Ruralis* 44, 301–316.

Guthman, J. (2004b). *Agrarian Dreams? The Paradox of Organic Farming in California*. Berkeley, CA: University of California Press.

Guthman, J. (2004c). Back to the land: the paradox of organic food standards. *Environment and Planning A* 36, 511–528.

Guthman, J., Morris, A.W. and Allen, P. (2006). Squaring farm security and food security in two types of alternative food institution. *Rural Sociology* 71 (4), 662–684.

Harwood, R.A. (1990). A history of sustainable agriculture. In C. Edwards (Ed.), *Sustainable Agricultural Systems* (pp. 3–19). Ankeny: Soil and Water Conservation Society.

James, A. (1993). Eating green(s): discourses of organic food. In K. Milton (Ed.), *Environmentalism: The View from Anthropology* (pp. 205–218). London: Routledge.

Klonsky, K. and Tourte, L. (1995). *Statistical Review of California's Organic Agriculture 1992–1993*. Davis, CA: University of California Cooperative Extension.

Kloppenberg, J.J., Henrickson, J. and Stevenson, G.W. (1996). Coming into the foodshed. *Agriculture and Human Values* 13, 33–42.

Kuh, P. (2001). *The Last Days of Haute Cuisine: The Coming of Age of American Restaurants*. New York: Penguin.

Leibman, E. (1983). *California Farmland: A History of Large Agricultural Land Holdings*. Totowa, NJ: Rowman and Allanheld.

Liebman, J. (1997). *Rising Toxic Tide: Pesticide Use in California 1991–1995*. San Francisco, CA: Californians for Pesticide Reform.

Levenstein, H.A. (1988). *Revolution at the Table: The Transformation of the American Diet*. New York: Oxford University Press.

Levenstein, H.A. (1993). *Paradox of Plenty: A Social History of Eating in Modern America*. New York: Oxford University Press.

Lockie, S. and Halpin, D. (2005). The 'conventionalisation' thesis reconsidered: structural and ideological transformation of Australian organic agriculture. *Sociologia Ruralis* 45, 284–307.

Lockie, S., Lyons, K., Lawrence, G. and Mummery, K. (2002). Eating 'green': motivations behind organic food consumption in Australia. *Sociologia Ruralis* 42, 5–22.

Luck, M.T. and Woodruff, S. (1931). *The Food of Twelve Families of the Professional Class*. Berkeley, CA: University of California Press.

Martin, P. (2004). *Promise Unfulfilled: Unions, Immigration, and the Farm Workers*. Ithaca, NY: Cornell University Press.

Mennell, S. (1986). *All Manners of Food: Eating and Taste in England and France from the Middle Ages to the Present*. New York: Basil Blackwell.

Michelsen, J. (2001). Recent development and political acceptance of organic farming in Europe. *Sociologia Ruralis* 41, 3–20.

Miele, M. and Murdoch, J. (2002). The practical aesthetics of traditional cuisines: slow food in Tuscany. *Sociologia Ruralis* 42, 312–328.

Morgan, K. and Murdoch, J. (2000). Organic vs. conventional agriculture: knowledge, power and innovation in the food chain. *Geoforum* 31, 159–173.

Peters, S. (1979). *The Land in Trust: A Social History of the Organic Farming Movement*. Unpublished PhD dissertation, McGill University.

Pisani, D.J. (1984). *From the Family Farm to Agribusiness: The Irrigation Crusade in California and the West 1850–1931*. Berkeley, CA: University of California Press.

Price, P.L. (2000). No pain, no gain: bordering the hungry new world order. *Environment and Planning D: Society and Space* 18, 91–110.

Schilling, E. (1995). Organic agriculture grows up. *California Journal*, May 1995, 21–25.

Schlosser, E. (2001). *Fast Food Nation: The Dark Side of the American Meal*. Boston, MA: Houghton Mifflin Company.

Shreck, A., Getz, C. and Feenstra, G. (2006). Social sustainability, farm labor, and organic agriculture: findings from an exploratory analysis. *Agriculture and Human Values* 23 (4), 439–449.

Thomas, R.J. (1985). *Citizenship, Gender, and Work*. Berkeley, CA: University of California Press.

Toussaint-Samat, M. (1994). *History of Food*. Cambridge, MA: Blackwell.

Unterman, P. (2000). Fresh off the farm. *San Francisco Examiner* August 20.

Walker, R. (1990). The playground of US capitalism? The political economy of the San Francisco Bay Area. In M. Davis, S. Hiatt, M. Kennedy, S. Ruddick and M. Sprinker (Eds), *Fire in the Hearth* (pp. 3–82). London: Verso.

Walker, R. (2004). *The Conquest of Bread: 150 Years of Agribusiness in California*. New York: The New Press.

Warde, A. and Martens, L. (2000). *Eating Out: Social Differentiation, Consumption, and Pleasure*. Cambridge: Cambridge University Press.

Worster, D. (1985). *Rivers of Empire*. Oxford: Oxford University Press.

Chapter 15

Buying into 'Buy Local': Engagements of United States Local Food Initiatives

Patricia Allen[1] and Clare Hinrichs[2]
[1]Center for Agroecology and Sustainable Food Systems, University of California, Santa Cruz, USA
[2]Department of Agricultural Economics and Rural Sociology, The Pennsylvania State University, USA

Introduction

Food system localisation has become a central goal and strategy of alternative agri-food movements in Europe and the United States. Instances of food system localisation include the emergence and growth of community supported agriculture (CSA), the spread of farmers' markets and the renaissance of small-scale and speciality food growers and processors. In addition, many alternative agri-food organisations and groups have emphasised food system localisation in their platforms and repertoires of activities. For example, more than half of the alternative agri-food organisations studied in a research project in California were involved in some fashion with the development of local food systems (see Allen et al. 2003). Farmer-based sustainable agriculture organisations, such as the Practical Farmers of Iowa, now also incorporate local food systems work into their activities, in part to build non-farmer support and membership (Bell 2004). At a larger scale, the US-based community food security movement places itself 'squarely within the anti-globalisation community', and groups within the movement are working to develop 'concrete alternatives that promote locally grown foods instead of globally sourced ones' (Fisher 2002: 5). In Europe, localisation has become central to the EU's efforts to improve rural livelihoods and preserve European heritage (DuPuis and Goodman 2005).

Beyond this, food system localisation is seen as a goal and strategy that can unite diverse groups in the alternative agri-food movement. According to Schwind (2005), for example, 'Fresh, local food is a vision that unites community food security activists, environmentalists, slow food enthusiasts, and small-scale farmers globally'. For Henderson (1998), localisation and decentralisation serve as key sources of vitality and strength for the alternative agri-food movement. Clearly, the need for developing local food systems is an idea that has captured the imaginations of those in many locations in this movement.

Buy Local Food campaigns have become a primary expression of local food system interests and concerns in the United States. They arguably constitute the most articulated and focused effort to date for promoting and developing local food systems. The greater systematisation and networking associated with Buy Local Food campaigns make it possible to examine the scope of alternativeness, not just in individual projects, but also in the larger institutionalisation of local food system efforts. Significantly, alternative food *networks* more so than specific *nodes* offer the greatest potential to provide a strong alternative to conventional food supply chains (Watts *et al.* 2005).

There is substantial evidence regarding the growth and expected promise of food system localisation efforts. However, the specific agendas, objectives and assumptions behind these efforts have received comparatively little investigation. As a consequence, we know relatively little about the core aspects of food system localisation. For instance, what concepts and relationships define and delimit food system localisation and resulting local food systems? What ideological constructions, politics and social relations prevail? To what extent do these efforts offer coherent alternatives to the conventional food system? Studying self-identified Buy Local Food campaigns offers one way to address these questions in the US context.

Our research therefore seeks to document and analyse the engagements, assertions and understandings evident in US Buy Local Food campaigns. We ask what are the motivations, understandings, visions, and assumptions behind 'local' as a strategy for developing sustainable food systems. This research involves, first, documenting the engagements and, second, learning about the subjective understandings of those involved in Buy Local Food campaigns through interviews. This chapter presents a preliminary exploration of our central question by examining the recent rise, discursive accents and dissemination of texts produced by Buy Local Food campaigns in the US. We begin by examining the origins and history of Buy Local Food campaigns, situating present efforts within prior patterns of 'selective patronage' consumer purchasing campaigns. This anchors our account of the rise and spread of Buy Local Food campaigns in the US and informs our analysis of patterns of thematic engagement evident in such campaigns. We conclude with observations about local food initiatives centred on the ambiguities and political tensions percolating within food system localisation in general and Buy Local Food campaigns in particular.

Origins and History of Buy Local Food Campaigns

Whilst Buy Local Food campaigns are of recent vintage in alternative agri-food movements, they derive from a long history of political (or ethical) consumerism through which people work to create change through the marketplace. 'Political consumers' engage deliberately with the power relations implicit in consumer products and make consumption choices informed by particular political values or ethical concerns (Micheletti 2003). Whilst political consumerism is not entirely new, it has expanded rapidly in size and scope in the past two decades. Political participation through consumer boycotts increased fourfold between 1974 and 1999, whilst other forms of political activity such as petition signing, demonstrations and occupations levelled off (Stolle *et al.* 2005).

Buy Local Food campaigns also link in instructive ways to historical initiatives intended to direct and mobilise American consumer purchasing. Campaigns based on the theme of 'Buy Local Food' have clear antecedents in the history of American boycotts and 'buycotts'. Whereas boycotts involve bringing pressure by eschewing purchase of certain products, a 'buycott' is more proactive. Here consumers deliberately and selectively purchase products with specific qualities or attributes or produced by certain people considered deserving of support. Aggrieved, disadvantaged and disgruntled groups have frequently sought to mobilise consumer purchasing to build (or rebuild) market presence and economic clout (Frank 2003). Indeed, local purchasing injunctions can be traced back to the non-importation movement of American colonists leading up to the Revolutionary War (Gabriel and Lang 1995), when the boycott of British goods necessitated reliance on colonial cloth and foodstuffs. Most famously expressed by the Boston Tea Party, the non-importation movement fundamentally protested unfair colonial tax laws, but it also involved wider assertions of cultural independence and distinct American national values through market development and self-reliance (Gabriel and Lang 1995). There is a distinct congruence here with Kloppenburg *et al.*'s (1996) advocacy of 'secession' from the dominant agri-food system (i.e. creating alternatives to it) followed by the 'succession' of these alternative forms (i.e. gradually transferring resources and commitments from dominant to alternative agri-food institutions).

In the early 2000s, Buy Local Food campaigns were launched in various parts of the US (see Figure 15.1, for example), exemplifying a 'buycott' tactic for achieving the goal of food system localisation. Groundwork for explicit campaigns to encourage the purchase of locally produced food took place in the 1970s and 1980s, as interest in alternative agriculture and an alternative whole foods sector grew. The influential private non-profit Rodale Institute conducted its Cornucopia Project in the 1980s, drawing attention to declining local and regional self-reliance for food supply in the US (Cornucopia Project 1981). Born of OPEC oil embargo era concerns about energy use, the Cornucopia Project anticipated the various arguments for local food to emerge in the 1990s and 2000s. Early networks of food cooperatives and natural foods groceries frequently sourced produce from local and regional growers and suppliers out of necessity, as national supply networks for low-input or organic foods were then weak or inconsistent. But no

Figure 15.1: US 'Buy Local' promotional food label (Source: authors' survey).

formal campaigns or initiatives centred on exhortations to Buy Local Food existed until the late 1990s.

In 1999, a regional grassroots organisation, Community Involved in Sustaining Agriculture (CISA), launched its 'Be a Local Hero/Buy Locally Grown' campaign in the Pioneer Valley of Massachusetts (specifically Hampshire and Franklin Counties). The consumer campaign was centred on coordinated promotional messages and advertising, using the 'Local Hero' logo, to urge consumers to find and patronise area farmers. As CISA declares on its website, 'Be a Local Hero/Buy Locally Grown' is currently the 'country's longest running and most comprehensive "buy local" program for farm products' (CISA n.d.). It has indeed been a pioneer in this arena, both in its concerted, consistent use of conventional advertising tools, its growing use of the Internet for 'farmer location' and in its efforts to track consumer recognition and response to the campaign. Widely viewed as a success, the CISA model has been disseminated throughout the US in both direct and indirect ways (CISA n.d.: para. 1).

Spread of Buy Local Food campaigns as a movement tactic hastened through CISA's partnership with another more nationally oriented non-governmental organisation, the FoodRoutes Network, which formed in 1997 with the goal of promoting 'sustainable food systems in critical regions of the United States' (FoodRoutes Network 2002). FoodRoutes positioned itself as a national player able to provide technical assistance in the form of strategic communications development and analysis and network facilitation. Working from CISA's success in involving local farmers and consumers in its Buy Locally Grown campaign, FoodRoutes organisers sought to extend that promotional model more widely in support of local food systems. They obtained funding from the W. K. Kellogg Foundation, long a supporter of alternative food and agriculture projects in the US, to work with other organisations around the country interested in starting their own Buy Local Food campaigns. FoodRoutes issued a request for proposals from organisations interested in participating in what one FoodRoutes organiser described as a 'learning community' for developing and implementing Buy Local Food campaigns.

In 2002, FoodRoutes announced Buy Local partnerships with 10 community-based groups from around the US (see Table 15.1). Groups were chosen based on their potential to build regional markets for local foods and their willingness to participate in FoodRoutes' collaborative learning process. Participating groups would gain access to advertising as well as promotional messages and materials developed based on market research commissioned by FoodRoutes. Through periodic retreats and meetings, group organisers would share insights about successes and challenges in conducting their campaigns, a process intended to provide support and spark creativity in adapting core campaign elements to particular circumstances. Beyond this more formal enrolment of organisations into a process for developing and implementing their own Buy Local Food campaigns, FoodRoutes and CISA co-published a manual titled, *Harvesting Support for Locally Grown Food: Lessons from the 'Be a Local Hero, Buy Locally Grown' Campaign* in 2002. This manual has been sold to numerous groups and organisations around the US and, in part, accounts for the increasing number of consumer campaigns focused on 'Buy Local Food'. These new campaigns and the campaigns that predate FoodRoutes comprise an additional eight Buy Local Food campaigns that we included in this research (Table 15.1).

Local food systems advocates and authors suggest that local food systems present a needed and desirable alternative to the conventional food system and make wide-ranging claims about their benefits. One of the earliest writers to lay out the argument

Table 15.1: Buy Local Food campaigns in the US

Project name	Lead organisation(s)	State	Date of inception
FoodRoutes projects			
Regional Food System Program	Land Stewardship Project	MN	2001
Buy Fresh–Buy Local PA	Pennsylvania Association for Sustainable Agriculture	PA	2002
Central Coast Buy Fresh–Buy Local	Community Alliance with Family Farmers	CA	2002
Food and Farms Program	Eco Trust	OR	2002
Buy Fresh–Buy Local	Baton Rouge Economic and Agricultural Development Alliance	LA	2003
Buy Fresh–Buy Local Iowa	Practical Farmers of Iowa; Center for Energy and Environmental Education	IA	2003
Buy Local–Buy Appalachian	Appalachian Sustainable Agriculture Project	NC	2003
Eat Local Foods (ELF) Coalition's Buy Local Campaign	Maine Organic Farmers and Gardeners Association	ME	2003
Select Michigan Local Food Program	Michigan Integrated Food & Farming Systems; Michigan Department of Agriculture	MI	2003
Buy Local Food and Farm Products Initiative	Alternative Energy Resources Organization; Mission Market of Ronan	MT	2005
Independent projects			
Be a Local Hero, Buy Locally Grown	Community Involved in Sustaining Agriculture	MA	1999
Buy Local – The 10% Difference	Vermont Agency of Agriculture, Food and Markets	VT	2002
Buy Local Campaign	Southeastern Massachusetts Agricultural Partnership	MA	2002
Farm Fresh Atlas	Research, Education, Action and Policy on Food	WI	2002
Princeton Borough's Buy Local Campaign	Princeton Future	NJ	2003
Think local – Buy local – Be Local	Business Alliance for Local Living Economies; Sustainable Connections	WA	2003
Buy Local Guide	Growing Small Farms; Chatham County Center; North Carolina Cooperative Extension	NC	2004
Inland Northwest Buy Fresh Buy Local Campaign	Inland Northwest Community Food Systems Association; Rural Roots	ID	2004

for localised food systems is Friedmann (1993: 228), who traces the increasingly distant and durable food regimes over the past centuries, and states that, 'the promising solution lies in locality and seasonality'. This duality – the virtues of food system localisation juxtaposed against the globalisation of the food system and its deleterious effects – captures the essence of the movement for local food systems (Hinrichs 2003). Feenstra (1997: 28) sums up the virtues of local food systems, writing that they 'use ecologically sound production and distribution practices, and enhance social equity and democracy for all members of the community'. Lyson and Green (1999) claim that, whilst global food systems are based on the economics of self-interest, the economic relationships in local food systems are consciously shaped by society and that ownership and control by local people means that they will have a stake in the long-term future of their communities.

In general, local food systems are considered less alienating than the conventional food system, because they reduce not only physical distance (as in the environmental argument), but also social distance between producers and consumers. They are seen as fostering direct democratic participation in the local food economy and cultivating caring relationships among people in a community. Kloppenburg et al. (1996) argue that in a local foodshed, taking collective responsibility for people becomes necessary rather than optional, as in the current food system. Writing from an anti-hunger perspective, Poppendieck (1997) states that locally based solutions are fundamental to solving hunger problems. A related thread is that a shared commitment to place, and the knowledge that you will be encountering your neighbours, will reduce behaviours that pursue individual interests at the expense of a community's common interests (Campbell 1997).

These types of assertions and claims have not, however, been examined in terms of the discursive and material alternatives to the present conventional agri-food system, including those that cohere with the movement's sustainability goals. The situation parallels that identified by Stolle et al. (2005) who observe that whilst there has been a significant rise in political consumption as a type of political participation, there is little information about the political nature of these activities. More systematic social science study of Buy Local Food campaigns, as explicit efforts to promote food system localisation, can fill these gaps. The next section begins to assemble this information for Buy Local Food campaigns in the US.

Patterns of Engagement Within Buy Local Food Campaigns

Drawing primarily on the study of textual materials – the web sites, brochures, reports, and other promotional literature of Buy Local Food campaigns – we document the engagements of Buy Local Food campaigns. Our account here is informed and supported by ongoing participant observation and informal interviews at various Buy Local Food events, which are part of the larger research project. Here we use content analysis to identify thematic patterns in the textual materials of Buy Local Food campaigns. Content analysis is a research method that examines words or phrases within texts, the goal of which is to compress many words of text into fewer content categories (Stemler 2001). We used an emergent coding method wherein categories were established based on preliminary examination of the data.

Through this process, we identified six themes as the primary patterns of engagement: aesthetics, community, economics, environment, equity and health, which we explain here. Into the category of aesthetics, we grouped factors that had to do with sensory pleasures such as taste and viewshed (an area of land, water and other environmental elements that is visible from a fixed vantage point). In our data, the category of aesthetics includes, for example, statements such as that local food is fresher and/or tastes better ('quality') and that purchasing local food preserves the rural landscape. In the community theme, we placed items such as that local food systems build relationships and community and connect farmers and consumers. Economics included issues such as strengthening the local economy by keeping dollars in the local community, increasing profits for local farmers, supporting local businesses and increasing connections between local producers and retailers. The environmental category included priorities such as preserving farmland, protecting genetic diversity and increasing awareness of organic or sustainably grown food. Only one statement fell into the equity category – supporting socially responsible farms. Finally, the health category included statements such as that local food is more nutritious and that buying locally helps with nutrition education.

Our first step was to determine the self-identified objectives of Buy Local Food campaigns, that is, how they frame what are they working to accomplish. To make this assessment, we reviewed the web sites of the 10 FoodRoutes Buy Local Food campaigns in the US and counted the programmatic objectives listed. Of the 10 campaigns, only seven contained material that we considered to be programmatic objectives, that is, mission statements or reasons they were conducting the campaign. We counted 22 different programmatic objectives (Table 15.2). Clearly, economic themes strongly dominate the objectives of these campaigns (73 per cent). By contrast, objectives of community, environment and health garnered only nine per cent each, whilst objectives focused on aesthetics and equity themes were not mentioned. Whilst these numbers should not be construed as definitive, they do illustrate general tendencies as to which objectives are most important and which may be more tangential to the *raisons d'être* of these campaigns overall.

The campaign's textual materials offered a separate type of discourse in how they promoted the value of buying locally to consumers and the wider public. That is, the programmatic objectives (organisation mission) and programmatic promotions (marketing of mission) were often distinct (Table 15.2). Two categories stand out in particular.

Table 15.2: Themes of programme objectives and promotions of FoodRoutes Buy Local Food campaigns

	Per cent of objectives ($n = 22$)	Per cent of promotions ($n = 37$)
Aesthetics	0	23
Community	9	5
Economics	73	36
Environment	9	23
Equity	0	5
Health	9	9
Total	100	101*

* Does not equal 100 due to rounding.

First, whilst aesthetic themes did not feature in programmatic objectives of the Buy Local Food campaigns, they comprise a quarter of the reasons given in publicly oriented, promotional material for why people should be buying local food. Taste is given even more prominence in Buy Local Food posters. For example, in the Santa Cruz, California campaign, the only reason listed on the poster is, 'Taste the difference!'. In a similar vein, whilst economics was the top category both for objectives and promotions, it was much more important for the programme objectives (73 per cent) than for the programme promotions (36 per cent). It is perhaps unsurprising that project objectives and project promotional rhetoric would differ in emphasis, since one is about strategic planning and the other is essentially about marketing. Nonetheless, this disjuncture points to a certain fungibility in what 'buy local' is really about. It suggests that Buy Local Food resonates in potentially quite different ways for those in nonequivalent economic positions, for example, for producers as compared with consumers. Whilst this may be unavoidable or perhaps even necessary, it points to potential ideological disjunctures – both presences and silences – in Buy Local Food campaigns that merit further inquiry.

We also identified the claims made about why buying local is considered important and looked at the rationales given to support these claims. We marked reasons cited for why local food systems are 'good' and counted the number of times a given item was mentioned. These reasons were collated and organised into thematic categories as above (Table 15.3). The figure in the right-hand column of this table is the percentage of total claims made in the programmatic literature of the eighteen local food campaigns, that is, the 10 FoodRoutes campaigns along with the eight Buy Local Food campaigns not explicitly affiliated with FoodRoutes. Table 15.3 shows the distribution of the claims among the thematic categories used in the analysis. Again, economics predominates (42 per cent), closely followed by environment (30 per cent). The remaining 28 per cent of claims are distributed among community (14 per cent) and health and aesthetics (7 per cent each). No claims were made about equity in the programmatic literature. These patterns differ from those made in the academic and popular literature about local food systems, where claims about social equity outcomes receive much greater attention (e.g. Kloppenburg *et al.* 1996; Feenstra 1997; Henderson 1998). Conversely, the programmatic literature of actual campaigns to advance food system localisation places greater emphasis on economics.

Table 15.3: Claims about buying local food in programmatic literature of 18 Buy Local Food campaigns in the US

Claim	Per cent of stated reasons ($n = 57$)
Aesthetics	7
Community	14
Economics	42
Environment	30
Equity	0
Health	7
Total	100

For more than a decade, many activists and academics have put forth the concept of local food systems and the process of food system localisation as a promising pathway for achieving environmental, economic and social change in the agri-food system, often under the umbrella of sustainable food systems. Yet, not all three sustainability priorities – environment, economics and equity – are given commensurate attention in local food system projects. As the data in the next section indicates, Buy Local Food campaigns accent economics and environment, comparatively subordinating social justice.

Shades of Alternatives in Buy Local Food Initiatives

This section discusses the alternativeness of local food systems from the vantage point of the three sustainability goals articulated within the alternative agri-food movement. The discussions of environment, economics and equity include the evidence provided by Buy Local proponents to support claims about benefits of buying locally (Table 15.4).

Environment

In this study, one-third of the claims made about the benefits of local food systems centre on environmental issues. Arguments include the idea of reduced food miles and therefore reduced energy use and pollution and greenhouse gas emissions. In *Bringing the Food Economy Home*, Norberg-Hodge et al. (2002) see this as one of the strongest arguments in favour of local food systems, a position common in the local food system literature. In addition, local food systems are considered to require less packaging, again reducing resource use and pollution from waste. Another often cited environmental benefit of local food systems is the preservation of farmland, the argument being that a vibrant local economy will reduce pressures from urbanisation. The drive for food system localisation is part of a broader orientation towards eco-localism and bioregionalism. For eco-localists such as Curtis (2003: 1), for example, 'The road to environmental sustainability lies in the creation of local, self-reliant, community economies'. Although he is not writing solely about food systems, he includes farmers' markets and CSA in his list of eco-localist strategies, along with local currency systems, co-housing and car-sharing schemes.

However, whilst there are numerous claims about the benefit of local food systems in terms of energy savings, recent empirical work suggests a more complicated picture. Indeed, one study comparing local and global food systems found that claims of energy savings could not be substantiated (Schlich and Fleissner 2005). Whilst acknowledging that their results directly apply only to the foods studied (fruit juices and lamb meat), the authors point out that small farmers require much more energy per unit of food than larger units to produce and distribute their products. They conclude that marketing distance itself has little to do with ecological quality and that, 'claims for regional food production and distribution instead of global process chains are not generally valid' (Schlich and Fleissner 2005: 223). Research in Sweden, for example, has focused on transport energy intensity and found that the positive impacts of short distances can be counteracted by the low loading capacity of vehicles typically used for transport to local food markets (Wallgren 2006). Other research has found positive environmental benefits

Table 15.4: Themes with associated claims about the benefits of buying local foods

Themes	Claim	Rationale
Aesthetics		
	Improves food quality	Ensures better tasting food
Community		
	Increases food supply dependability	Revitalizes communities
	Is tailored to community needs	Enhances culture
	Is developed by local citizens	Links consumers and producers
	Enhances community/individual well being	Builds trust between producers and consumers
	Increases self-reliance	Builds community
	Improves bonds in community	Enhances social health
Economics		
	Reduces food costs	'Keeps taxes in check'
	Increases grower income/profitability	Preserves local businesses
	Increases employment	Improves customer service
	Helps local economy	Offers wider choices to consumers
	Helps develop alternative commerce	Lowers public costs
	Strengthens economies	Encourages local investment
	Supports local farm families	Adds/supports local jobs
Environment		
	Improves farming practices	Is adapted to local environments
	Saves energy	Preserves genetic diversity
	Preserves open space	Supports a clean environment
	Enables harmonious relationship with nature	Uses ecologically sound production practices
	Reduces emissions/pollution	Protects wildlife
	Leads to accumulation of renewable assets	Protects the environment
Equity		
	Is built on justice	Women play key roles
	Puts people first	Creates moral economy
	Enhances social equity	Can catalyze positive local transformations
	Accessible to everyone	Is socially just
	Improves wages and working conditions	Includes everyone
	Meets people's needs	Reduces health inequalities
	Improves food security	Increases consumer influence
	Distributes benefits fairly	Enhances democracy
Health		
	Makes food more nutritious	Safeguards your health
	Improves food safety	Improves health

of food system localisation. For example, Pretty *et al.* (2005), who take a broader view of food system externalities, estimate that if all food items sold in UK retail outlets were sourced from local food systems, it would result in £2119 million of avoided annual external costs per year related to factors such as pesticides and fertilizers in water, adverse effects to human health due to pesticides, microorganisms and BSE and the environmental and health costs related to transporting food items to retail centres. Resolving these contradictory conclusions requires further research to understand more about environmental claims and the evidence upon which they are based and would provide important clarification for consumers and for the campaigns themselves.

Economy

From an economic perspective, the discourse in these campaigns centres on supporting local farmers and other businesses and keeping money in the community. Local food systems are seen as a way to create or expand local economic activity – generating jobs for workers and profits for businesses. Another goal is to reverse the decline in the number of family farms. In local food systems, it is reasoned, more of the money spent on food accrues to the farmer rather than to corporate middlemen (Norberg-Hodge *et al.* 2002). As Halweil (2004: 18) writes, 'money spent on local produce at farmers' markets and locally owned shops stays in the community, cycling through to create jobs and raise incomes'. Kloppenburg *et al.* (1996) write that secessionist and successionist alternatives should be built around small- and mid-sized enterprises that are proximate to each other.

Yet a key tension revealed through our examination of Buy Local Food campaigns is the place and role of capitalism (and its auxiliary institutions) in these efforts. On one hand, food system localisation efforts have emerged in reaction to the juggernaut of global capital in agri-food systems and concern about the resulting harms for society and the environment (Hinrichs 2000; Allen 2004). But on the other, and particularly so with Buy Local Food campaigns, local food initiatives still strongly privilege economic discourse and justifications by accenting entrepreneurship and stressing small business development. Furthermore, the campaigns appropriate and deploy fundamental tools and techniques of conventional consumer marketing and advertising, albeit in service of 'local food'. For example, FoodRoutes has developed a Buy Fresh, Buy Local 'toolbox' that contains marketing and campaign materials for local organisations. Perhaps this is simply 'fighting fire with fire', but it raises important questions about the implications of fighting capitalism with capitalism. Buy Local Food proponents speak eloquently in meetings and workshops about the importance of knowledgeable and aware consumers, yet catchy slogans and familiar, though appealing, visual images can become proxies for substantive engagement with the extra-economic complexities of the food system.

Still, efforts to conceptualise alternative economic structures and behaviours may point us towards more variegated notions of capitalism. In theorising possibilities for a local economic politics, for example, Gibson-Graham (2002) delineates a 'diverse economy' typology that corrects for the assumed primacy of capitalist wage and market relations and recognises both difference and viability in 'alternative capitalist' and 'non-capitalist' forms of organisation. Within this more diverse economic landscape fit increasingly differentiated accounts of consumerism. Economic consumers may still shop for 'good buys', whilst political consumers shop attuned to the politics of products (Micheletti

2003) and green consumers shop carefully to reduce environmental harms (Gabriel and Lang 1995). But broad questions remain about the transformative potential of political consumerism and the extent to which it alters capitalism or can transcend capitalist laws of motion over the longer term (see Allen and Kovach 2000).

Just as Buy American and other 'selective patronage' consumer campaigns have tended to emerge when other avenues for social change have been difficult or blocked (Frank 1999), Buy Local Food campaigns can be seen as a logical and appropriate tactic, given the current backdrop and constraints of neoliberalism in the US. Emphasising various economic rationales for local food initiatives also reflects shrewd, realistic recognition by agri-food activists and practitioners of the shifting political-economic landscape and its current possibilities in the US. The current neoliberal environment and its emphasis on choice has played a key role in the development of organic certification legislation in the US (Allen and Kovach 2000) and is similarly shaping farm-to-school programmes in California (Allen and Guthman 2006). As it happens, Buy Local campaigns are able to manoeuvre within this comfortable zone of individual consumer choice, even as they endeavour to guide consumers in preferred local directions. In addition, when cast as economic populism or alternative capitalism (and an underdog at that), local food initiatives pose little obvious threat to anyone and therefore find ready and widespread popularity. In the US context, these initiatives are also consonant with a longstanding cultural privileging of the individual and a national preoccupation with self-reliance (Allen 2004).

What may also explain the predominance of economic objectives in these campaigns may be that they are not only economic *means* for change, but they are simultaneously economic *ends*. In contrast, most forms of political consumerism involve boycotts or buycotts to achieve an objective outside of the marketplace – for social or environmental improvement – with the idea that the strategy will be abandoned once the objectives are met. For example, the boycott of table grapes organised by the United Farm Workers was to force improvements in the workplace for farm labourers; it was not intended to last forever. It is this kind of social justice goal that has often been featured in selective patronage efforts, but, as we discuss in the next section, social justice is much less prevalent in Buy Local Food campaigns.

Equity

Historically, both boycotts and 'buycotts' have been employed as methods for creating social change through consumer actions to help socially or economically disadvantaged groups. For example, the 'Don't Buy Where You Can't Work' movement emerged in African-American neighbourhoods in Chicago in the 1920s and 1930s and spread throughout the US Midwest (Frank 2003). Here the consumer boycott was designed to counter racial discrimination and open up job opportunities for black citizens. Other, more recent, efforts focus on harnessing the purchasing power of women to support other women economically. In October 2004, the online network, '85 Broads' organised a variation on a traditional boycott – a day where women were to buy nothing in order to demonstrate the enormous buying power of women and call attention to the fact that this has not translated into economic power in the workplace (Carroll 2004).

In this framework, Buy Local Food campaigns can be seen as efforts to provide economic support to the disadvantaged group of small-scale farmers and independently owned businesses. However, rarely are benefits for other disadvantaged groups such as women, low-income people or oppressed minorities considered, although such groups are certainly affected. Historically, boycotts and buycotts promoted through union campaigns both extended and complicated the work of household shopping done by working-class wives and mothers (Frank 2003). Similarly, the more involved tasks of buying and preparing fresh, local food, as advocated in Buy Local Food campaigns, can actually reinforce traditional gender roles within households (Allen 1999; Sokolofski 2004). In addition, our textual analysis of Buy Local Food campaigns found no reports of their working directly with public food assistance programmes (e.g. Women, Infants and Children or senior nutrition programmes) designed to assist low-income consumers. These programmes are employed in farmers' markets in an effort to create 'win-win' economic solutions for both farmers and low-income consumers (Guthman *et al.* 2006). To the extent that Buy Local Food campaigns encourage the patronage of farmers' markets, there is an oblique connection to social justice, but it does not appear as an overt goal in the campaigns.

Aside from scarce attention to social justice issues, some aspects of the goals and language of buying locally may to an extent be contrary to the achievement of social justice. For example, there is a type of 'othering' at the core of food system localisation ideology that is troubling. Protecting from amorphous threats runs the risk of creating collateral 'others', in the same way that Buy American purchasing campaigns included hostility, insults and sometimes physical harm to racial-ethnic minorities.

American consumer purchasing campaigns launched by worker interests in the nineteenth and twentieth centuries, for example, fostered racism. During the late 1970s and 1980s, unions in the US, notably automakers and garment makers, implemented 'Buy American' campaigns in reaction to their declining market share in increasingly globalised consumer goods sectors (Frank 1999). Such campaigns urged purchases of American-made cars or clothing as an expression of patriotism and a way to ensure good US jobs. The accompanying expressions of hostility, however, for example, when Detroit autoworkers smashed Toyotas with sledge hammers, reveal the deep currents of anti-'foreigner' and racist sentiment possible in consumer purchasing campaigns (Frank 2003). In the nineteenth century, union labels applied by white-dominated labour unions often served also as a signal to working-class consumers that Asian workers had had no hand in producing the goods, thereby stoking existing racist and exclusionary impulses (Frank 2003).

Rather than promoting the interests of special groups (other than farmers) or explicit social justice concerns, arguments in favour of buying locally tend to focus on the value of community writ large. As much as clarifying what comprises the community that is being protected, it is necessary also to reflect on from what the community is being protected. Exhortations to Buy Local Food are often accompanied by undertones of defensive localism (Allen 1999, 2004; Hinrichs 2003; Winter 2003). But just as there is little direct specification of who is being protected, the external threat to be protected from is also sometimes vague or even disquieting, coming as it does from a putatively progressive movement. In the post 9–11 era, for example, some local food initiatives have tied Buy Local Food messages to homeland security and risks posed by terrorism to the food supply, demonising the 'other' from whom we need to protect ourselves.

Other equity considerations include the role of large, multilocation corporations versus small, local businesses in buy local efforts. For example, large companies, such as

Sodexho, have begun to source local food for institutional dining and provision. Halweil, a strong proponent of food system localisation, celebrates this saying, 'There's no reason that big companies can't source local ingredients' (Worldwatch Institute 2004: 1). Is the incorporation of local food into a vast corporate food enterprise a victory and if so, for whom? Probably not for small, local businesses that compete with these enterprises. Whole Foods Market, often considered a bane of small, independent natural foods markets, is now actively promoting its efforts to buy local and regional foods (Ness 2006) to defuse criticism about its business practices. Similarly, fast food chains, such as Burgerville operating in the Pacific Northwest, have enjoyed great success sourcing local food ingredients and marketing that angle to consumers. On one hand, this represents a canny, innovative adaptation of the entrenched American tradition of 'fast food'. But on another, it may encourage the 'McDonaldisation' of local food, along with its attendant labour relations, sloganeering and drive toward cultural homogeneity. These examples raise questions about the social and economic goals and methods of 'buy local' efforts. As history shows us, the Buy American campaigns failed to save 'good' American jobs. At least in part this is because by casting foreign workers as threatening 'others', they failed to focus on the role of transnational corporations that were restructuring global economy in ways that affected workers throughout the world, regardless of ethnicity or country of origin. Such examples underscore potential perils in too readily equating 'local' with 'alternative'.

Open Questions and Future Directions

As a specific and increasingly popular tactic within broader efforts to localise the food system, Buy Local Food campaigns highlight a number of issues important for thinking about 'alternative food geographies', the subject of this volume. Foremost among these is the very slippery notion of 'alternativeness'. Our account demonstrates the variations in and mutability of 'alternativeness' both in conceptualising and in representing local food systems. This results in a number of potential tensions and inconsistencies in the practice and dissemination of US Buy Local Food campaigns.

The agendas of Buy Local Food campaigns operate at multiple levels, which, under scrutiny, can reveal mixed messages. On one hand, many campaigns take an 'everything-but-the-kitchen-sink' approach to recounting the benefits of local food systems for prospective consumers. Benefits may emphasise economic considerations, but they also invoke environmental and aesthetic appeals, and to a lesser degree health, community and equity rationales. Schwind (2005), for example, writing for the NGO Food First, advocates for local food systems on the grounds that they can simultaneously reduce global climate change and rural poverty. Pretty (2001) suggests that local food systems increase jobs in the local area, keep more money in the local economy, reduce unnecessary transportation of food and therefore greenhouse gas emissions, use more sustainable production systems, provide better incomes for farmers and create greater trust and connectedness within and between consumer and producer groups.

Aside from the general absence of clearly articulated reasoning and evidence as to how and to what extent such benefits are achieved, this array of benefits merits pause on other counts. A lengthy benefits list may have a greater likelihood of touching on at least one issue salient to most consumers, but it glosses potential tensions and trade-offs between some objectives and thus may encourage unrealistic performance expectations of more localised food systems. And whilst the broad range of alleged benefits may widen appeal, in practice, certain goals and values receive greater emphasis, and, more subtly, legitimation.

Our analysis of Buy Local Food campaigns as a specific tactic within a broader strategy of food system localisation points to broader, but still unanswered questions about what exactly is being protected within the 'local' and from what it is being protected. 'Sustainable' is often considered too ambiguous or polarising, whilst 'organic', some fear, is increasingly captured by large corporations marketing processed organic lines and pressing now for looser, less exacting US national organic standards. Frustrated producers and confused consumers can sidestep the ambiguities of 'sustainable' and the disappointments of 'organic' by valorising resonant (and still unregulated) attributes such as 'local' (Planck 2005). Yet the ambiguity of what 'local' references – a place, certain people, particular practices? – allows it to be about anything and, at the margin, perhaps very little at all. Current enthusiasm notwithstanding, *local* food systems may not offer any greater clarity than terms such as sustainable and organic. Furthermore, many of the claims about local food systems seem to be offered on faith. Our examination of print and electronic campaign materials revealed strikingly little factual evidence to build the case for the suggested outcomes.

The disjuncture between the reasons campaigns emphasise in promotions directed to consumers and objectives the sponsoring organisations stress in their organisational statements is instructive here. Organisational statements stress the economic benefits of Buy Local Food campaigns in building local food systems, commonly touting improving farm business viability and profitability, the elimination of middlemen and capturing value added. This attention to the economic rejuvenation and entrepreneurial development of family farms emerges, in part, from organisational imperatives driven by the expectations of agencies, foundations and sometimes also local governments that provide grants to start up or support such local food initiatives (see also Allen *et al.* 2003). Given that Buy Local Food campaigns involve printing fliers, brochures and signage, buying advertising space and airtime, updating websites and local farm and food business directories, not to mention the labour to coordinate it all, they generally require more money than they are likely to take in from fees or licensing paid by participating producers or businesses. Selling their Buy Local Food campaigns to prospective funding sources, as well as participants, requires organisations and groups to 'buy in' on some level to the not necessarily entirely 'alternative' agendas of those with available dollars. Given the exigencies of organising and running such campaigns, actual practitioners and on-the-ground promoters of local food systems may be inclined to be less circumscribed about the logic and benefits of local food systems in order to have the mental freedom to get their work done.

All this suggests we should continue to probe and locate the centre of gravity in local food. What moral considerations and social relations matter most? What factors influence the presences and absences of particular priorities? Our ongoing research in this area will address several related issues. First, we seek also to understand current consumer and citizen understandings of 'local' as descriptor, attribute and qualification within agri-food systems. Various studies, often based on survey research, have offered useful clarifications of public response to 'local' food as opposed to 'sustainable' or 'organic' (Wilkins 1995; Weatherell *et al.* 2003; Smith and Sharp 2004; Tegtmeier and Duffy 2005). More qualitative research with local food customers and consumers will be important for deepening the understanding of the current appeal of 'local' food, its symbolic and ideological contours and for probing tacit assumptions, potential contradictions and unconsidered opportunities within local food initiatives explored in this chapter. Why do certain themes receive emphasis and in particular ways, whilst others are effectively bracketed out?

Second, we will examine how some of the more ironic manifestations of food system localisation unfold over time. For example, in promoting local food production and consumption, US Buy Local Food campaigns can be seen as making a stand for local differentiation and autonomy before the steamroller of globalisation. Yet the very homogeneity of message and promotional techniques across so many of the individual campaigns – from Pennsylvania to California – potentially obscures much of what might be distinct or particular about either the localities or the food itself. What does it mean when 'local food' assumes the trappings of a standardised brand? Buy Local Food campaigns in the US gained traction through fairly sophisticated application of stock consumer research and marketing techniques. It remains to be seen how the resulting packaged approaches to promoting 'local' food will fare over time, to what extent they evolve or become embellished in distinctive ways and also how they complement or compete with more explicit territorial markers and place-of-origin labelling coming into favour with some US alternative agri-food activists and practitioners (Barham and Hinrichs 2005).

The primary appeal of and justification for food system localisation as a strategy for achieving sustainability are that local food systems are considered *ipso facto* to embody the qualities of sustainable food systems, that is they are environmentally sound, economically viable and socially just. Yet despite their tremendous appeal, these representations of the initiatives we studied have emphases and omissions, key notes and silences. Deeper understanding of the reasons behind these is crucial before we fully embrace the rhetoric and practice of food system localisation.

References

Allen, P. (1999). Reweaving the food security net: mediating entitlement and entrepreneurship. *Agriculture and Human Values* 16, 117–129.
Allen, P. (2004). *Together at the Table: Sustainability and Sustenance in the US Agrifood System*. University Park, PA: Penn State University Press.
Allen, P., FitzSimmons, M., Goodman, M. and Warner, K. (2003). Shifting plates in the agrifood landscape: the tectonics of alternative agrifood initiatives in California. *Journal of Rural Studies* 19 (1), 61–75.
Allen, P. and Guthman, J. (2006). From "old school" to "farm-to-school": neoliberalization from the ground up. *Agriculture and Human Values* 23 (4), 401–415.
Allen, P. and Kovach, M. (2000). The capitalist composition of organic: the potential of markets in fulfilling the promise of sustainable agriculture. *Agriculture and Human Values* 17, 221–232.
Barham, E. and Hinrichs, C. (2005). *Regionally Identified Foods in the United States*. Paper presented at Joint annual meetings of Agriculture, Food and Human Values Society and Association for the Study of Food and Society, June, at Portland, OR.
Bell, M. (2004). *Farming for Us All: Practical Agriculture and the Cultivation of Sustainability*. University Park, PA: The Pennsylvania State University Press.

Campbell, D. (1997). Community-controlled economic development as a strategic vision for the sustainable agriculture movement. *American Journal of Alternative Agriculture* 12, 37–44.

Carroll, J. (2004). *Don't Shop 'til You Drop During Buycott.* http://www.news8austin.com/content/top_stories/default.asp?ArID=122253. Accessed on 4 January 2004.

CISA (Community Involved in Sustainable Agriculture) (n.d.). *CISA's Local Hero Campaign.* http://www.buylocalfood.com/Local%20Hero.htm. Accessed on 5 December 2005.

Cornucopia Project (1981). *Empty Breadbasket?* Emmaus, PA: Rodale Press.

Curtis, F. (2003). Eco-localism and sustainability. *Ecological Economics* 46, 83–102.

DuPuis, E.M. and Goodman, D. (2005). Should we go "home" to eat?: toward a reflexive politics of localism. *Journal of Rural Studies* 21, 359–371.

Feenstra, G.W. (1997). Local food systems and sustainable communities. *American Journal of Alternative Agriculture* 12, 28–36.

Fisher, A. (2002). Community food security: a promising alternative to the global food system. *Community Food Security News,* 5. Venice, CA: Community Food Security Coalition.

FoodRoutes Network (2002). *Building Support for Buying Local.* http://www.foodroutes.org/doclib/52/BuildingSupport.pdf. Accessed on 6 December 2002.

Frank, D. (1999). *Buy American: The Untold Story of Economic Nationalism.* Boston, MA: Beacon Press.

Frank, D. (2003). Where are the workers in consumer–worker alliances? Class dynamics and the history of consumer-labor campaigns. *Politics and Society* 31, 363–379.

Friedmann, H. (1993). After Midas' feast: alternative food regimes for the future. In P. Allen (Ed), *Food For The Future: Conditions And Contradictions of Sustainability* (pp. 213–233). New York: John Wiley & Sons.

Gabriel, Y. and Lang, T (1995). *The Unmanageable Consumer: Contemporary Consumption and Its Fragmentations.* London and Thousand Oaks: Sage.

Gibson-Graham, J.K. (2002). Beyond global v.s local: economic politics outsides the binary frame. In A. Herod and M.W. Wright (Eds), *Geographies of Power: Placing Scale.* Oxford: Blackwell.

Guthman, J., Morris, A.W. and Allen, P. (2006). Squaring farm security and food security in two types of alternative food institutions. *Rural Sociology* 71 (4), 662–684.

Halweil, B. (2004). *Eat Here: Reclaiming Homegrown Pleasures In A Global Supermarket.* New York: W. W. Norton & Company.

Henderson, E. (1998). Rebuilding local food systems from the grassroots up. *Monthly Review* 50, 112–124.

Hinrichs, C.C. (2000). Embeddedness and local food systems: notes on two types of direct agricultural market. *Journal of Rural Studies* 16, 295–303.

Hinrichs, C.C. (2003). The practice and politics of food system localization. *Journal of Rural Studies* 19, 33–45.

Kloppenburg Jr., J., Hendrickson, J. and Stevenson, G.W. (1996). Coming in to the foodshed. *Agriculture and Human Values* 13, 33–42.

Lyson, T.A. and Green, J. (1999). The agricultural marketscape: a framework for sustaining agriculture and communities in the northeast. *Journal of Sustainable Agriculture* 15, 133–150.

Micheletti, M. (2003). *Political Virtue and Shopping: Individuals, Consumerism, And Collective Action.* New York: Palgrave Macmillan.

Ness, C. (2006). Whole foods, taking flak, thinks local. *San Francisco Chronicle,* 26 July 2006. San Francisco.

Norberg-Hodge, H., Merrifield, T. and Gorelick, S. (2002). *Bringing the Food Economy Home: Local Alternatives to Global Agribusiness.* London: Zed Books.

Planck, N. (2005). Beyond USDA organic: buying local, organic, & grass-fed. *The New York Times,* 23 November 2005. New York, NY. http://www.organicconsumers.org/btc/beyond112805.cfm.

Poppendieck, J. (1997). The USA: hunger in the land of plenty. In G. Riches (Ed), *First world Hunger: Food Security and Welfare Politics* (pp. 134–64). London: Macmillan.

Pretty, J. (2001). *Some Benefits and Drawbacks Of Local Food Systems.* Briefing note for TVU/Sustain Agrifood Network, November 2nd 2001. http://www.sustainweb.org/pdf/afn_m1_p2.pdf. Accessed on 21 May 2003.

Pretty, J.N., Ball, A.S., Lang, T. and Morison, J.I.L. (2005). Farm costs and food miles: an assessment of the full cost of the UK weekly food basket. *Land Use Policy* 30, 1–19.

Schlich, E.H. and Fleissner, U. (2005). The ecology of scale: assessment of regional energy turnover and comparison with global food. *International Journal of Life Cycle Assessment* 10, 219–223.

Schwind, K. (2005). Going local on a global scale: rethinking food trade in the era of climate change, dumping, and rural poverty. *Backgrounder* 11(2):1–4.

Smith, M.B. and Sharp J.S. (2004). *Factors Associated With Support for Local and Organic Foods in Ohio.* Paper presented at 67th annual meeting of the Rural Sociological Society, August, Sacramento, CA.

Sokolofski, L. (2004). *Managing Household Food and Feeding: Gender, Consumption and Citizenship among Community Supported Agriculture Members.* Unpublished MS thesis, Iowa State University, Ames, IA.

Stemler, S. (2001). An overview of content analysis. In *Practical Assessment, Research & Evaluation.* http://PAREonline.net/getvn.asp?v=7&n=17.

Stolle, S., Hooghe, M. and Micheletti, M. (2005). Politics in the supermarket: political consumerism as a form of political participation. *International Political Science Review* 26, 245–269.

Tegtmeier, E. and Duffy, M. (2005). Community Supported Agriculture (CSA) In The Midwest United States: A Regional Characterization. Ames, IA: Leopold Center for Sustainable Agriculture, Iowa State University. http://www.leopold.iastate.edu/pubs/staff/files/csa_0105.pdf.

Wallgren, C. (2006). Local or global food markets: a comparison of energy use for transport. *Local Environment* 11, 233–251.

Watts, D.C.H., Ilbery, B. and Maye, D. (2005). Making reconnections in agro-food geography: alternative systems of food provision. *Progress in Human Geography* 29, 22–40.

Weatherell, C., Tregear, A. and Allinson, J. (2003). In search of the concerned consumer: UK public perceptions of food, farming and buying local. *Journal of Rural Studies* 19, 233–244.

Wilkins, J.L. (1995). Seasonal and local diets: consumers' role in achieving a sustainable food system. *Research in Rural Sociology and Development* 6, 149–166.

Winter, M. (2003). Embeddedness, the new food economy and defensive localism. *Journal of Rural Studies* 19, 23–32.

Worldwatch Institute (2004). *Local Food: A Holiday Recipe That's Better For You, For Farmers, And For Homeland Security.* http://www.worldwatch.org/press/news/2004/10/06/. Accessed on 1 December 2004.

Chapter 16

Manufacturing Fear: The Role of Food Processors and Retailers in Constructing Alternative Food Geographies in Toronto, Canada

Alison Blay-Palmer and Betsy Donald
Department of Geography, Queen's University, Kingston, Ontario, Canada

Introduction

One of the frustrations for those working in alternative agro-food systems is the inability to move significantly beyond its 'alternative' status and become the dominant, agro-food paradigm in North America (Kloppenberg et al. 1996; Allen 2004). One reason for the failure of the alternative food system to move past the margins is the confusion that exists as to what defines an alternative food system. For some, alternative food has come to mean any type of food that is labelled 'organic', 'local' or 'fair trade'. In these cases, the alternative responds to specific consumer concerns created by the industrial food system. Problems range from fears about food safety, such as the 'Mad Cow' crisis to concerns about social equity issues, for example Third World working conditions (Kneafsey et al. 2004; Ilbery et al. 2005). Although all of these concerns and anxieties shape and define the alternative food industry, in this chapter we focus on issues of food safety and how they mould the alternative industry. Essentially, we argue that by reacting to specific food safety issues created by the industrialisation of food processing, the 'alternative' plays into the hands of the industrial. Our research and analysis confirms that the expression of 'alternative' in specific, codified terms enables the co-option of alternative market segments by dominant agro-food players. As a result, alternative food networks (AFN) are repeatedly undermined as industrial food cannibalises bits of the alternative and absorbs them as branches of the mainstream food economy. This is problematic as it does not create the fundamental shift needed to produce a genuinely alternative food system. Rather, the mainstream reconfigures aspects of the alternative in ways that create similar problems to those the alternative sought to address. Guthman (2004a: 179) applies this to organics and observes that, 'the unfortunate confluence of regulation-driven rents with existing mechanisms of intensification contributes to some of the ecological problems that organic farming is supposed to alleviate'. So by adopting alternative food system goals in a disjointed and piecemeal manner, contradictions arise in the 'new' version of the industrial food system.

Given this history then, the question that interests us is what needs to shift so the alternative becomes the mainstream North American food? Like Allen (2004), we posit that the alternative food system will remain marginal until we address some of the long-term structural as well as social and cultural patterns prevalent in our food:

'Much work still remains to be done. Now that the ideas and priorities of alternative food movements have taken hold, it is time for the next—even more challenging—step. Alternative agro-food systems must acknowledge and address the deeper structural and cultural patterns that constrain the long-term resolution of social and environmental problems in the agro-food system' (2004: i).

In particular, there is merit in building on the work of Guthman (2004a) and others who acknowledge the challenges posed by codification (e.g., the creation of USDA organic standards in 2002). Deconstructing the way fear is translated by the actors in the food network provides insights into the neoliberal discourse that creates spaces for AFN and then co-opts them (Lee and Leyshon 2003; Watts *et al.* 2005).

In this chapter, we present the case that AFN are weakened in part as a result of socio-cultural factors. More specifically, we explore: (1) the response of the alternative food industry to food safety issues; and (2) food scares as a defining feature of alternative food systems. Of particular interest for this chapter is the relatively unexplored middle space between the producer and the consumer. Our focus then is on food processors and retailers as the intermediary nexus in the alternative food system. Although we have a developing and thoughtful literature on the production and consumption ends of the food chain, little is known about alternative food retailers, and even less about processors. As processors and retailers interpret, mediate and translate the food that passes from producer to consumer, they are a critical but poorly understood point in the food chain. Further, the research that has been done on these intermediate spaces tends to focus on rural-based processing and/or retail links (Hinrichs 2000; Holloway and Kneafsey 2000; Ilbery and Maye 2005a, 2005b; Ilbery *et al.* 2005). This chapter takes us instead into an *urban* setting by exploring alternative food systems in Toronto, Canada. By addressing the opportunities and challenges that emerge for alternative food systems in large cities, we begin to address the urban and retailer/processor gap in the alternative food literature (Donald and Blay-Palmer 2006). Accordingly, we investigate urban food systems through the lens of food safety and the way that North American consumer fears are translated into products and shelf-space by processors and retailers. The chapter builds on Goodman's (2003) conceptualisation of North American AFN as reflexive social constructions, so that AFN build shortened supply chains[1] based on trust and quality.

We begin with a brief overview of the North American history of food fear and food regulation to set the context for our discussion of processor and retailer roles in AFN. We then report research results from a 3-year study of AFN in Toronto, Canada. As one of the most multicultural cities in the world, Toronto is an interesting case study site for food (UNDP 2004). It offers dense material for the study of the alternative food industry and the role that food processors and retailers play as they respond to food fears and define alternative products and networks to address food safety concerns. Our interviews with food processors, retailers, distributors and food experts provide empirical evidence that: (1) food processing and retail firms define themselves in response to consumer food fears about personal health and about more pervasive social concerns; (2) by reacting to these specific concerns, alternative food businesses define and promote their products in clearly codified terms; and (3) these codified solutions to food scares provide the mechanisms for the conventional system to co-opt elements of alternative food, and ultimately undermine the integrity of the 'alternative'. Our working definition of 'alternative' began as oppositional in the sense that we took alternative to be anything outside the mainstream, conventional food system. To understand this system, we included interviews with multinational corporations (MNC) engaged in food processing in the region (Blay-Palmer and Donald 2006; Donald and Blay-Palmer 2006). These results helped us refine our understanding of 'alternative' to include organic, ethnic and/or fusion food produced by Small and Medium-sized Enterprises (SMEs). For the purposes of this chapter, the results reported have been refined to report only on organic processors, retailers and distributors. Their priority is the delivery of quality food from the local foodshed for local consumption. Our findings lead us to conclude that AFN need to take stock of what they can offer as a system and define themselves in these terms to create strong, independent, resilient and local food systems (Watts et al. 2005).

The Industrialisation of Food

The industrialisation of food is well documented. Since the mid-1800s, there has been a steady transformation of the food system as it has shifted from being the 'intimate commodity' to the 'industrial commodity' (Winson 1993), and from local consumption to mass consumption (Goodman and Watts 1997; Murdoch and Miele 1999). As the food system evolved, production–consumption connections weakened, and eaters were increasingly separated from their food. This process necessitated intermediaries to process, transport, distribute and sell food. Shifts from rural- to urban-based industrialised populations facilitated this change.[2] As a result, fewer people knew where or how their

---

[1] Either a physical shortening as in the case of direct sale farmer–eater relations in a local community or fair trade when a conceptual shortening occurs and consumers develop a deeper appreciation for farmer needs (Goodman 2003).
[2] With the advent of the industrial revolution, population shifted to urban centres (e.g. in Iowa by 1900, the rural share of the population had dropped to 74.4 per cent (State Library of Iowa 2005); in Canada the rural population dropped to 63 per cent by 1901 (Statistics Canada 2005)].

food was being produced or processed. At the same time, science was developing a better understanding of germs, bacteria and their relationships to food. In the 1880s, connections were established between infective organisms and food poisoning (Draper and Green 2002). Accordingly, science and technology were applied enthusiastically to food processing and handling to make it as safe and efficient as possible. As the links were being made between disease, bacteria and food at the end of the nineteenth century, there was a genuine sense of urgency to impose scientific approaches and government regulations onto the food preparation process. These 'watershed' food regulations were important and necessary as they eliminated food-borne illnesses as the leading cause of death in the US (Bobrow-Strain 2005: 5). Food regulation was seen as a way to generate a healthier, more productive America. This sentiment is aptly reflected in a 1905 Good Housekeeping article:

> 'National virility... depends upon individual health to such an extent, and this in turn is so largely governed by our food, that the healthfulness of foods is a matter of the most serious consequence to the nation' (from Bobrow-Strain 2005: 12).

By the turn of the twentieth century, food no longer came in bins from the local store – this was denounced as unhygienic. Food needed to be packaged and processed to be safe. Crackers, cereals and other processed foods were now sealed and marketed as Post, Coca-Cola and Kellogg's became household names. And so, food was on its way to becoming industrialised, and people moved another step further away from the source of their food (Stacey 1994). Shifts in meat processing provide an excellent example of the alterations occurring in the early part of the 1900s.

At the beginning of the twentieth century, Chicago was the centre of the meat-processing industry where the one square mile meat district employed over 40 000 people. Animals were processed in four- to five-storey slaughterhouses where cows went in the top floor and sides of beef came out on the ground floor to be shipped across the US and to Europe. The workers that were employed were skilled, but poorly treated, so that they commonly suffered lacerations, amputations and other injuries (Cronon 1991). In an attempt to expose the terrible working conditions, Upton Sinclair wrote a novel based on in-depth research about the state of the industry. Entitled *The Jungle*, the book's account was so horrific that it sparked a public investigation into working and sanitary conditions in the slaughterhouses. The public report that was produced in 1906 from the investigation connected industrial food processing practices with food safety and raised public awareness about the health threats stemming from the industrial food system. This report led President Theodore Roosevelt to enact 'The Pure Food and Drug Act' to mitigate public health concerns.

The creation of food standards paved the way for the post-World War II era shift to an emphasis on modern, sanitary food as embodied in Andy Warhol's famous Campbell's Soup tin. This period in North American food production valued competence, scientific inquiry and the consolidation of mass production. Efficiency and hygiene were of prime importance, so through the mid-1900s the emphasis moved to promoting convenience and 'snap, crackle, pop' (Stacey 1994: 15). Speed and efficiency were the bottom line as food was further transformed into a processed commodity, capitalising on the view that food was fuel, and not a source of pleasure:

'Processed foods are often felt to be more healthful, and in this way more desirable, than foods consumed in a natural state, particularly since such foods are often presented with some sort of covering. Packaged, canned, wrapped foods, and bottled liquids at present have become endowed with the connotation of the *pure, the sanitary, and the healthful*' (Cussler and deGive 1952: 35, emphasis added).

This was an era when people revered plastic and bowed down at the altar of perceived cleanliness. It was also the time when bland, processed food became the standard and people moved another step away from natural, fresh food.

By the 1980s, food processors fashioned themselves as promoters of healthy food that fed the American babyboomer desire for immortality. For example, in 1984, Kellogg's launched All-Bran cereal as an anti-cancer weapon. In so doing, they reflected back the growing public desire to control their bodies, possibly beat death, and, at the very least, manage risk (Stacey 1994). The characterisation of food as fuel and the emphasis on calories as a way to value food continues today and creates the ultimate fear of food – food phobias that surface as eating disorders such as anorexia and bulimia.

By the end of the twentieth century, industrial food companies used science and technology to give consumers what they wanted. And, food regulators continued to supervise the processing of food to reassure consumers that the food produced by the global, industrialised food system was safe to eat. As food historian Harvey Levenstein (2004: 6) has put it, 'when you have a culture in which food is the object of fear and loathing as well as love, there are people who are going to discover innumerable creative and inventive ways of exploiting these fears'. Jolly Time candy apple flavoured popcorn is a case in point. This microwavable snack product promotes itself as a fat-free healthy option – one that contains artificial sweeteners and is artificially candy apple flavoured. The company president explains in a press release for the product launch that, 'JOLLY TIME is banking on the concept that *more choice equals more sales* from the growing number of health-and weight-conscious consumers' (Jolly Time 2004, our emphasis).[3] In this case, corporate food processors capitalise on fears of poor health to carve up the food market into ever-smaller niches. Beginning in the 1980s, a series of food scares, however, once again raised concerns about food safety and shook public confidence in the reliability of industrial food.

Fear, Food and AFN

Food fears present an interesting intersection between science and society, risk and regulation, externality and corporeality (Beck 1992, 1999; Whatmore 2002; Kneafsey *et al.* 2004). In the previous section, we discussed how the fear of contamination and adulteration at the inception of the industrial food system necessitated a central role for government regulations, science and technology. As food became increasingly processed and people were unsure about where their food came from, their uncertainty fed the

---

[3] According to Marion Nestle, in 1998 11,037 new food products were launched. Of these, 25 per cent were marketed as 'nutritionally enhanced' (2004: 25).

need to regulate food and supported an increased reliance on the power of science to guarantee safe food.

Today, food processors diminish the perceived distance between field and fork through packaging, labelling and certification programmes (Kneafsey et al. 2004; Watts et al. 2005). These relationships intend in part to take the fear out of food and offer eaters quality food they can trust (Hinrichs 2000). In some cases, the search for safe food has contributed to an emphasis on food 'quality', trust, shorter food chains and direct buying–selling relationships (Murdoch et al. 2000; Goodman 2003; Hinrichs 2003; Whatmore et al. 2003) – in short the de-industrialisation of the food economy. AFN, however, are rife with contradictions (Ilbery and Maye 2005). Growing inequalities within AFN have elite consumers eating ' "designer" organic vegetables that get shipped around the world in a sophisticated "cool chain"' (Goodman and Watts 1997: 3). Guthman (2004a, 2004b) is very vocal about the co-option of organic food by agro-business in California as it undermines the alternative food agenda. This line of research raises questions about the complexity and hybrid nature of AFN and has attracted more attention about how to define 'alternative' food practices.

With a view to tackling these definitional challenges, Watts et al. (2005) contrast strong chains – characterised as short, trust-based, personal systems with direct links between producer and consumer – with weak ones. For Watts and his colleagues, the difference between the two types of food system rests on the fact that 'weaker' alternatives focus on qualities inherent in the food, while 'stronger' alternatives define their alternativeness through their network. Another important distinction in the literature between different types of food systems is the recognition of complex, nuanced, hybrid food systems with no clear lines between alternative and conventional. Accordingly, foods are characterised as positioned at locations along continuums between de-localised/re-localised (Sonnino and Marsden 2006); strong/weak/hybrid alternative systems/chains of food provision (Watts et al. 2005); or local/global scales (Hinrichs 2003; Winter 2003).

However, a missing dimension from these discussions is the role of retailers and processors participating in the 'alternative' food market and the way they mediate and translate social concerns. Although the literature has explored in depth the rural context for producer–consumer links, as well as direct sell relationships, farmers' markets and labelling/certification schemes (Hinrichs 2003; Ilbery and Maye 2005a, 2005b; Watts et al. 2005) little is known about how food processors and retailers interpret social cues from AFN consumers. The void is especially notable in urban settings. Given the growing importance of urban populations and the clear role that processors and retailers play in mediating food systems, understanding the intersection of these two areas would be useful. To this end, we turn our attention to our interviews with food processors and retailers in Toronto, Canada.

Fear in the Toronto Food Case Study

We originally became interested in trying to understand the role that fear might be playing in the rise of AFN (especially the increased North American sales of foods labelled organic) when we noticed a pattern of answers in our interviews with various food actors in the Toronto area.[4] In response to questions about the personal and business motivations for starting 'alternative' food companies, key informants told us that *fear* played a role in shaping their ways of life and businesses. Although not initially a theme in our research, a re-review of our interviews led us to deconstruct the Toronto food system in terms of the social and emotive drivers that act to create spaces for alternative food companies. These observations led us to re-cast our empirical findings in the light of fear and the way it shapes food provision.

The responses were an unexpected by-product of our research on the innovative dimensions of food in the North American urban economy (particularly the rise of ethnic, organic and fusion foods). Our project on the speciality food economy in Greater Toronto was part of a larger body of work that sought to understand regional development opportunities and challenges (Wolfe and Gertler 2001).[5] Results from this work raised interesting questions about the role of alternative small and medium food enterprises as drivers of innovation (Blay-Palmer and Donald 2006). It also pointed to the way that food facilitates cultural inclusion for new Canadians. It became clear from our research that accessing culturally appropriate food helps people to feel at home in Canada (Donald and Blay-Palmer 2006). In the same way that processors of ethnic food bring the comfort of home to new Canadians through food, organic and other alternative processors, distributors and retailers respond to social needs. In this way, alternative food provides hybrid spaces for processors, distributors and retailers (Ilbery *et al.* 2005; Watts *et al.* 2005). Of particular interest for present purposes is how intermediaries in the food chain translate the need for safety into food products.

The results reported in this chapter are based on 65 interviews. These discussions took place between January 2003 and June 2005 with food producers, distributors, processors and retailers; non-governmental organisations including food security experts and consultants; restaurateurs and chefs; educational institutions; the media; and, municipal, provincial and federal government officials. While the focus of the interviews was on alternative food processing SMEs, 15 per cent of our interviews were with MNC and institutions in the industry. This established the context from the MNC perspective for the larger food processing industry in Toronto. The results discussed in this chapter report on a subset of 25 interviews with five retailers, 14 processors, four distributor/retailers and two producer/processors. While the majority of interviews were with retailers and intermediaries, two participants were farmers engaged in production, processing and retail activities. Seventeen of the businesses interviewed are certified organic, with the remainder focusing on local food for niche markets. In the rest of this section, we present our findings under two sub-headings: defining alternative food processing and retailing; and, codifying the message.

---

[4] Food anxiety was also an unintended result remarked upon by Kneafsey *et al.* (2004) in their research on the (re)connection of consumers and producers through 'alternative' food networks.
[5] *Innovation Systems and Economic Development: the role of local and regional clusters in Canada* was a five-year study examining how local networks of firms and supporting infrastructure of institutions, businesses and people interacted to facilitate economic growth.

Defining Alternative Food Processing and Retailing

Our analysis revealed that fear emerged as defining features of the products and businesses of our key informants. Interview subjects identified both personal fear and a more pervasive societal fear as stimulating consumer interest in alternative food. Retailers and food processors stated that these fears served as focal points for their firms. As one retailer commented, 'We position ourselves against two of the largest bodies of power—the agro-chemical companies and pharmaceutical companies. We are the antithesis of them. People's focus on health food scares, and the need for alternative foods and medicine allows us to offer a choice' (Retailer, April 2003).

Clearly, a key emphasis for this retailer is on alleviating consumer fears. These retailers describe the interaction between personal and societal spaces and fears as a powerful force driving growth in AFN. Personal and societal forces feed off each other so that consumers turn to AFN to assuage their concerns about the corporate-dominated food system. This fear creates gaps for alternative retailers where they can provide a safe and trusted space where agro-chemical and pharmaceutical companies are not.

Personal fears also provide the motivation for entrepreneurs to develop alternative food products. A processor of organic jams and condiments offered her personal rationale for engaging with AFN:

> 'North America is a *giant Petri dish*. [Our] children are sick. How much is being caused by the food we eat? Our air, our water... we are worried about terrorists. This is nothing to worry about in comparison to what we are doing to ourselves' (Organic processor, February 2003, emphasis added).

In this case, the key informant started her business as an alternative in every sense. As a result of personal health issues and an ethical crisis about the financial industry she was working in, she searched out an alternative business opportunity. In this case, her concern for societal well being, as well as her own health, motivated her to make drastic lifestyle changes.

The desire to reconnect with nature and bridge the gap between nature and society spurred this farm family to switch from conventional to organic farming and processing:

> 'We found it [organic farming] was a safe environment to work in. We don't want chemicals for ourselves and our family, for the soil and our present/future capability to produce food. We were tired of what big industry was telling us to do, to be part of their agenda. We were a tool for them to make money; this was not to our benefit. We want to be independent, and support the natural world to enhance diversity' (Organic producer and processor, March 2003).

Indeed, the impacts of outside forces and shocks on the food industry such as severe acute respiratory syndrome (SARS) and bovine spongiform encephalopathy (BSE) have generated consumer fear about personal health. This fear is compounded by the related perception that science and institutions have failed to regulate, control and protect the integrity of North American food (Klint-Jensen 2004) or to properly represent and protect the public interest (Concentration of GE corporations: Harhoff *et al.* 2001; Phillips 2002). The discovery of BSE in Canada in 2003 and in the US in 2005 created a demand for naturally raised and organic meats. As one organic meat producer, processor and retailer explains, the fall-out from SARS and BSE in Canada has been a boon to his organic meat business:

> 'In the first two weeks of SARS home delivery doubled... Mad Cow disease did wonders for the industry, it was inevitable that consumers were going to start asking more questions, Mad Cow just gave it a prod... In organics we have audit trails, accountability... we can 100% guarantee our product' (Organic Meat producer, processor and retailer, April 2003).

Clearly shocks within the mainstream food system and the fear that these events create push consumers to seek out alternative food provision systems.

The increasing demand for trust-based, quality relationships and products are the fastest growing sector in the food industry (Minou and Willer 2003). But we also know that agro-food companies from the industrialised food system are filling these spaces as they profit from food fears (Guthman 2004a, 2004b). Recalling the example of Jolly Time candy apple popcorn, the industrialised food system plays on these fears as a way to carve up the market and create new, niche opportunities (Schlosser 2002). When AFN translate food fears into codified lists that can be replicated, they provide a directory of factors that the mainstream food system can – and does – address. In Watts et al.'s (2005) terms, they rely on properties of the food and not the network and are more easily copied. These weaker food networks lack the resilience of stronger food networks.

Codifying Fear

Clearly, food retailers and processors understand and capitalise on the fear that consumers have about their food. The space that these intermediaries occupy in the food chain privileges them to act as translators of consumer fears. From this vantage point, food processors and retailers are able to interpret, focus and consolidate consumer concerns into market opportunities. Preoccupations about food safety are fed back to consumers in codified forms through a myriad of techniques including product labelling, advertising, use of certified standards and website development. As one food distributor describes, the organic food network guarantees that:

> 'Organically grown foods are produced without the use of synthetic fertilizers, pesticides, herbicides, hormones or antibiotics. Organic certification is the consumer's guarantee that foods are grown and handled according to strict standards that are verified by independent organisations. Certified Organic foods are not irradiated and do not contain genetically modified organisms' (March 2003).

This sample description of the benefits of organics is a check list that mainstream food producers can use to fit themselves into this market stream (Guthman 2004a). Ironically, the power of these labels is bolstered by a conventional network that has taught consumers to trust labels and standards. Sophisticated marketing and labelling in concert with government food safety regulations work to create an image of food safety. And, the alternative network uses certification standards and labels to confirm the authenticity of their products. The existence of standards in both food networks provides a way into the AFN for mainstream food processors as they can more easily co-opt this codified knowledge. Another retailer in describing their role in the community explains,

> 'We work tirelessly, advocating fewer and safer pesticides in non-organic foods, in educating our customers about the value of foods produced without harmful or questionable food additives, and we have worked with manufacturers to supply our stores with foods that meet our strict quality standards... We educate our customers about the importance of food safety measures and techniques, including our concerns about irradiation, food borne illnesses, food handling, and material safety' (Retailer, April 2003).

One food processor has been determined to offer her products as an alternative to industrialised food and describes them as:

> 'using no chemicals, no additives, no preservatives, no fillers, no food colouring, no food flavouring—all our condiments are GLUTEN FREE, WHEAT FREE, DAIRY FREE, EGG FREE, YEAST FREE, NO SUGAR ADDED, SOY FREE and VEGAN' (emphasis in original) (Organic processor, February 2003).

The products are defined in terms of what they do not contain. At the same time, however, she is not completely secluded from the mainstream; and in fact one could argue that she is still deeply part of it. Since we interviewed her 2 years ago, her business has entered a new phase, growing to the extent that she is considering outside investors and larger retailers to broaden product exposure and increase economic viability. The evolution of this business speaks to the conflicted nature of alternative food network and the up-scaling versus down-scaling tensions that exist as the industry expands and changes. The personal fears that led her into the business are reflected in the societal fears that help her business expand and industrialise. By translating those fears into product characteristics, processors and retailers facilitate the co-option process.

As Kneafsey *et al.* (2004) make clear, there are two sides to the way these concerns are expressed. A review and classification of the language used on processor and retailer websites exposes the duality of the fear approach (see Table 16.1). On the one hand, processors and retailers use fear-based language to capture the market and remind consumers about why they need to buy alternative food. On the other hand, they offer solutions in terms of human, animal and environmental health. On the negative side, products are described as not containing a range of chemicals, additives and food qualities. On the positive side, the alternative food products and stores offer health and well being to consumers and the world they live in. By emphasising these dualities, the alternative network offers consumers a way to recover some of the power they have lost to the industrial food system. It gives people a way to control what they are putting into their bodies (Whatmore 2002).

Tacit knowledge, or 'knowledge that has not been documented' (Lundvall 2003: 6), is another matter. This leads us to our final point – that having identified fear as a defining factor of AFN, it is possible to circumnavigate fear to embrace hope. We suggest by clarifying the non-codified strengths of the alternative network, we can reinforce its success. It is through hope that we see the greatest potential for a re-framing of the food system.

Table 16.1: Fear-based and positive descriptors from key informant advertising and websites.

Firm type	Fear descriptors	Positive descriptors
Processors	Free of pesticides and herbicides	We are committed to providing all you need to live a healthy life
	No synthetic fertilisers, pesticides, herbicides, fungicides or insecticides	Biodynamic, ecological
	Livestock cannot be fed or treated with synthetic antibiotics, growth hormones, genetically modified organisms, colouring agents, animal by-products or medicated feed	Ethical environmental responsibility Community health and well-being
	Wheat-free	Delicious and convenient
	No preservatives	Fresh, flavourful, seasonal
	Cruelty-free personal care products	Local, authentic, ethnic
	Additive free products	We use only the best, all-natural ingredients
	Personal health and safety	Quality
	No chemicals, genetically modified organisms, or artificial fertilisers . . . no antibiotics or growth promoters are used	We fill the box with the best locally grown seasonal items. Healthy, nourishing and full of flavour; health and well being of the earth

Manufacturing Hope in AFN

By exploring a nuanced and complex set of macro-relationships that include personal and social constructs, science and safety, and the industrialisation/de-industrialisation of food, our research points to the influence fear exerts on the food industry. In our case study, we found that the intermediaries in the food network 'envisage' food for consumers as part of the global-local food system. By capturing and then codifying consumer concerns, processors and retailers translate food fear to food product. We suggest that this process facilitates the co-option of alternative by mainstream food producers. We argue below that if we are to 'mainstream' AFN to create an alternative food system, there needs to be a shift in public focus from fear to hope.

Food fears have been a useful starting point for AFN as a stimulus propelling consumers to search out 'safe' food. The quest for safe food links people in search of the same reassurance and leads to 'alternative visions' (Kneafsey et al. 2004: 8) embedded in shorter food connections, trust- and quality-based food (Goodman 2003). However, to move beyond the marginal and to become the mainstream, our findings suggest the need for an alternative food system that offers benefits that cannot be copied by industrial food retailers and processors. Our research has begun to fill the gap that existed in our understanding of how alternative urban food processors co-opt power from consumers and consolidate it in their own hands. This has the effect of displacing the construction of AFN beyond the reach of consumers and producers. Given these constraints on the nature, scale and structure of current North American food networks, changing the way food is provided requires fundamental institutional and structural change at multiple dimensions and scales.

One way to overcome fear as an obstacle to growing the alternative food system is to go local, scale down and establish reflexive food relationships that emphasise quality and freshness. That is, to eat from our 'foodshed' (Friedmann 1993; Kloppenberg *et al.* 1996). Shorter food chains and local producer–consumer connections signal an important shift in focus from farm production-based sustainability initiatives to consumer-driven quality initiatives offering the opportunity to re-empower the consumer. Trust-based, intimate food relationships between farmers and eaters mitigate fear that emerges from large-scale, anonymous industrial food systems and the food scares they precipitate. Initiatives at the macrolevel, such as the EU Public Sector Directive 26 (Sassatelli 2006), and at smaller scales, such as student-driven university buy-local programmes in the US and Canada (for example, Blay-Palmer *et al.* 2006; Yale 2006), provide templates to re-design the way people connect with their food. Hope can be constructed within the alternative system as it tries to deconstruct the old to produce new institutions and new ways of delivering food – the surge in farmers' markets, community shared agricultural projects and the increasing demand for local food are examples. These initiatives offer the chance to re-localise power through food and present practical solutions for local economic development challenges (see Kirwan and Foster's chapter on the Cornwall local food initiative in this volume). Enticott's (2003) work on unpasteurised milk (UPM) in the UK is useful in this context as it points to the way that tensions between dystopian and utopian consumer visions of food systems are resolved. Other research shows that consumers who buy food boxes balance out acceptable levels of availability with desired food 'production practices, freshness and origin' (Lamine 2005: 341). In these cases, consumers decide on levels of quality they can tolerate, find their balance between many conflicting priorities and seek out the best alternative foods to meet their needs. This process works as it is reflexive and iterative.

On a practical level, there is also the problem of 'label fatigue' as consumers must constantly read and educate themselves to engage with the food system (Goodman 2004: 10; Ilbery *et al.* 2005). Information overload can lead to confusion and the dilution of standards to the lowest possible denominator. McDonald's, for example, now has fair trade coffee in the north eastern United States and Wal-Mart recently introduced highly processed and packaged organic products, underscoring the contradictory and ambiguous nature of alternative foods and the spaces that labelling opens up for AFN. The presence of global players in the AFN can expand awareness about the merits of alternative food. However, the long distances of global industrial food supply chains and the risks these food production systems open up for contamination through large-scale farming compromise the essence of AFN as quality, trust-based, shortened supply chains. Further, if we are to understand the potential for the alternative to become mainstream, we need to quantify the willingness of consumers to engage in re-localised consumption networks (Holloway and Kneafsey 2000). We also need to determine what local and alternative means for consumers. Caution is needed though so that 'the ways in which such spaces [farmers' markets] become increasingly regulated, and the relationships between these spaces and the spaces of "conventional" food production and retail, should also be examined in recognition of tendencies towards bureaucratic and capitalist appropriation of what might become alternative economic spaces' (Holloway and Kneafsey 2000: 298; see also DuPuis and Goodman 2005; Donald and Blay-Palmer 2006).

However, scaling down does not necessarily require us to abandon top-down or scaled up policy (Morgan *et al.* 2006; Sassatelli 2006). On the contrary, it is in fact also necessary for national governments and international bodies to become involved in promoting more localised food procurement policies as well as national and international health and labelling laws that facilitate healthier eating and food re-localisation (Barling *et al.* 2002; Morgan and Morley 2003; Jackson 2004). National, regional, state or provincial government and non-governmental organisations can act as resource repositories and sources of funding, gathering expertise to help alternative food systems get established, standardised and labelled (Ilbery *et al.* 2005). Caution and vigilance are needed though to ensure that the institutions supporting inclusive activity are encouraged (Dupuis and Goodman 2005). Multiple-scaled public education policies are also important for re-framing the food system in North America. These revolutions in food systems provide a vehicle for public education that feeds change on a broader scale and could contribute to a sea change in the North American approach to eating and food emerging from the local scale. Public education needs to emphasise the 'difference' (Ilbery *et al.* 2005) so that consumers understand the benefits of buying local food for the long term – healthier food, animals and people; more robust and resilient local economies; and improved local ecologies. By building on the re-localisation of food, public education programs can provoke a lasting socio-cultural shift from food fear to food celebration. In keeping with DuPuis and Goodman (2005), to move in a truly alternative direction, the alternative food system needs to move away from fear and embrace democratic, reflexive localism. Multi-scaled changes from institutional change and public education can address problems head on through a respectful, inclusive process of food system reform.

There is also the opportunity to broaden the scope of food theory to engage more with ecological perspectives (Goodman 1999; Murdoch 2006). The validation of nature as an actor in food networks could shift our thinking from conceptualising food as a commodity to thinking about food as an essential part of our culture, to be celebrated and protected. In this way, we offer a hopeful vision for robust alternative food systems.

References

Allen, P. (2004). *Together at the Table: Sustainability and Sustenance in the American Agro-food System.* University Park, PA: Penn State University Press.
Barling, D., Lang, T. and Caraher, M. (2002). Joined–up food policy? The trials of governance, public policy and the food system. *Social Policy and Administration* 36, 556–574.
Beck, U. (1992). *Risk Society: Towards a New Modernity.* London: Sage Publications.
Beck, U. (1999). World Risk Society. Cambridge: Polity Press.
Blay-Palmer, A. and Donald, B. (2006). A tale of three tomatoes: the new city food economy in Toronto, Ontario. *Economic Geography* 82 (4), 383–400.
Blay-Palmer, A., Dwyer, M. and Miller, J. (2006). *Sustainable Communities: Building Local Foodshed Capacity in Frontenac and Lennox-Addington Counties through Improved Farm to Fork Links.* Report prepared for Frontenac and Lennox-Addington Community Futures Development Corporations. Available at: http://www.frontenaccfdc.com/downloads/Sustainablecommunities.pdf, accessed 17 November 2006.
Bobrow-Strain, A. (2005). *Since Sliced Bread: Purity, Hygiene and the Making of Modern Bread.* Paper presented at the American Association of Geographers Annual Conference, Denver.
Cronon, W. (1991). *Nature's Metropolis: Chicago and the Great West.* New York: W. W. Norton.

Cussler, M. and deGive, M. (1952). *Twixt the Cup and the Lip: Psychological and Socio-cultural Factors Affecting Food Habits*. New York: Twayne Publishers.

Donald, B. and Blay-Palmer, A. (2006). The urban creative-food economy: producing food for the urban elite or social inclusion opportunity? *Environment and Planning A* 38, 1901–1920.

Draper, A. and Green, J. (2002). Food safety and consumers: Constructions of risk and choice. *Social Policy and Administration* 36, 610–625.

DuPuis, M. and Goodman, M. (2005). Should we "go home" to eat?: Towards a reflexive politics of localism. *Journal of Rural Studies* 21, 359–371.

Enticott, G. (2003). Risking the rural: nature, morality and the consumption of un-pasteurised milk. *Journal of Rural Studies* 19, 411–424.

Friedmann, H. (1993). The political economy of food: a global crisis. *New Left Review* 197, 29–57.

Goodman, D. (1999). Agro-food studies in the 'age of ecology': nature, corporeality, bio-politics. *Sociologia Ruralis* 39, 17–38.

Goodman, D. (2003). Rural Europe redux? Reflections on alternative agro-food networks and paradigm change. *Sociologia Ruralis* 44, 3–16.

Goodman, D. and Watts, M. (1997). *Globalising Food: Agrarian Questions and Global Restructuring*. New York: Routledge Press.

Guthman, J. (2004a). *Agrarian Dreams: The Paradox of Organic Farming in California* Berkeley: University of California Press.

Guthman, J. (2004b). The trouble with 'organic lite' in California: a rejoinder to the 'conventionalisation' debate. *Sociologia Ruralis* 44, 301–316.

Harhoff, D., Regibeau, P. and Rockett, K. (2001). Genetically modified food: evaluating the economic risk. *Economic Policy* 6, 264–299.

Hinrichs, C. (2000). Embeddedness and local food systems: notes on two types of direct agricultural market. *Journal of Rural Studies* 16, 295–303.

Hinrichs, C. (2003). The practice and politics of food system localization. *Journal of Rural Studies* 19, 33–45.

Holloway, L. and Kneafsey, M. (2000). Reading the space of the farmers' market: a preliminary investigation from the UK. *Sociologia Ruralis* 40, 285–299.

Ilbery, B. and Maye, D. (2005a). Food supply chains and sustainability: evidence from specialist food producers and in the Scottish/English borders. *Land Use Policy* 22, 331–344.

Ilbery, B. and Maye, D. (2005b). Alternative (shorter) food supply chains and specialist products in the Scottish–English borders. *Environment and Planning A* 37, 823–844.

Ilbery, B., Morris, C., Buller, H., Maye, D. and Kneafsey, M. (2005). Product, process and place: an examination of food marketing and labeling schemes in Europe and North America. *European Urban and Regional Studies* 12, 116–132.

Jackson, T. (2004). *Motivating Sustainable Consumption: A Review of Evidence on Consumer Behaviour and Behavioural Change*. Report prepared for DEFRA, London, Policy Studies Institute. Available online at: http://portal.surrey.ac.uk/pls/portal/docs/PAGE/ENG/STAFF/STAFFAC/JACKSONT/PUBLICATREVIEW.PDF, accessed 16 November 2006.

Jolly Time Popcorn. (2004). *Press Release: JOLLY TIME Pop Corn Expands "Smart Snack" Offerings With Healthy Pop 94% Fat Free Caramel Apple Microwave Pop Corn*. Available online at: http://www.jollytime.com/newsroom/releases/caramel_apple.asp, accessed 17 November 2006.

Klint-Jensen, K. (2004). BSE in the UK: why the risk communication strategy failed. *Journal of Agricultural and Environmental Ethics* 17, 405–423.

Kloppenberg, J., Jr., Hendrickson, J. and Stevenson, G. (1996). Coming in to the foodshed. *Agriculture and Human Values* 13, 33–41.

Kneafsey, M., Holloway, L., Venn, L., Cox, R., Dowler, E. and Tuomainen, H. (2004). *Consumers and Producers: Coping with Food Anxieties through 'Reconnection'?* Cultures of Consumption and ESRC-AHRB Research Programme Working Paper Series, Working Paper No. 19, Available online at: http://www.consume.bbk.ac.uk, accessed 28 November 2005.

Lee, R. and Leyshon, A. (2003). Conclusions: re-making geographies and the construction of 'spaces of hope'. In A. Leyshon, R. Lee and C. Williams (Eds), *Alternative Economic Spaces* (pp. 193-198). London: Sage Publications.

Levenstein, H. (2004). *Fear of Food*. CBC Ideas Program Transcripts. Developed by Jill Eisen. Produced by Alison Moss with the assistance of Liz Nagy. Archival research by Debra Lindsay. Technical operations by Dave Field.

Lamine, C. (2005). Settling shared uncertainties: local partnerships between producers and consumers. *Sociologia Ruralis* 45, 324-345.

Lundvall B. (2003). *The Economics of Knowledge and Learning*. Department of Business, Aalborg University. Available online at: http://www.globelicsacademy.net/pdf/BengtAkeLundvall_1.pdf, accessed 17 November 2006.

Minou, Y. and Willer, H. (2003). *The World of Organic Agriculture 2003 – Statistics and Future Prospects*. Available online at: http://www.ifoam.org/, accessed 28 November 2005.

Morgan, K. and Morley, A. (2003). *School Meals: Healthy Eating & Sustainable Food Chains*. Candiff: The Regeneration Institute, Cardiff University.

Morgan, K., Marsden, T. and Murdoch, J. (2006). *Worlds of Food: Place, Power and Provenance in the Food Chain*. Oxford: Oxford University Press.

Murdoch, J. (2006). *Post-structuralist Geography*. London: Sage Publications.

Murdoch, J. and Miele, M. (1999). 'Back to nature': Changing 'worlds of production' in the food sector. *Sociologia Ruralis* 39, 465-483.

Murdoch, J., Marsden, T. and Banks, J. (2000). Quality, nature, and embeddedness: some theoretical considerations in the context of the food sector. *Economic Geography* 76, 107-125.

Nestle, M. (2004). *Food Politics: How the Food Industry Influences Nutrition and Health*. Berkeley, CA: University of California Press.

Phillips, P. (2002). Biotechnology in the global agro-food system. *Trends in Biotechnology* 20, 375-381.

Sassatelli, R. (2006). *Empowering Consumers: The Creative Procurement of School Meals in Italy and the UK*. Available at: http://www.matforsk.no/web/wakt.nsf/ee079377855e855ec1256dbb002f1e6b/e5c31b2792458b3ec12571640033ffbe/$FILE/Empowering%20Consumers.doc, accessed 27 August 2006.

Schlosser, E. (2002). *Fast Food Nation: The Dark Side of the All American Meal*. New York: Perennial Press.

Sonnino, R. and Marsden, T. (2006). Beyond the divide: rethinking relationships between alternative and conventional food networks in Europe. *Journal of Economic Geography* 6, 181-199.

Stacey, M. (1994). *Consumed: Why Americans Love, Hate and Fear Food*. New York: Simon and Schuster.

Statistics Canada (2005). *Canadian Statistics: Population urban and rural, by province and territory (Canada)*. Available at: http://www40.statcan.ca/l01/cst01/demo62a.htm?search strdisabled=1951&filename=demo62a.htm&lan=eng, accessed 21 May 2006.

State Library of Iowa (2005). *Urban and Rural Population (1850-2000) and Metropolitan and Non-metropolitan Population (1950-2003)*. Available online at: http://data.iowadatacenter.org/browse/urbanruralareas.html#Population, accessed 15 September 2006.

United Nations Development Program (UNDP) (2004). *Human Development Report 2004. Cultural Liberty in Today's Diverse World*. London: Oxford University Press.

Watts, D., Ilbery, B. and Maye, D. (2005). Making reconnections in agro-food geography: alternative systems of provision. *Progress in Human Geography* 29, 22-40.

Whatmore, S. (2002). *Hybrid Geographies: Natures, Cultures, Spaces*. London: Sage Publications.

Whatmore, S., Stassart, P. and Renting, H. (2003). Guest editorial: what's alternative about alternative food? *Environment and Planning A* 35, 389-391.

Winson, A. (1993). *The Intimate Commodity: Food and the Development of the Agro-Industrial Complex in Canada*. Guelph, Canada: Garamond Press.

Winter, M. (2003). Embeddedness, the new food economy and defensive localism. *Journal of Rural Studies* 19, 23–32.

Yale. (2006). *Organic Options Growing Across Yale*. Available online at: http://www.yale.edu/sustainability/foodproject.htm, accessed 10 August 2006.

Chapter 17

Networking Practices among 'Alternative' Food Producers in England's West Midlands Region

David Watts[1], Brian Ilbery[2] and Gareth Jones[3]
[1]Institute for Transport and Rural Research, School of Geosciences, University of Aberdeen, UK
[2]Countryside and Community Research Unit, University of Gloucestershire, UK
[3]National Farmers' Retail and Markets Association (FARMA), Southampton, UK

Introduction

This chapter examines the networking practices of a sample of 'alternative' food producers located in England's West Midlands (WM) region (which is shown in Figure 17.1). It aims to contribute to the development of a more nuanced understanding of so-called alternative food networks (AFN) in an advanced industrial country and to reflect on their possible spatial characteristics. The setting of these aims was guided by recent studies that appear to problematise both the 'alternativeness' of AFN and their relationship with particular geographical scales, notably the 'local'.

A useful starting point for reflecting on the development of AFN is provided by Whatmore *et al.* (2003).

> 'Far from disappearing, those diverse and dynamic food networks that had been cast as remnant or marginal in the shadow of productivism have strengthened and proliferated. This unexpected turn of events has garnered unprecedented interest from researchers and policymakers in, variously, 'alternative' and/or 'quality' and/or 'local' food networks... These overlapping but nonidentical collective nouns... share in common... their constitution as/of food markets that redistribute value through the network against the logic of bulk commodity production;

290 *Practising Alternative Food Geographies*

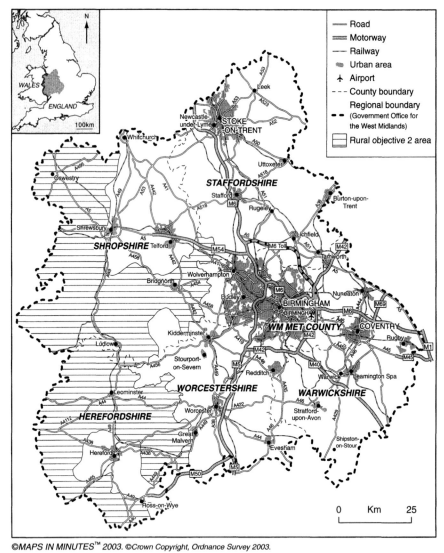

©*MAPS IN MINUTES*™ *2003.* ©*Crown Copyright, Ordnance Survey 2003.*

Figure 17.1: The West Midlands region: selected administrative zones, places and transport links.

that reconvene 'trust' between food producers and consumers; and that articulate new forms of political association and market governance' (Whatmore *et al.* 2003: 389).

That quotation sets the context for, and shows the limitations of, this chapter. The key limitation is that this chapter does not engage with consumers' uses of AFN.[1] It means

---

[1] Although, in mitigation, it should be noted that the empirical data on which this chapter is based were collected as part of a larger research project that did examine the role of consumers in AFN.

that the chapter cannot comment on what factors are sufficient for the success of AFN, which would of course fail if nobody bought and ate the food produced (cf. Lee 2000: 140). Instead, it is premised on the argument that a focus on the networking practices of 'alternative' food producers is necessary in order to investigate both the 'alternativeness' of the 'supply side' of AFN, and their possible spatial outcomes.

The discussion is structured as follows. The following section reflects on some key debates in the academic literature that are relevant to recent studies of AFN in the European Union (EU) and the US. The third section summarises the methodology for a study of 'alternative' food producers in the WM, the data from which are analysed in the fourth. Finally, the authors, draw together the empirical and theoretical sections and reflect on their contribution to the development of a more nuanced understanding of the 'alternativeness' of the 'supply side' of AFN and their possible spatial outcomes.

Some Key Debates Informing Recent Studies of AFN

Of the three 'types' of AFN identified by Whatmore *et al.* (2003) – 'alternative', 'local' and 'quality' – the last seems to have generated most interest among researchers. Indeed, the volume of studies and policy initiatives devoted to them has led Goodman (2004) to argue that there has been a 'quality turn', which constitutes a new rural development paradigm in the EU. This 'quality turn' has been influential, particularly among those seeking to foster endogenous economic development in 'lagging' (e.g. EU Objectives 1 and 2)[2] rural areas. Food has been considered important to this process, both as part of a 'regional culture' that can be commodified to boost tourism (e.g. Ray 1998; Kneafsey *et al.* 2001) and as a means of trying to retain value added in rural areas by protecting and promoting foods that have a link to them. Examples of this are the EU's protected designation of origin (PDO) and protected geographical indication (PGI) labels, awarded to food that either originates from, or is processed in, a specific area (Ilbery and Kneafsey 2000a; Parrott *et al.* 2002).

The production of 'quality' food products can be interpreted as an alternative to agricultural productivism. Indeed, it has been presented as part of a post-productivist transition (Ilbery and Kneafsey 1999). However, it is doubtful whether the 'quality turn' is 'alternative' in the sense that the term has recently been used in economic geography (e.g. by Leyshon *et al.* 2003; Hughes 2005). This work discusses practices that seek to 'perform the economy otherwise' (Leyshon and Lee 2003: 4), thereby breaking down 'the traditional association between economy and capitalism' (Hughes 2005: 499; c.f. Lee 2000: 138). This consideration of 'alternativeness' has found its way into some studies of land-based economic activity: for example, through work on relations of 'regard' by Lee (2000) and Sage (2003).

The main reason why the production of 'quality' food products is unlikely, in and of itself, to 'perform the economy otherwise' is because most (e.g. those protected by

---

[2]For more detail on EU Objectives 1 and 2 areas, see http://ec.europa.eu/comm/regional_policy/objective1/index_en.htm and http://ec.europa.eu/comm/regional_policy/objective2/index_en.htm respectively (sites accessed 12 May 2006).

PDO or PGI) are locality foods.[3] As Watts et al. (2005) argue, the production of locality foods is analogous to flexible specialisation, a hallmark of the transition to post-Fordism. And, as Potter and Tilzey (2005) argue, it is just such a transition that EU agriculture is currently experiencing. Potter and Tilzey (2005: 589) also argue that debate over agricultural restructuring in the EU is dominated increasingly by a neoliberal agenda. This, in turn, stems from the global hegemony of the neoliberal project (Peck and Tickell 2002: 381), manifested in the World Trade Organisation's Uruguay round of negotiations aimed at 'liberalising' trade in agricultural products. It can be argued, therefore, that the 'quality turn' is unlikely to give rise to new forms of market governance. For, not only is it compatible with the continued hegemony of the neoliberal project, it can also be interpreted as a manifestation of the transition to a post-Fordist European agriculture and could, therefore, be said to be part of it.

This is why Watts et al. (2005) regard those AFN that are concerned with 'quality' foods as presenting 'weaker' alternatives to currently dominant food supply chains (FSC). For, while they present an alternative to agricultural productivism, they are often distributed through 'conventional' FSC and, in a number of cases, are produced by large companies subject to capitalist modes of evaluation (notably in the UK – see Ilbery and Kneafsey 2000a).

Instead, Watts et al. (2005) argue that it is AFN that prioritise the networks through which food products move that are more likely to present 'stronger' alternatives to currently dominant FSC. The type of AFN with which they make this argument is concerned with the production and sale of 'local' food: that is, food consumed within a limited radius (commonly 50–80 km or within a given local administrative district) of its site of production. However, a number of authors have expressed concerns over the valorisation of the 'local'. Winter's (2003) critique of 'defensive' localism – the potential for local food networks to exclude remote areas and those with poorer quality agricultural land, and the danger that focusing on the 'local' could truncate political action[4] – has made an important contribution here and has received wide coverage. In addition, DuPuis and Goodman (2005) warn about the dangers of 'unreflexive localism', whereby:

'a small, unrepresentative group decides what is "best" for everyone and then attempts to change the world by converting everyone to accept their utopian ideal' (DuPuis and Goodman 2005: 361).

One of the studies with which they develop this critique is Hendrickson and Heffernan's analysis of Kansas City's Food Circle, whose aims – the health of the community and the environment; local control over the food system; and a 'dense' social network – are represented as being preferable to the current state of affairs (Hendrickson and Heffernan 2002: 362). This, DuPuis and Goodman argue, is problematic for two reasons.

---

[3] This term is drawn from the report of the Policy Commission on Farming and Food (2002: 43). 'Locality' foods have a specific geographical provenance but circulate nationally and possibly internationally. They are therefore distinct from 'local' food, which is retailed within a relatively short distance – commonly 50–80 km – of its site of production.

[4] Based on Holloway and Kneafsey's (2000) critique of English farmers' markets as being potentially socially conservative, parochial and nationalistic.

'First, it can deny the politics of the local, with potentially problematic social justice consequences. Second, it can lead to proposed solutions, based on alternative standards of purity and perfection, that are vulnerable to corporate cooptation' (DuPuis and Goodman 2005: 360).

Such criticisms do not, as DuPuis and Goodman (2005: 368) emphasise, apply equally to all 'local' food networks. Instead, they demonstrate that it is necessary to pay attention to the social construction of the 'local', by and through food networks that purport to offer an 'alternative', before coming to a judgement as to whether they are 'deployed for or against global forces' (DuPuis and Goodman 2005: 368).

Work that engages with these issues in the UK has yielded further difficulties. For example, Morris and Buller (2003: 361) argue that the local food sector in England is 'empirically contestable and spatially indeterminate'. Ilbery and Maye's (2005a, 2006) work with 'alternative' food producers and food retailers in the Scottish–English borders found that individual enterprises are often involved in 'local' and conventional FSC simultaneously. They suggest that 'local' food networks are not, in practice, separate from the 'the rest of the agro-food system' (Ilbery and Maye 2006: 355), and that it may be better to 'think instead about *hybrid food spaces*' (Ilbery and Maye 2006, original emphasis). Such critiques dovetail with more widespread concerns about 'the continually fraught relationship between the alternative and the conventional' (Hughes 2005: 501).

Running parallel to this problematisation of the functional and spatial distinctiveness of 'local' food networks are recent contributions to debate over the 'relational turn' in economic geography (e.g. Hess 2004; Yeung 2005). Spatial relations are central to this 'relational turn'.

'Territorial development is theorised to be significantly embedded in networks of relational assets and spatial proximity, particularly at the local and regional scales, such that "territorialization is often tied to specific interdependencies in economic life"' (Yeung 2005: 39, embedded quote from Storper 1997: 20).

However, Hess argues that much of this work has used: 'an "overterritorialized" concept of embeddedness by proposing "local" networks and localized social relationships as the spatial logic of embeddedness' (Hess 2004: 174). Hess is particularly critical of an often posited, but theoretically unjustified, 'mutual determination of spatial proximity and trust' (Hess 2004: 175) in business networks, and questions the attribution of trust to one particular geographical scale. Indeed, after reviewing the theoretical literature, Hess (2004: 180) goes on to argue that there is no necessary relationship between network and territorial embeddedness.

While Hess' (2004) argument is convincing, it leaves itself vulnerable to critique by not defining what it means by 'trust'. This is important, because trust is widely considered to be an important component of AFN. For example, the growth of AFN has been attributed, in part, to an erosion of trust in conventional FSC (Ilbery and Kneafsey 2000a: 317; Renting *et al*. 2003: 396). By contrast, the presence of trust-based relationships is considered important to endogenous rural development in general (Knickel and Renting 2000; Murdoch 2000; Ventura and Milone 2000; Kneafsey *et al*. 2001; Marsden *et al*. 2002) and to AFN in particular (e.g. Hendrickson and Heffernan 2002: 361; Whatmore *et al*. 2003: 389). Research in England has found high levels of trust in local food among its consumers (Dürrschmidt 1999: 149–150; Archer *et al.* 2003: 492; Youngs 2003: 538). It is not surprising, therefore, that farmers in the English county of Gloucestershire, interviewed by Morris and Buller (2003: 564), claimed that one of the benefits of becoming involved with the local food sector is that it has helped re-establish trust between consumers and producers.

However, as none of these studies define what they mean by 'trust', they do not provide a convincing counter-argument to that developed by Hess (2004). Indeed, the plethora of definitions of 'trust' used in studies of supply-chain relationships (Sahay 2003: 554) means that making such a counter-argument would be almost impossible. Moreover, Blois (2003: 191) warns that it may not always be appropriate for producers to 'trust' their customers. It is arguable, therefore, that what many researchers have done is naturalise their own definition of trust by treating it as fixed and as signifying something that is intrinsically good (DuPuis and Goodman 2005: 364).

This does not mean, however, that the concept of 'trust' can (or should) be dismissed as an ideological construct, used to assert a sufficient link between what Hess (2004: 180) terms network and territorial embeddedness. Rather, what is required, with regard to 'local' and 'alternative' food networks, is the development of a more nuanced understanding of how their constituent producers interact, and whether there is anything 'spatial' about that interaction. For, as Yeung (2005: 42) argues, what is often lacking, when considering concepts such as 'relations' and 'networks' (to which could be added 'trust') is any consideration of how 'concrete/spatial outcomes are produced through them'.

Specifically, for Yeung (2005), what are lacking are a focus on the practices that (re)produce actors' inter-relations and networks and an emphasis on the role of power (defined as an actor's exercise of their capacity to influence others) on those practices. Although, as Yeung (2005: 46–47) acknowledges, the 'concrete' outcomes of past practices will, in turn, influence them in the future, this focus on practice seems to offer a way forward. Indeed, a concern with the practice of networking is another 'current' of recent debate that appears to being having an influence on studies of rural development (e.g. Jones 2003; Clark 2005; Lee *et al.* 2005).

Of these, the study that is most relevant here is Clark (2005), which examines agri-food diversification in England's East Midlands region. Three findings are pertinent. First, Clark (2005) demonstrates that individual farmers are involved simultaneously in 'conventional' and 'alternative' food networks (cf. Ilbery and Maye 2006). Second, a strong case is made for the influence of geographical factors:

'ninety-three per cent of all respondents [$n = 118$] characterised agro-food diversification as a "locally-driven" activity and claimed that features of embeddedness, including importance of natural and non-natural capitals, the needs of local economies, distance to market, access to labour, and existence of territorial infrastructure, directly affected agro-food diversification outcomes' (Clark 2005: 484).

Third, Clark (2005) emphasises the importance of both practice and what Yeung (2005: 46) terms 'complementary' power in the building and maintenance of networks. For instance, the design of collaborative projects enables network members 'to reposition their production–consumption activities outside conventional agro-FSC, and to recast existing power relations established under the agro-industrial model' (Clark 2005: 494).

However, Clark's (2005) main focus is on the outcomes, not the practices, of networking. This limits its applicability here in three ways. First, although Clark (2005: 494) lists a number of important network components, little is said of the practice of mobilising them other than that 'networking needs substantial investment of time and social and economic capitals' (Clark 2005). Second, there is little discussion of what Yeung calls 'relational' power: that is, power relations specific to the network in question, 'whereby dedicated commitment is enforced' (Yeung 2005: 46). Last, although Clark (2005: 490) considers 'trust' to be an important characteristic of a successful network, no definition is offered.

The purpose of setting out these limitations is not to attempt to undermine Clark's (2005) work. Rather, they set the context for the following sections. Thus, the remainder of this chapter will focus on three issues: the networking practices of a sample of 'alternative' food producers in the WM and the deployment of trust and relational power in those practices.

Research Methodology

This section describes the conduct of the research that forms the empirical basis for this chapter and explains how the debates in the previous section influenced the analysis of the data gathered. Research data were collected from interviews with 20 local and locality food producers in England's WM region. The main characteristics of each interviewee's enterprise are listed in Table 17.1. All interviewees produce and/or sell local and/or locality foods, and sixteen process or sell produce from their own farm or from one belonging to a family member or colleague. Most defined themselves as local or locality food producers through advertisements in local, regional or national food directories. Others were successful applicants for food-related grants from the England Rural Development Plan whose business activities, on enquiry, positioned them as producers of local or locality foods. The choice of discussion topics was guided by a review of the relevant literature (Watts *et al.* 2005) and covered interviewees' food production and/or retailing activities; their judgement of the quality of their produce; and their relationships with those from whom they buy and to whom they sell. As the aim was to elicit interview subjects' qualitative descriptions of these phenomena, questions were semistructured and open-ended (Kvale 1996: 124), allowing issues to be explored at a level of detail appropriate to each interviewee. Interviews were recorded, transcribed and coded according to the discussion topics. To generate additional data about their networking practices, interviewees were also asked to list the sources of their main inputs and the destinations of their outputs. This information was used to draw business network diagrams (cf. Blundel 2002; Ilbery and Maye 2005a, 2006).

Analysis of the coded transcripts highlighted the importance of networking, trust and integrity to the interviewees. These are discussed in the next section, within the context of the chapter's aims. However, a necessary prelude to that discussion, given the concerns raised above about the imprecise use of concepts such as 'network' and 'trust', is a clarification of their use here. As the terms 'networking' and 'trust' have already received considerable attention, the following discussion of them will focus on aspects of each that have received little coverage to date. The term 'integrity' will also be clarified, as it has been little used in agri-food studies.

Table 17.1: Characteristics of interviewees' agricultural and non-agricultural enterprises.

Interviewee code	Characteristics of non-agricultural business	Farm type (farm size in hectares)
a	Abattoir and butcher; local wholesale; retail and mobile shops	Not a farmer
b	Village shop, post office and delivery office	Conventional; sheep (32.4)
c[a]	Dairy processor with shop on site; regional trade deliveries; attends farmers' markets	Organic; dairy (89)
d	Cider and perry production; national distribution; farm shop and restaurant	Organic; orchards and beef cattle (12.1)
e	Fruit juice production; local distribution; farm gate sales	Conventional; arable, potatoes, fruit, honey and beef cattle (47)
f	On-farm butchery and shop; some local delivery	Conventional; sheep, cattle and pigs (32.4)
g	Farm shop	Conventional; arable, sheep, fruit and vegetables (164)
h	On-farm butchery and shop	Conventional; beef, sheep, pigs and arable (145.7)
i	Milk and cream processing; local doorstep and trade deliveries; farm gate sales	Conventional; dairy (72.9)
k	Farm gate and local sales	'Almost organic'; pigs, cattle and sheep (8.9)
l	Cheese, cream and yoghurt production; farm gate sales; attends farmers' markets	Organic; dairy (28.3)
m[a]	Two butchers' shops; local (own van) and national delivery	Conventional; beef and arable (647.5)
n	On-farm butchery and shop; some local delivery	'Natural'; cattle and fodder crops (9.3)
p	Local lamb sales and delivery	Conventional; sheep, fodder crops and some cattle (140)
r	Conventional and high-quality cider production; national trade distribution; farm shop	Conventional; apple orchards (80.94) plus 20.24 hectares grazing land rented out
s	Monthly farmers' market, touring to regional food fairs	Organic; fruit and vegetables, plus some sheep and pigs (4.1)

Table 17.1: Continued

Interviewee code	Characteristics of non-agricultural business	Farm type (farm size in hectares)
t	Makes preserves; farm gate sales; local delivery; sells through local shops	Organic; sheep, cattle, fruit and vegetables (32)
u	Ice cream producer; direct sales at events, festivals, etc.; mail order; regional and national trade distribution	No longer involved in farming
w	Cider and apple juice production; pick-your-own; farm gate and local trade sales	Conventional; orchards plus arable and vegetables (31.2)
x	Makes fruit syrups; national distribution; attends farmers' markets	Organic (not registered); orchard (1.2)

a In these two cases, the farm listed is a part-owner of the non-agricultural enterprise.

One of the present authors has considerable experience in dealing with farmers who have diversified into retailing, and from this deduced that their networking practices tend to be of three main types. When applied to the interview data, this typology was found, with the addition of a further type of highly informal networking, to be a useful heuristic basis for distinguishing between different types of networking practice.[5] Thus, this typology is used to structure the discussion of the business networking practices of these 20 'alternative' food producers. Its four types of networking are informal, non-material and often untraded exchanges with other enterprises; collaborative stocking of shops by producers local to one another; informal cooperation, which could involve carrying another's goods or livestock on a journey that one had to make oneself; and joint investment. Although this typology can be considered hierarchical, as the different types of networking are divided according to the degree of resource commitment required, it is not proposed as a developmental model of networking practice, 'up' which 'alternative' food producers should be encouraged to progress. The data analysed do not support such a hypothesis. Rather, it should be considered as an initial, heuristic attempt to build a more nuanced understanding of the types of networking practices that might constitute AFN. Its validity, therefore, cannot be assumed, but requires testing through further research.

It was noted in the previous section that, although the concept of 'trust' occurs repeatedly in studies of rural development and AFN, most seem to treat it as little more than an ideological construct. For 'trust' to be a credible (i.e. empirically contestable) concept, some attempt must be made to provide a working definition of it. The one that seems most applicable here is Blois' (1999). In commercial relationships, Blois (1999) distinguishes between 'trust' and 'reliance'.[6] Reliance is the belief of one party that another party will fulfil their side of a bargain. Trust encompasses, but goes beyond,

---

[5] Following Lee et al. (2005: 269), practice may be defined as the 'actualities of the processes' that constitute a particular phenomenon.
[6] In a more recent article, Blois (2003) uses the terms 'weak' and 'strong' trust instead. However, for the sake of clarity, 'trust' and 'reliance' are used here.

reliance by embodying an expectation of benign intent on the part of the other: a belief that they will consider and act in the interests of both parties (Blois 1999: 199–200).

Last, the term 'integrity' is used here because it captures something of the beliefs and behaviour discussed by interviewees. Its use could be considered problematic, as 'integrity' is a complex notion with a moral dimension (Pascalev 2003). These complexities cannot be explored in any detail here, so its use is restricted to its lay definition, denoting 'wholeness; the unimpaired state of anything; uprightness; honesty' (*Chambers Dictionary* 1993: 869). Its moral aspects are considered insofar as they relate to Yeung's (2005: 46) concept of 'relational' power in the context of interviewees' networking practices.

Empirical Reflections on Networking, Trust and Integrity

Several interviewees produce, or are closely connected (by ties of family or co-ownership) to those who produce the key raw material for at least some of what they sell and/or process (see Table 17.1). However, few grow or rear all that they sell, and even fewer do all of any processing required to make their produce saleable. The exceptions are two fruit growers who sell or process all their produce (interviewees r and w), but even they buy in some ingredients (e.g. cultures and sugar). Thus, all the interviewees rely, to varying extents, on external actors in order to make their produce marketable.

The business network diagrams show that, in most cases, a significant proportion of interviewees' inputs and outputs travel less than about 80 km to and from their premises (e.g. Figure 17.2). However, over a third of interviewees (7) sell a significant proportion of their produce through conventional channels, so its final destinations are unknown. Indeed, four livestock producers dispose of the great majority of their animals through conventional markets, while selling a small proportion direct to consumers (interviewees g, h, m and p). Although one interviewee pointed out that the animals for their shop are hand-picked (interviewee m), the fact remains that similar beasts from these farms find their way into different FSC. This supports Ilbery and Maye's (2005b, 2006) finding that there is no straightforward division, at the level of the individual enterprise, between production for local and non-local markets. Similar caution must be exercised regarding interviewees' inputs. While most draw a significant proportion of these either from their own farm (or that on which they are located) or from within about 80 km, these may not originate locally (cf. Ilbery and Maye 2006: 362).

Nevertheless, it remains the case that most interviewees obtain a significant proportion of their inputs from within about 80 km. There are some familiar explanations as to why this is so. The transport and time costs of obtaining inputs seem likely to favour the use of local suppliers. In addition, the potential to exploit economies of scope (Marsden *et al.* 2002) helps to explain the use of inputs produced on interviewees' farms (or those where they are located). However, interviewees highlighted other reasons why obtaining local inputs is important to them. These are discussed in the context of the typology of producer networking outlined in the previous section.

The least resource-intensive networking activity mentioned by interviewees concerned the informal exchange of views and information that can occur when talking to suppliers, customers or fellow entrepreneurs. This may take the form of mutual commiseration and morale boosting (interviewee f). Another interviewee enjoys visiting a particular supplier because they face similar challenges:

Networking Practices Among 'Alternative' Food Producers 299

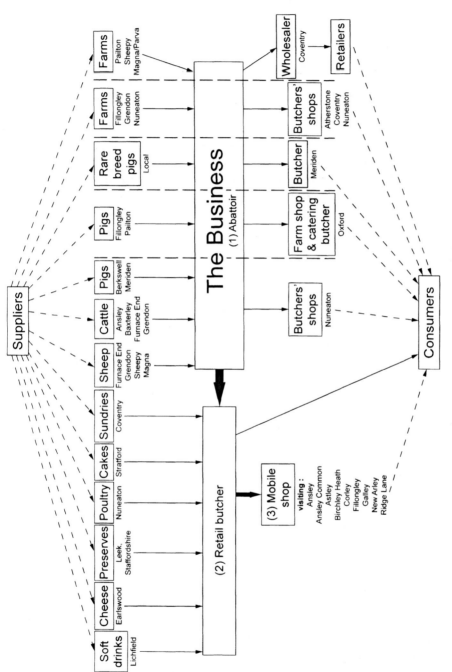

Figure 17.2: Business Network Diagram for interviewee 'a'.

'It's so informal, it's friendly, you don't have to dress up, you just go down and it's not... I don't phone them up and ask for advice but I get a pick-me-up and a pep talk every time I go' (interviewee b).

Such informal interaction, whether with contacts from individuals' social or business networks, can have material benefits. For example, two interviewees reported that informal networking generates word-of-mouth advertising (interviewees g and l). Another, who subsequently put in a successful grant application, found out about the scheme from a friend who had seen a newspaper article about a successful applicant and who passed the details on in the hope that they would be useful (interviewee a). Yet another, who was minded to apply for a grant, was able to get advice from a friend, who happened to be an agricultural consultant, on the most appropriate scheme to apply to (interviewee h).

At a higher level of resource commitment are relationships based on mutual compatibility: the second type of networking. Two interviewees, for example, benefit from neighbouring food producers selling some of their (the interviewees') produce through their (the neighbours') retail outlets. In both cases, however, this small-scale collaboration occurs within the context of large-scale economic relationships. However, some such small-scale relationships appear to be free-standing. One interviewee, for example, runs a village shop that sells fresh produce from local farmers, which can account for 40 per cent of their fresh food sales depending on the season (interviewee b). The quantities of food involved are not large, so the arrangement is 'not making anyone huge amounts of money' (interviewee b). However, it is to everyone's advantage because it keeps money circulating locally (interviewee b). A similar rationale was put forward by a retail butcher for supplying small quantities of packaged meat to nearby village shops:

'It's not making me any money but it's helping them. Because, I think this is something else that we have to be aware of, we have to help one another. As a small retailer I don't always look at the profit margin with things like that' (interviewee n).

For these two interviewees, such networking practices serve three purposes. They generate a moderate income for both parties; they can secure additional custom (interviewee n); and they support the local economy.

Several other interviewees also reported informal relationships based on mutual compatibility. Two (interviewees m and p) always pass enquiries for services and produce that they do not supply to the same local business. In both cases, this is reciprocated, either in kind (interviewee p) or by the purchase of meat for their catering service (interviewee m). Another interviewee, who had just started trading from their new farm shop, noted that their supply chain is 'fairly informal so far' (interviewee h), with some produce being bought from farming friends in the area. For others, trading opportunities arose from informal or chance meetings. One farmer occasionally supplies produce to box schemes run by people they have met socially through a regional organic growers group (interviewee t). At least two others have traded with other stall holders at farmers' markets (interviewees c and l).

Such trading can be mutually beneficial not only by bringing buyers and sellers together, but also by allowing them to trade with one another at the same place and time as they are trading as farmers' market stallholders (interviewee l). It can, therefore, reduce both the financial and opportunity costs of transporting produce to buyers. Thus, such practices equate to the third type of networking: mutual cooperation to save resources. One interviewee reported that they have, on occasion, had the same retail outlets to deliver to as another small producer, and have met midway between their premises, exchanged goods and delivered both parties' consignments to those customers located between the meeting point and their own premises (interviewee x). Such arrangements help to 'cut down that awful [cost of] carriage' (interviewee x).

However, local networking practices extend beyond the relatively small-scale and/or occasional mutual assistance and advice exemplified above. They can have a profound influence on interviewees' businesses. As one farmer remarked:

> 'Everybody we deal with is local, basically: my brother-in-law is a grain merchant who does fertilisers as well. You scratch my back, I will scratch yours, I suppose. You know, it does have its advantages' (interviewee g).

Such studied vagueness suggests a degree of informality not fit for strangers' ears. However, other interviewees were more forthcoming about the importance of local networks to their businesses. For example, the co-owner of a small abattoir explained that:

> 'we are very very committed to buying locally, sourcing the products locally, because you have got a bit more control over what you are getting: you can get what you want and keep the quality' (interviewee a).

Similarly, interviewee f (a butcher) buys livestock locally, visiting farms and choosing them on the hoof, in order to know the quality of the animals they are getting.

Interviewees a and f obtain their meat directly from local farmers and, as meat is their stock-in-trade, both rely on these local networking practices for their livelihoods. They also rely on their suppliers to ensure that they get the livestock they have chosen. Of course, most food producers are obliged to rely on others providing inputs of the quality specified. However, it can be argued that some of the relationships discussed by interviewees are based not simply on reliance, but on trust.

The operation of trust-based relationships, in the context of trade that is of critical importance to the enterprise, was exemplified by several interviewees, none more so than the abattoir co-owner (interviewee a). As their business network diagram (Figure 17.2) shows, this enterprise, which has been located on the same premises since the 1910s, takes in livestock from a considerable number of farms, almost all of which are within 30 km. Although some of these relationships were probably formed before long-distance livestock transport became commercially viable, the key issue here is the rationale for their continuance. As noted above, one reason is the interviewee's belief that, by buying locally, they can exercise greater control over the quality of their inputs. However, when asked whether they visit farms to select livestock, they replied:

> 'Well . . . not so much now because, I mean, basically the farmers that we deal with we have dealt with them so long they know what we want. And, alright, if they have got a lamb or a beast that is a little bit on the fat side I mean they will let us know . . . you know, and we try and do something with it . . . We just kind of work with each other, it's as simple as that. But in the main there is no need to

[visit] because they know what we are looking for. We want them to keep going, they want us to keep going, that's the bottom line' (interviewee a).

As this interviewee does not directly oversee the selection of livestock for their abattoir, they do not just rely on their suppliers to send them the livestock that they have specified, they trust them to select it and to tell them of any problems, even when doing so may result in a cost to the supplier. Similarly, a processor, who relies on a farmer to pick fruit for them at a specific state of ripeness, told of an instance when the farmer contacted them to say that an entire crop had, on a hot day just before it was due to be harvested, become over-ripe and that, as a result, they would not be supplying them with any that year (interviewee x). Clearly, this was to the detriment of both parties, as the farmer was depriving the processor of a key ingredient, if only until they could locate an alternative supply. Of course, the fruit may have been rejected by the processor if the supplier had tried to deliver it as originally planned. The farmer, therefore, may have had little to lose by this action. However, their promptness appears to have been intended to minimise an unavoidable inconvenience to the processor and could, therefore, be interpreted as putting their client's interests ahead of their own, with a view to maintaining mutually beneficial relations with them in the future.

Trust, therefore, appears to function as an informal quality assurance mechanism. Take, for example, the attitude to joining a quality assurance scheme of a small-scale farmer and retailer of rare-breed livestock:

'If I thought that it would have a major impact on my market then I would go for it but the people who I sell to aren't interested in that. They meet me—it's very much a direct selling operation—and they know me. Once they have tasted my products they come back for more' (interviewee k).

When asked how they police the quality of the apples that they buy in order to supplement their own crop, two members of a fruit processing enterprise stated that it was a combination of picking dates and 'confidence in the person who is doing it' (interviewee e). One of them went on to observe that:

'quality assurance doesn't assure quality at all, it assures box ticking. Whereas we are buying from, let's say [names farmer], and we know perfectly well that . . . what he says is what he does' (interviewee e).

The other added that: '[w]e are talking to them about what is available and the rest of it right through the year pretty well' (interviewee e). Thus, the trust that exists between these interviewees and their key suppliers and customers would appear to be based on at least two things. First, confidence in their ability to produce what is wanted to the required quality. Second, a belief in their integrity: that they will alert one to any problems or opportunities that arise. These, in turn, may rely on regular contact.

Indeed, it can be argued, from these examples, that the integrity (honesty) of their suppliers, which is maintained partly through regular contact, helps to ensure the integrity (unimpaired state) of their own processing enterprises, and hence the quality of their products. However, this argument requires qualification. For it can also be argued that the more important the quality of the ingredient is to the end product, the more important is the relationship with the supplier. It is notable, for example, that for some ingredients interviewees are content to use the cheapest source. One fruit processor, who trusts to the integrity of the farmer from whom they buy their fruit, buys the sugar that they use in processing it from the cheapest supplier (interviewee x). However, for the defining ingredients of their products, the quality of which can be highly variable, such price-driven, reliance-based relationships are inadequate. It could be argued, therefore, that the quality (excellence) of the end product, as defined 'objectively' (Ilbery and Kneafsey, 2000b) by the producer, depends not only on their own integrity and on that of the processing, but also on the integrity of those who supply their key ingredients.

It is clear, therefore, that for those producers who must buy in key or defining ingredients, the integrity of their suppliers is vital to the quality of the end product. That integrity forms the basis of the trust that they bestow on their suppliers. This, in turn, would appear to depend, at least in part, on a mutual knowledge and understanding of one another maintained through regular contact. For some interviewees, this gives a spatial dimension to relationships with their key suppliers: 'the main thing is to source locally' (interviewee a).

Of course, it can be argued that the cost of transporting bulky inputs will also tend to favour local suppliers. However, this is not always the case. The economies of scale achieved by some large producers and suppliers can, along with lower costs of production in many other countries, make it more expensive to buy locally (interviewee x). Other factors, therefore, are likely to be as important, if not more so, than transport costs in any decision to obtain supplies locally. One, as argued above, is the relative ease with which regular contact can be maintained. And although people can maintain contact without meeting, face-to-face interaction is often considered preferable (interviewee b).

Propinquity can also facilitate the operation of surveillance and social sanctions. One interviewee noted that: 'the other thing we found was, from using organisations, the closer they are to us the higher standard we have had, the most honest answers, the more realistic picture' (interviewee b). Another appears to rely on an implicit threat of social sanctions to regulate the farmers' market they run:

> 'because we are small and autonomous we are able to police ourselves, as it were... because everybody knows everybody and you know when people are not obeying the rules and so forth... [O]ne of the things about farmers' markets, you have to keep the integrity going, you can't let your standards slip at all' (interviewee s).

Concluding Remarks

Comments such as those made in the previous paragraph suggest that the sorts of networking practices described above can be interpreted as displaying 'relational geometry' (Yeung 2005: 46). Yeung argues that the capacity for relational geometry to produce 'concrete' (including spatial) outcomes is proportional to their 'relational specificity': that is, to the extent to which network actors are 'dependent on and "locked-in" to their ongoing power relations for resources and information' (Yeung 2005: 47). It is certainly the case that producers who sell through the farmers' market run by interviewee s are 'locked-in' to a set of power relations: their continued attendance depends on maintaining certain standards.

It would also seem that relational geometry influences the networking practices of interviewees who process others' agricultural produce. A striking example of this was provided by an organic cider producer (interviewee d) who, to supplement their own crop, buys apples grown in orchards that they have 'annexed' from local farmers. This interviewee used the verb in its colloquial sense (i.e. to appropriate – *Chambers Dictionary* 1993: 59), indicating that they exercise a high degree of control over how the orchards from which they buy are managed. There would appear to be a high degree of relational specificity in this relationship, as the farmers in question might find it difficult to locate another buyer for their cider apples in a market dominated by large producers who frequently use imported dessert apple concentrate in order to keep costs down (noted by interviewee r).

However, the presence of two additional factors suggests that the relational power exhibited in interviewees' networking practices may not be so one-sided as these examples imply. The first factor is that these networking practices are attempts to construct 'accumulation and reproduction strategies' (Clark 2005: 495) that are alternative to the 'dominant agro-industrial logics in English agriculture' (Clark 2005). They can therefore be interpreted as attempts to build 'relational complementarity—a relational advantage defined and practiced through their [i.e. the network actors'] co-operative relations' (Yeung 2005: 46).

Although this makes them alternative to currently dominant FSC, the difference would appear to be quantitative rather than qualitative. As Lee and Leyshon (2003: 193) point out, 'all economies are irreducibly material'. This means that, at some point, calculations of equivalence (exchange values) must be made (Hughes 2005: 499). And, as economic activity is the 'means through which people sustain their lives' (Lee 2000: 140), sufficient gross profit must be generated in order to enable a network's constituent actors to sustain themselves financially. This may enable them to exist '*within* the market but *outside* the norms of capitalist evaluation' (Lee 2000: 138, emphasis in the original) but it does not exempt them from the requirement to generate a surplus. Indeed, for smaller 'alternative' enterprises, unable to exploit economies of scale, gross profit on individual transactions may need to be high, relative to larger 'conventional' enterprises, for this to occur. On this basis, therefore, it could be argued that what separate 'alternative' from 'conventional' producers are the extent and destination of any net profit (i.e. any surplus over and above that required to sustain the networking practices concerned). In and of itself, therefore, this particular aspect of the 'alternativeness' of AFN does not appear any more likely to produce a particular spatial outcome, such as a local food production and processing network,[7] than conventional FSC.

The second factor that may limit the potential one-sidedness of the relational power exhibited in interviewees' networking practices appears to be related to the trust that they place in key networking partners. Not all of those on whom interviewees rely require

---

[7]The extent to which it may produce a local food supply network must, of course, await a discussion of the role of consumers in AFN.

trusting (recall the example of the fruit processor who buys sugar from the cheapest source), but for key and defining inputs trust in the supplier appears vital. Trust, in turn, would appear to be related to regular contact, which may have a spatial component. This is not to argue that trusted suppliers must be local to the producer, although it transpired that the great majority of suppliers trusted by interviewees were. However, it does suggest that there may be something about certain types of trust-based relationship that make a particular spatial outcome (in this case a localised production network) more likely.

What that 'something' might be remains uncertain. However, it may be related to interviewees' deployment of the concept of 'integrity'. As Pascalev (2003: 586) argues: 'With respect to institutions or practices, the notion of integrity signifies their unity and internal coherence and mandates that those be preserved and protected'. Thus, a shared belief in the 'right-ness' of certain institutions and practices can underpin joint efforts to create and maintain them. For several interviewees, a shared belief in the 'right-ness' of trying to create and maintain 'local' economic activity appears to underlie at least some of their networking practices (notably a, b, n and t). Thus, such networking practices could be interpreted as being 'alternative' precisely because they attempt to create and maintain a 'local' scale through their economic activities.

However, there appears to be another set of shared beliefs in the 'right-ness' of certain 'institutions and practices' at work in the networking practices of several interviewees. It was noted above that there may be something about certain types of trust-based relationship that make a particular spatial outcome (in this case a localised production network) more likely. It was also noted that, for key and defining inputs, trust in the supplier seems vital. It may be, therefore, that those inputs possess characteristics which, when taken in combination with certain other factors, make a particular spatial outcome more likely. Thus, it may be necessary to qualify Watts *et al.*'s (2005) argument that 'stronger' alternatives to currently dominant FSC are more likely to be those wherein attention is focused primarily on the networks through which food moves. This should not be taken as implying that AFN are only 'weakly' spatial. It is, instead, to suggest that their spatial outcomes may depend on more than the networking practices of 'alternative' food producers. It may be that such spatial outcomes are the result of a combination of certain networking practices with certain types of food input.

Acknowledgements

This chapter is based on a paper presented at the Annual Conference of the Royal Geographical Society with the Institute of British Geographers in September 2005. We are grateful to the organisers of that session (the editors of this volume) for encouraging us to write it and for providing constructive criticism on a previous draft. We also thank an anonymous referee for their comments. The research on which this chapter draws was funded by the Economic and Social Research Council (grant R000239980), to whom the authors are grateful for their support. We are also grateful to those who gave up their time to be interviewed; to Andrew Gilg, Jo Little and Sue Simpson; and to Stuart Gill for drawing the figures.

References

Archer, G.P., García Sánchez, J., Vignali, G. and Challiot, A. (2003). Latent consumers' attitude (sic) to farmers' markets in North West England. *British Food Journal* 105, 487–97.
Blois, K.J. (1999). Trust in business to business relationships: an evaluation of its status. *Journal of Management Studies* 36, 197–215.
Blois, K. (2003). Is it commercially irresponsible to trust? *Journal of Business Ethics* 45, 183–93.
Blundel, R. (2002). Network evolution and the growth of artisanal firms: a tale of two regional cheese makers. *Entrepreneurship and Regional Development* 14, 1–30.
Chambers Dictionary, The (1993) Edinburgh: Chambers.
Clark, J.R.A. (2005). The 'new associationalism' in agriculture: agro-food diversification and multifunctional production logics. *Journal of Economic Geography* 5, 475–98.
DuPuis, E.M. and Goodman, D. (2005). Should we go "home" to eat?: toward a reflexive politics of localism. *Journal of Rural Studies* 21, 359–71.
Dürrschmidt, J. (1999). The 'local' versus the 'global'? 'Individualised milieux' in a complex 'risk society'. The case of organic food box schemes in the South West. *Explorations in Sociology* 55, 131–52.
Goodman, D. (2004). Rural Europe redux? Reflections on alternative agro-food networks and paradigm change. *Sociologia Ruralis* 44, 3–16.
Hendrickson, M.K. and Heffernan, W.D. (2002). Opening spaces through relocalization: locating potential resistance in the weaknesses of the global food system. *Sociologia Ruralis* 42, 347–69.
Hess, M. (2004). 'Spatial' relationships? Towards a reconceptualization of embeddedness. *Progress in Human Geography* 28, 165–86.
Holloway, L. and Kneafsey, M. (2000). Reading the space of the farmers' market: a preliminary investigation from the UK. *Sociologia Ruralis* 40, 285–99.
Hughes, A. (2005). Geographies of exchange and circulation: alternative trading spaces. *Progress in Human Geography* 29, 496–504.
Ilbery, B. and Kneafsey, M. (1999). Niche markets and regional speciality food products in Europe: towards a research agenda. *Environment and Planning A* 31, 2207–22.
Ilbery, B. and Kneafsey, M. (2000a). Registering regional speciality food and drink products in the United Kingdom: the case of PDOs and PGIs. *Area* 32, 317–25.
Ilbery, B. and Kneafsey, M. (2000b). Producer constructions of quality in regional speciality food production: a case study from south west England. *Journal of Rural Studies* 16, 217–30.
Ilbery, B. and Maye, D. (2005a). Alternative (shorter) food supply chains and specialist livestock products in the Scottish–English Borders. *Environment and Planning A* 37, 823–44.
Ilbery, B. and Maye, D. (2005b). Food supply chains and sustainability: evidence from specialist food producers in the Scottish/English borders. *Land Use Policy* 22, 331–44.
Ilbery, B. and Maye, D. (2006). Retailing local food in the Scottish-English Borders: a supply chain perspective. *Geoforum* 37, 352–67.
Jones, A. (2003). 'Power in place': viticultural spatialities of globalization and community empowerment in Languedoc. *Transactions of the Institute of British Geographers* NS 28, 367–82.
Kneafsey, M., Ilbery, B. and Jenkins, T. (2001). Exploring the dimensions of culture economies in rural West Wales. *Sociologia Ruralis* 41, 296–310.
Knickel, K. and Renting, H. (2000). Methodological and conceptual issues in the study of multifunctionality and rural development. *Sociologia Ruralis* 40, 512–28.
Kvale, S. (1996). *InterViews. An Introduction to Qualitative Research Interviewing*. London: Sage.
Lee, R. (2000). Shelter from the storm? Geographies of regard in the worlds of horticultural production and consumption. *Geoforum* 31, 137–57.
Lee, J., Árnason, A., Nightingale, A. and Shucksmith, M. (2005). Networking: social capital and identities in European development. *Sociologia Ruralis* 45, 269–83.

Lee, R. and Leyshon, A. (2003). Conclusions: re-making geographies and the construction of 'spaces of hope'. In A. Leyshon, R. Lee and C.C. Williams (Eds), *Alternative Economic Spaces* (pp. 193–98). London: Sage.

Leyshon, A. and Lee, R. (2003). Introduction. In A. Leyshon, R. Lee and C.C. Williams (Eds), *Alternative Economic Spaces* (pp. 1–26). London: Sage.

Leyshon, A., Lee, R. and Williams, C.C. (Eds) (2003). *Alternative Economic Spaces*. London: Sage.

Marsden, T., Banks, J. and Bristow, G. (2002). The social management of rural nature: understanding agrarian-based rural development. *Environment and Planning A* 34, 809–25.

Morris, C. and Buller, H. (2003). The local food sector. A preliminary assessment of its form and impact in Gloucestershire. *British Food Journal* 105, 559–66.

Murdoch, J. (2000). Networks – a new paradigm of rural development? *Journal of Rural Studies* 16, 407–19.

Parrott, N., Wilson, N. and Murdoch, J. (2002). Spatializing quality: regional protection and the alternative geography of food. *European Urban and Regional Studies* 9, 241–61.

Pascalev, A. (2003). You are what you eat: genetically modified foods, integrity, and society. *Journal of Agricultural and Environmental Ethics* 16, 583–94.

Peck, J. and Tickell, A. (2002). Neoliberalizing space. *Antipode* 34, 380–404.

Policy Commission on Farming and Food (2002). *Farming and Food: a Sustainable Future*. Norwich: The Stationery Office.

Potter, C. and Tilzey, M. (2005). Agricultural policy discourses in the European post-Fordist transition: neoliberalism, neomercantilism and multifunctionality. *Progress in Human Geography* 29, 581–600.

Ray, C. (1998). Culture, intellectual property and territorial rural development. *Sociologia Ruralis* 38, 3–20.

Renting, H., Marsden, T.K. and Banks, J. (2003). Understanding alternative food networks: exploring the role of short food supply chains in rural development. *Environment and Planning A* 35, 393–411.

Sage, C. (2003). Social embeddedness and relations of regard: alternative 'good food' networks in south-west Ireland. *Journal of Rural Studies* 19, 47–60.

Sahay, B.S. (2003). Understanding trust in supply chain relationships. *Industrial Management and Data Systems* 103, 553–63.

Storper, M. (1997). *The Regional World*. New York: Guildford.

Ventura, F. and Milone, P. (2000). Theory and practice of multi-product farms: farm butcheries in Umbria. *Sociologia Ruralis* 40, 452–65.

Watts, D.C.H., Ilbery, B. and Maye, D. (2005). Making reconnections in agro-food geography: alternative systems of food provision. *Progress in Human Geography* 29, 22–40.

Whatmore, S., Stassart, P. and Renting, H. (2003). What's alternative about alternative food networks? *Environment and Planning A* 35, 389–91.

Winter, M. (2003). Embeddedness, the new food economy and defensive localism. *Journal of Rural Studies* 19, 23–32.

Yeung, H.W.-C. (2005). Rethinking relational economic geography. *Transactions of the Institute of British Geographers* NS 30, 37–51.

Youngs, J. (2003). Consumer direct initiatives in North West England farmers' markets. *British Food Journal* 105, 498–530.

Chapter 18

The Appropriation of 'Alternative' Discourses by 'Mainstream' Food Retailers

Peter Jackson[1], Polly Russell[2] and Neil Ward[3]
[1]Department of Geography, University of Sheffield, UK
[2]The National Sound Archive, The British Library, London, UK
[3]Centre for Rural Economy, University of Newcastle, UK

Introduction

In social science analyses of the production and consumption of food, there has been a tendency to employ binary notions of 'conventional' and 'alternative' food systems. This has raised questions about the nature of 'alternativeness' (Whatmore *et al.* 2003) which, it has been argued, has remained relatively free from critical questioning. Alternative food systems are generally taken to imply some notion of oppositionality to the conventional, although Goodman (2003) identifies two distinct bodies of literature on 'alternative food networks'. The first is dominated by European scholars and casts alternative within the context of European policy imperatives to improve farming incomes and the prospects for farm survival in increasingly competitive conditions through developing niche market opportunities. Here, the conventional mass food supply system is not threatened by the emergence of new niche markets, and research has focused on constructions of food quality and the embeddedness of more localised food production and consumption systems (Gilg and Battershill, 1998; Murdoch *et al.* 2000). Holloway *et al.* (2005: 7) see the use of 'alternative' in this literature as problematic because "it seems to have to represent a collection of other terms and senses, and is used in a polarised manner as part of a conventional-alternative dualism". The second literature is characterised more strongly by North American scholars and associates 'alternative' more explicitly with a political discourse of oppositional activism. Here, alternative food networks are more about wresting control over food production and consumption away from the corporate-controlled industrial mass food supply system. While there are risks in such research of what Holloway *et al.* (2005: 9) call "romanticising the radicalised 'alternative' ", the North American literature at least opens up the prospect of a more relational understanding of 'alternativeness' (see Freidberg, 2004).

Building on these debates, this chapter traces the way that 'mainstream' (supermarket) retailers have begun to appropriate 'alternative' discourses in their product development and marketing strategies. Our particular contribution is to challenge the all-too-common assumption that 'alternative' and 'mainstream' producers are diametrically opposed to one another in their production methods and marketing strategies. Rather, we join forces with those like Susanne Freidberg (2004) whose work focuses on the 'ethical complex' of corporate food power where non-governmental organisations (NGOs) and the popular media are campaigning for changes in the global supply chains of Britain's high-street retailers. While Freidberg explores the tensions between the NGOs' international focus and the more localised concerns of many of their constituencies, our work focuses on the appropriation of 'alternative' discourses by 'mainstream' retailers. Specifically, we examine the strategy adopted by one British high-street retailer (Marks & Spencer) in the development and marketing of a new brand of chicken (Oakham White). In a highly competitive marketplace which saw Marks & Spencer's profits falling in the late 1990s, the development of the Oakham White brand allowed the company to charge premium prices for a staple product through their emphasis on quality, provenance and taste. The 'Britishness' of the brand was also crucial to their marketing strategy as were other points of difference such as improved animal welfare. The strategy was, we argue, as much about *manufacturing meaning* as it was about the agricultural, technical or economic development of a new product.[1]

In developing the product and in their marketing strategy, we argue, Marks & Spencer borrowed the language of 'alternative' food producers and adapted it for their own commercial ends. We are not suggesting that the company misled consumers by adopting this approach and we make an explicit comparison between their marketing strategy and that of a small-scale producer who sells free-range chickens via a local farmers' market. Our evidence suggests that the relationship between 'mainstream' and 'alternative' producers is complex and that the outcome of different modes of production in terms of advancing the consumer interest is by no means clear.

A Narrative Approach

In this chapter, we take a narrative approach to understanding the process of brand development and marketing, seeking to record the views of those most closely involved in their own words. Our approach relies on the recording of life-history interviews with the key actors involved in the development of the brand, moving along the supply chain from chicken hatcheries, growers and processors through to the agricultural and processing technologists, buyers and category managers who work for retailers.[2]

---

[1] This paper derives from a project on "Manufacturing meaning along the food commodity chain", funded by the Arts and Humanities Research Council and the Economic and Social Research Council as part of the *Cultures of Consumption* programme (award no. RES-143-25-0026).
[2] Interviews were tape-recorded and transcribed in full. The tapes, tape summaries and transcripts are deposited with The British Library Sound Archive as part of the National Life Stories' project, *Food: From Source to SalesPoint*. Subject to any restrictions imposed by the interviewees, they can be accessed via the Library's online catalogue at http://www.bl.uk.collections/sound-archive/cat.html.

The life-history method allows us to place these individual stories within a wider set of narratives concerning the institutional and technological changes that are driving the food industry. As will become clear, these individual accounts are embedded within a range of other 'stories', including the industrialisation of British agriculture (where broiler chickens were particularly amenable to the process of intensification), the globalisation of food production and the response of the British food industry to increasing foreign competition, the recent crises affecting British agriculture and the growth in consumer demand for 'alternative' (free-range, organic and GM-free) food. All of these narratives played a part in the development and marketing of the Oakham White brand. Primarily though, we argue, our evidence reveals the way 'mainstream' food retailers like Marks & Spencer are responding to these challenges via a selective appropriation of the discourses that are normally associated with 'alternative' producers.

In recording and analysing these life histories, we are interested in their narrative style (how people 'tell the tale') as well as in their narrative content (see Perks and Thomson (2006) for an overview of oral history approaches and Russell (2003) for a life-history analysis of recent developments in British culinary culture). Underpinning this approach is a relational understanding of the narrative constitution of identity, where each individual's life story is subject to particular rules and conventions. For, in Margaret Somers' words:

> Narrative identities are constituted by a person's temporally and spatially variable place in culturally constructed stories composed of (breakable) rules, (variable) practices, binding (and unbinding) institutions, and the multiple plots of family, nation, or economic life. Most importantly, however, narratives are not incorporated into the self in any direct way; rather, they are mediated through the enormous spectrum of social and political institutions and practices that constitute our social world (1994: 635).

The life-history approach situates individuals within a biographical context and provides an opportunity to examine how work places and practices are negotiated, implemented and understood by individuals. A typical life-history account includes a detailed biography, starting with earliest memories and moving on to describe family background, housing, education, social life, friendships and work expectations and experiences. In this project, interviews ranged in length from 2 to 11 hours, with the average recording being 5 hours. The lengthy nature of life-story interviews and their emphasis on recording a narrative account of 'the life' provides a vehicle for examining biographical dimensions of individuals' descriptions, reflections and explanations. A typical life-history account includes both detailed descriptions of events or practices and opportunities for individuals to reflect on how these connect to other areas of their lives. The focus of this paper is on how the practices and narratives adopted by one corporate retailer are shaped by the critiques of intensive food production advanced by various NGOs, food campaigners and advocates of 'alternative' food systems. While the focus is not on individual biographies and their intersections with the processes of food production, we do argue that the examples cited here, and in particular the degree of reflective analysis on the part of interviewees, were facilitated by the life-story approach (see Jackson and Russell, in press, for a more detailed account of the life-story method).

In what follows, we apply this approach to the development and marketing of the Oakham White brand as described to us in life-story recordings of key players in the Marks & Spencer poultry department. Primarily, this paper focuses on an analysis of the relationship between 'mainstream' retail brand development and the discourses and practices associated with 'alternative' food systems. We are keen, however, to draw attention to the individual biographies and subjectivities accounted for through the life-history method, particularly where these provide insights into the tensions and ambivalences at work in the production of the Oakham White brand. These tensions, we believe, shed light on the discursive function of the Oakham White brand and on the constitutive relationship between 'mainstream' and 'alternative' food discourses. In the analysis that follows, we start by describing the impetus for and characteristics of the Marks & Spencer Oakham White brand, using a combination of company information and the words of Marks & Spencer employees. This description suggests that the Oakham White brand borrows discourses and practices associated with 'alternative' production methods in order to differentiate products in a mass market. In the second section, we describe how the brand was marketed and the different external and internal constraints at work in determining the brand, drawing in particular on the reflections of three key individuals.

Developing the Brand

The Oakham White brand took 18 months to develop and was part of Marks & Spencer's attempt to re-position the firm in relation to their main competitors (Asda-Walmart, Morrisons, J. Sainsbury and Tesco). Although recent financial results for Marks & Spencer show a 35% rise in profits,[3] the Oakham White brand emerged in a different commercial context. After decades of commercial success, when Marks & Spencer had enjoyed 'legendary and iconic status' as one of the world's greatest retailers (Mellahi *et al.* 2002), the company's sales and pre-tax profits started to fall in 1998. The company's share price collapsed from a high of £6.60 on 3 October 1997 to a low of £1.70 on 20 October 2000 (Burt *et al.* 2002: 192). There are many possible explanations for what one journalist describes as the 'rise and fall of Marks & Spencer' (Bevan, 2001). Critics have pointed to the company's over-reliance on their own brand (St Michael); their reluctance to accept credit and debit cards; their continued attachment to high-street outlets in preference to out-of-town locations; their reluctance to advertise; and the increasing level of competition from other 'quality' brands such as *Tesco's Finest* and Sainsbury's *Taste the Difference* ranges. The company also engaged in some ill-fated overseas acquisitions where many of their traditional ways of doing business, including their 'buy British' policy, were hard to adapt to a different retail environment (Burt *et al.* 2002).

This was the commercial context in which Marks & Spencer developed the Oakham White brand. The company had been a pioneer in the development of the cold-chain in which poultry was a key driver.[4] This involved comprehensive changes to distribution,

---

[3]Marks & Spencer press release (23 May 2006), http://www2.marksandspencer.com/thecompany/mediacentre/pressreleases/2006/fin2006-05-23-00.pdf.
[4]Seth and Randall (1999: 125–6) describe the development of chilled prepared food as 'the breakthrough' for Marks & Spencer as a food retailer.

transportation, refrigeration and hygiene, with chicken as a major driver of technological change within the industry.[5] The introduction of the Oakham White brand was the latest development in this long history of product innovation. According to Catherine Lee, poultry buyer at Marks & Spencer, Oakham White was introduced after a major review of their poultry lines and a decision to rationalise the poultry supply base.[6] The 'fresh' (i.e. not prepared) poultry business was put out to tender with all their existing poultry suppliers. As part of the tender, process suppliers were invited to propose long-term strategies for improving poultry sales, profitability and quality. The strategy proposed by Two Sisters Food Group included a plan to introduce a poultry brand to differentiate Marks & Spencer products from their rivals and this was a key point in the company winning the contract to supply.

What, then, sets Oakham White – the name eventually given to the Two Sisters Marks & Spencer chickens – apart from the chickens that are sold by Marks & Spencer's high-street rivals? According to several of our interviewees, Oakham White was designed to provide Marks & Spencer's customers with 'chicken the way it used to be', in between a free-range and a standard (intensively reared) chicken. According to Mark Ranson, Marks & Spencer's Agricultural Technologist responsible for animal welfare and a former Royal Society for the Prevention of Cruelty to Animals (RSPCA) Scientific Inspector, the product development process had to start with what customers wanted:

> I mean this intensively produced chicken, is a commodity, or was a commodity, and what we needed to do was to differentiate our chicken offer compared to the rest of the high street. So it was about looking at what our customers ... starting off at the other end, whereas normally we go at the farm end and work upwards towards the customer, speaking to customers about what they wanted. So they wanted chickens to taste like chicken used to taste years ago. So we looked at various different factors: we've got a named, a bespoke breed of bird, or strain of bird which is exclusive to Marks & Spencer; we've got a bespoke diet which we've formulated with nutritionists ... It's a slower growing breed ... we've set certain husbandry standards ... And then probably the other thing is that we put the name of the farmer on the pack of the birds ... So that in itself, rather than saying as a point of difference and a key selling 'we have 100% traceability', just by saying we are so proud of our farmers we put the name of the farmer on the pack.[7]

---

[5] Recalling the technological innovation required to develop the poultry cold-chain, David Gregory (Head of Food Technology at Marks & Spencer) referred to Marks & Spencer's partnership with BOC distribution, the development of polar stream nitrogen-chilled vehicles and the need for piped refrigeration to ensure consistent chilling capacity in stores (interviewed by Polly Russell, 26 April 2004, C821/130/05, Transcript pp. 90–1).
[6] Catherine Lee interviewed by Polly Russell, 14 May 2004, C821/129/05, Transcript pp. 100–2.
[7] Mark Ranson interviewed by Polly Russell, 30 January 2004, C821/121/04, Transcript p. 98. The labelling is actually more complicated than this. Each label has the name of the farmer stamped on it. But it also quotes 'a typical Marks & Spencer farmer', who gives their further endorsement, saying that they are proud to produce quality chicken for Marks & Spencer. The label also bears the photograph of this 'typical' farmer, not that of the farmer who reared this particular chicken.

Mark Ranson's description suggests a shift from a production- to a market-determined supply chain. Market-driven supply chains are characterised, among other things, by deriving profits from control of marketing and research rather than relying entirely on economies of scale (Gereffi, 1994). Understanding customer concerns, applying new agricultural standards and putting these practices on display via packaging and marketing are integral to the Oakham White branding exercise. For Mark Ranson, whose role is to oversee and implement Marks & Spencer's standards of animal welfare, a more transparent production system provides a commercial opportunity to differentiate the company's poultry business as well as to implement improved practices on an intensive scale.

Although the Oakham White brand was developed as a 'bespoke' bird, in practice, Oakham White chickens are a standard breed, the Ross 508, known for its high meat-yield and good feed-conversion ratio.[8] Marks & Spencer worked with the breeder company, Aviagen, to select the ideal bird for their needs.[9] As Mark Ranson explained:

Well, [Aviagen] had this bird available and it was, it was... it wasn't being used anywhere else because again it had been originally developed for the fast food industry. What they wanted was a bird with lots of breast meat with very little leg so they could use the breast meat for cutting up into portions and nuggets and God knows what else. So it was a slower growing bird overall, but you got this bigger, larger breast meat from it. If we had gone to them and said 'Right, OK... this is our checklist of all the different points that we want, can you start developing this bird?' I mean M&S must be, in terms of their overall business, must be point something of a point nought or something a percentage of the total, you know, world population of poultry... I mean, I think, in an ideal world... we would look at it from a different perspective in terms of looking at, you know, starting from scratch, but you're probably talking about 10–15 year development, but having something very exclusive and the cost implications of that would be absolutely astronomical.[10]

The difference with Oakham White, then, is not predominantly about breed but about how the birds are reared: "It's not about the bird itself but it's a package including breed,

---

[8]The Ross brand is owned by Aviagen who also owns three other poultry breeders: Arbor Acres, L.I.R. and Nicholas. According to the company's website, 'The Ross 508 parent stock has been developed to maximise breeder performance without compromising the efficiency of meat production' (http://www.aviagen.com, accessed on 28 July 2005).
[9]Aviagen is one of only a handful of breeder companies worldwide and Marks & Spencer represent just a small part of their business.
[10]Mark Ranson interviewed by Polly Russell, 30 January 2004, C821/121/04, Transcript pp. 105–6. Mark Ranson felt that breeder companies were only interested in feed conversion ratios, breast meat yield and growth speeds. Succulence, tenderness and flavour were not on their agenda at all (ibid., p. 105).

diet, living conditions and farm".[11] According to a Marks & Spencer press release, Oakham White chickens are fed on "a nutritious feed ration which is non-GM, cereal-based and free from antibiotic growth promoters and encourages health and welfare too" (23 November 2003). Besides this special feed regime, Oakham White chicken also have access to 'natural behaviour enablers' (like bales of straw that the chickens can pick, claw and peck at), lower stocking densities and a slower growing cycle.[12] Mark Ranson described the typical life of an Oakham White chicken. The breeder farm produces the eggs which are then sent to the hatchery where they stay for 21 days. Once hatched, they are transported to the farm. The chickens spend 40–45 days on the farm, depending on the weight required. The birds are then caught and driven to the processor where they are killed, prepared and packaged for distribution and retail sale. As Mark Ranson explained, "Oakham is a package of various different points which we can then... market and sell to the customer. So Oakham, or the Oakham concept is the bird, the diet, the living conditions, the name of farmer on pack... so it's all of those things together".[13]

The Oakham White brand offers solutions to a range of issues associated with intensive poultry production. Mark Ranson emphasised, for example, the importance of a longer growing cycle to determining taste and texture, compared to the 40-day industry standard:

> One of the things about chicken is the way, the length of time it's grown for will greatly affect the flavour and the texture of the chicken. So if you compared, from today's product on the shelf, if you compared an intensely produced bird which could be typically 39 days of age when it was, when it's been slaughtered compared to an organic chicken or a La Belle Rouge could be anything from 91 days plus and the flavour's much greater in the organic bird and the texture is much denser and there is more bite... I mean some of the chicken you get intensively produced today is very, well pappy, the only flavour to it is whatever sauce is being added onto it. It's just a protein, you know in many regards you could almost be producing tofu or something like Soya, Quorn or something like that because it's the same, the actual meat itself is giving some texture to the recipe dish that you are eating but not chickeny, if you see what I mean.[14]

---

[11] Later in the interview, Mark Ranson concluded that "my work with the product developers is about product differentiation in terms of the agriculture, what can we say agriculturally? And I think because over the years we've seen changes in sales of chicken, meat or whole birds and portions, and it's been much more and more... seen as a commodity, that we've needed to differentiate" (Mark Ranson interviewed by Polly Russell, 30 January 2005, C821/121/05, Transcript p. 138.).

[12] At 34 kg per square metre, Oakham White chickens are reared at lower stocking densities than the industry standard of 38 kg m^2 but still above the RSPCA's (2005) recommended maximum density of 30 kg m^2. The growing cycle is, on average, 4 or 5 days longer than a normal broiler chicken with obvious implications in terms of the increased cost of feed and other inputs.

[13] Ibid., C821/121/04, Transcript p. 102.

[14] Ibid., C821/121/02, p. 63. With the intensification of the industry and improvements in feed-weight ratios, a chicken that would have taken 70 or 80 days to grow will now take just 39 days.

Within living memory, the intensification and acceleration of poultry production has reduced the time taken to grow a chicken to slaughter weight by half. Broiler chicks are reared to their slaughter weight of 2 kg within about 40 days of being hatched. Similarly, the amount of feed needed to achieve this weight gain has been reduced by almost 40% since 1976 (Compassion in World Farming, 2003: 7). The length of time taken for organic or specialist label chickens to reach slaughter weight is reflected in its price and, according to Mark Ranson, its taste and texture. The 10% longer growing cycle therefore allows for claims of improved texture and flavour, even though the growing time of an Oakham White chicken falls well short of the production ideals of La Belle Rouge and 'organic' chicken cited by Mark Ranson.

The Oakham White brand can also claim to be more welfare-friendly than standard intensively produced birds. Summarising the main points of difference between Oakham White chickens and those sold by their rival companies, Marks & Spencer's Protein Group Agricultural Technologist Paul Wilgos argued:

> I think a lot of our points of difference are about how the birds are grown, the animal welfare standards, the stocking density that we would use, the use of medicines or not, the fact that we won't allow the use of antibiotic growth promoters while other people will, and the way the birds are treated in terms of transport, the way that they're settled before they're killed, the way they're killed and so on and so forth... our points of difference in poultry is more about the way we produce the chickens rather than what we're doing to the meat, and then it's about recipe, and you know, excellence of products.[15]

It is significant that each of the points of difference highlighted by Paul Wilgos pertains to an aspect of poultry welfare that has been the subject of focus from NGOs including the RSPCA and Sustain (the campaign for better food and farming).[16] As David Gregory, Head of Food Technology at Marks & Spencer explained, the relationship between food retailers, NGOs and customers has altered in recent years, with NGOs playing an important part in influencing customer opinion and product sales:

> And whereas my predecessors would really never have thought about dealing with press or Non Government Organisations, NGOs, or whatever... I mean today, probably twenty per cent of my life is spent dealing with that type of activity... Food safety is on everybody's agenda, because people worry, is this piece of beef I'm eating safe, or is there an issue around BSE? Is this piece of produce contaminated in some way with pesticides that might impact upon me and my life? And then, so that's the sort of big central area of trust and my role in this is a key role in the company, to ensure that we have strategies that anticipate the issues our customers will be concerned about.[17]

---

[15] Paul Wilgos interviewed by Polly Russell, 20 February 2004, C821/123/01, Transcript p. 40.
[16] See, for example, the campaign reports by Sustain (1999) and RSPCA (2001, 2005) which focus on the consequences of genetic selection, excessive stocking densities and inappropriate lighting.
[17] David Gregory interviewed by Polly Russell, 26 March 2004, C821/130/03, Transcript p. 54.

In this extract, David Gregory slides from an explanation about how his time, as compared with his predecessors, is spent communicating with NGOs and the media, to asserting that anticipating customer 'issues' is central to his job. Clearly, these two activities are inextricably linked: communicating with customers is as much about engaging with NGOs and the media as it is about providing actual customers with the marketing and messaging necessary to promote and protect the brand. Paul Wilgos's earlier description of the welfare and health standards of Oakham White is evidence of the ways conversations and pressure from NGOs and the media can influence food production (cf. Freidberg, 2004). In the case of poultry at Marks & Spencer, the Oakham White brand provided a vehicle for implementing some improved environmental and welfare practices without unduly sacrificing intensive production methods, high volume and low costs. This compromise is something that individuals reflected upon in some detail and to which we return in the next section.

Having established the difference between Oakham White poultry production and standard intensively reared broiler chickens, we turn now to the business of how the brand was marketed. In this, we follow Celia Lury's argument that branding is the process through which the meaning of goods is "made manageable" (2004: 47).

Marketing the Brand

Marks & Spencer introduced the Oakham White brand in March 2003. From October that year, all of Marks & Spencer's fresh chicken (including whole birds, breasts and drumsticks) as well as the chicken used in some of their prepared recipe dishes and other products is sold under the Oakham label. The brand was launched as part of the company's commitment to bringing customers "the best tasting, freshest products" (press release: 4 November 2003). Oakham White, Marks & Spencer maintain, has been specially chosen because it is slower growing than conventional chickens. This, the company promise, "combined with its diet, which is bespoke and promotes better animal health and welfare, delivers exceptionally flavoured meat". In the same press release, Mark Ranson (described here as 'Marks & Spencer's livestock and animal welfare specialist') was quoted as saying that

> The introduction of the Oakham chicken is based around listening to our customers and their concerns about issues such as the welfare of animals and what they are fed... Our customers told us they wanted chicken to 'taste like it used to taste'. So, responding to this, we've introduced birds that are grown in enriched environments, with the Oakham White grown in barns with straw bales enabling the birds to perch and rest, and as a result, the Oakham is a slower growing bird (ibid.).

Similar information was included on the Marks & Spencer website, in a section entitled 'All about our food' (http://www2.marksandspencer.com, accessed 28 February 2005). Here, the company insists: "We inspect everything from the hatchery where the chicks are hatched to the farm where birds are grown to ensure they meet our own high standards". The website continues:

Our new Oakham Gold chicken [the free-range version], launched in March 2003 and exclusive to Marks & Spencer, is fed on a nutritious non-GM wheat and maize based diet, including oils and herb extracts. We ensure birds have constant daytime access to pastures where they're free to roam, encouraged by the availability of conifer 'wigwams' which provide them shade and shelter if they wish. Indoors, they have plenty of space to move around and act normally, with bales of straw for them to perch and roost. You'll find the name of the farmer who grew the bird on the pack.

The website concludes: "We maintain these high standards of diet and welfare for our Oakham White brand . . . a breed of bird which grows more slowly, providing you a fuller, succulent flavour".

While the food industry is driven by the constant demand for acceleration and intensification, these are issues about which consumers are increasingly concerned, particularly in the context of recent 'food scares' (from *Escherichia coli* and salmonella to BSE and Food and Mouth Disease). Oakham White chicken was developed precisely to allay these fears and to address changing customer tastes and perception. In talking about Oakham White's brand values, Paul Wilgos made direct reference to problems associated with intensive poultry production.

It is a *product rather than a breed*, which we call Oakham White . . . the brand values around Oakham White are more than just having a unique breed. It's a slower growing bird and the reason for that is because of the welfare concerns that faster and faster growing birds have a problem with leg strength and the ability for the birds to mature properly and not, I think the phrase in the industry is 'come off its legs', which is not a nice phrase, but I think it's pretty descriptive [laughs].[18]

Having described the advantages of the Oakham White production practices as compared with standard broilers, Paul Wilgos then went on to reflect upon the appeal of the brand and its value to the company and customer. In this, he revealed his feelings towards poultry production and a pragmatic recognition of how commercial interests weigh against ideal practices. Towards the end of this extract, Paul Wilgos draws from the memory of his childhood to describe chickens roaming around farmyards to explain his preference for free-range poultry production, an example of how individual memories and subjectivities are bound up with commercial practice. Memory serves here as an explanatory framework and as a device to legitimise the distance between ideal and reality. Drawing from memory enables Paul Wilgos to align himself with the policies he implements, while simultaneously setting himself apart from them:

I mean politically I go back to, we're on a journey, you've got to start somewhere. I actually think that with all of these things, there's a cost to it. And in a commercial business you have to be able to demonstrate to the customer that what you're charging them for a product represents value for money. You can take them so far, but I go back to the chicken tikka masala example, about that customer,

---

[18] Ibid., p. 41 (emphasis added).

isn't actually at this stage – and I say at this stage – isn't actually interested in whether it's roosting or whether it's crowding up to a straw bale or not. That's not to say that it's, and I think there is an element of the birds can exhibit natural behaviour in a standard production facility, so they can scratch on the floor and they can peck and so on and so forth, so they can actually you know, the shavings on the floor or the straw on the floor which will be a couple of inches or so, two or three inches deep, they can actually rummage through that and they can exhibit some natural behaviour in that perspective. I think what we're trying to do is, there is an element of customer messaging in it, of course there is, because you have to get the benefit back for everybody otherwise if you just make your supply chain perfect from a welfare and from a principle perspective, but nobody buys your chicken because it's so expensive they can't understand it, then you've actually, it's worked detrimentally to the welfare of the animals and it's worked detrimentally to the producer base and so on and so forth, so there is a degree of pragmatism within the whole thing. Do I want us to have Oakham White across our business as a minimum standard? Yes. I do, because I actually think that that's what, that's who we are, that's what we should do and as a principle I personally believe that's where we should be. Would I like us to be all free range? Yes. And, but I'm being very personal now and I don't think, we may do it on our fresh chicken at some point in the future, but I wouldn't be surprised if it takes us five years. I wouldn't be surprised if it was twenty years before we saw it across the whole of our chicken business.

And you would want to be free range – is that to do with welfare, or is that to do with the point of difference and how you can sell it?

No. I think no, if you go and see, just go into a broiler shed, even the Oakham White ones where they're a little bit more, the birds are a little bit more attentive and so on and then go and see a flock of free range birds, and they're sparky and flighty, they're excited, they're running in and out, they're having a great time. They still die, but ... And you go into a broiler shed and they are lethargic and whilst I think ours are probably less lethargic and particularly the Oakhams would be, I still think that at the end of the day maybe it's something that goes right back to my childhood of walking round farms and friends and whatever, who had free range, you know chickens just scrabbling around for eggs, production, you know, not commercially but ... And chickens roam around and they roost and they go and you know, that's how chickens are. Now that might be just me being mad, but that's what I think, you know, I don't know, I think the animals just, it's more interesting and I think the meat's got more character as a consequence....[19]

Moving from description to explanation to evaluation and reflection, this extract suggests the ways that individuals negotiate commercial structures and situate their own, at times differing, views alongside the procedures and practices they design and perpetuate as company employees. What starts out as an explanation about the relative merits of the Oakham White brand becomes a reflection grounded in individual memory.

---

[19] Ibid., pp. 46–7.

The past in this example provides an explanatory framework for the individual but also offers a framework through which the brand can be focused and marketed.

The search for an appropriate brand name, with the right kind of nostalgic connotations ('like chicken used to taste'), took a long time. Poultry buyer, Catherine Lee explained how the poultry team decided upon the Oakham White name: "We were months at it, not working tirelessly, but you know...".[20] Market research with customers revealed that they wanted to be reassured that "from an M&S perspective everything's OK, but they don't want to know the details. So, saying it's from Marks & Spencer is great, reassuring, saying it's exclusive to M&S [as is the case with Oakham Gold] is even better".[21] The Oakham White brand was originally to have been called 'Suffolk White' because the processing plant was located in Flixton, Suffolk, and because the bird is white in colour. Much debate followed, however, as it was acknowledged that Marks & Spencer might be able to sell a 'Suffolk' chicken in the South-East of England but not in Scotland. According to Catherine Lee, the ideal name would be quite generic and would include "some provenance type of imagery".[22] 'Gold' was considered to signal 'top tier' in terms of quality, but 'Norfolk Gold' and 'Suffolk Gold' were already trade-marked. Oakham White beer (from Rutland) had won several awards at national CAMRA festivals (the Campaign for Real Ale) and the name was liked because of its positive associations with 'countryside imagery and nice places'.[23] To quote Catherine Lee again:

> It's more about an image than it is a place and provenance. It's, I suppose, it does [sound] British, there is a Britishness to it, because it sounds like – I don't know where it is – but it sounds like it's a place in Britain, you know, and there's that kind of provenance feel, a bit like Aberdeen Angus, because that's effectively what we were looking for... the Aberdeen Angus of the poultry world.[24]

This explanation about the name Oakham White highlights the recognition by food producers that food consumption is an emotionally laden process and one that can be managed and appealed to via nostalgic narratives of practice and place. As Catherine Lee explains, this is about a 'feeling' as much as about a tangible place or practice. The Oakham White name was selected because of the positive geographical, national and historical associations that the name conjures up.

Moreover, the Britishness of Marks & Spencer's chicken was also significant in the light of other companies who were increasing their imports of chicken from Thailand and Brazil. In 2000, Marks & Spencer took the decision not to buy poultry from anywhere

---

[20] Ibid., p. 114. According to Mark Ranson, simply calling the product 'Ross 508' was not thought appropriate, conjuring up images of test tubes and laboratories rather than a living creature (Mark Ranson interviewed by Polly Russell, 30 January 2004, C821/121/04, Transcript p. 102).
[21] Catherine Lee interviewed by Polly Russell, 14 May 2004, C821/129/06, Transcript p. 114.
[22] Ibid., p. 113.
[23] Ibid., p. 113.
[24] Ibid., p. 113. Unlike Oakham White, however, Aberdeen Angus refers to a specific provenance (Aberdeenshire in Scotland) and to a specific breed. Later in the interview, Catherine Lee reaffirmed that the 'Britishness' of the brand was very important as the company has a long-established 'buy British' policy and Marks & Spencer are 'all about British' (Ibid., p. 118).

but Britain and the EU because they wanted to maintain their policy of using only fresh chicken in their recipe dishes. They put up signs in their stores to advertise their procurement policy when chicken from the Far East became infected with avian flu. But the decision was a complex one and Paul Wilgos spoke about the importance of not "disadvantaging our business" and not "taking an opportunity away from our customers", some of whom might prefer cheaper imported poultry.[25] In taking the decision about sourcing from overseas, he indicated that the company would be driven by what their customers wanted, accepting that there was a "complex dynamic" involving a "balance between value, provenance and quality".[26] Marks & Spencer are currently investigating how best to protect their 'point of difference' if they follow other companies and start sourcing from Asia or South America. At present, however, the 'Britishness' of the Oakham White brand is a critical difference that the company proudly asserts through their promotion of the chicken's provenance down to the level of the particular farm.

While their competitors, such as Tesco and Sainsbury, sell chicken at a cheaper year-round price, Marks & Spencer have traditionally emphasised the quality of their produce. According to Catherine Lee, Oakham White was designed to address this market and to offer "a unique proposition on the high street".[27] Catherine Lee further distinguished the company from its commercial rivals by insisting that Marks & Spencer is not about economy fillets or buy-one-get-one-free promotions. Such strategies "don't fit anything else in our proposition"[28] and are not "what an M&S customer expects".[29] While their competitors were introducing everyday low prices and cost-cutting promotions, Marks & Spencer's 'proposition' emphasised quality and value via the discursive alignment of the brand with practices associated with 'alternative' free-range, organic and farmers' market poultry production.[30] The commercial difference between selling poultry on the basis of 'quality' rather than 'value' was reflected upon in detail by Catherine Lee. Her previous employment as a category analyst and then buyer for Safeway in the poultry department during the mid- to late 1990s gave her particular insight into the approaches and practices of different retailers and, in particular, on the difficulties of marketing the Oakham White brand.[31] Talking about Safeway, Catherine Lee explained how volume and price were the main drivers of the poultry business:

---

[25] Paul Wilgos interviewed by Polly Russell, 20 February 2004, C821/123/01, Transcript p. 37.

[26] Ibid., p. 37. A similar point was made by Mark Ranson who argued that 'A lot of people buy purely on price rather than quality or flavour... If their retailer started selling whole chickens from Thailand or Brazil [customers] wouldn't buy them, but they're quite happy buying a ready-meal which is being made with ingredients from Thailand or Brazil and standard-wise there may be no difference and in some areas the standards may be as good if not better' (Mark Ranson interviewed by Polly Russell, 15 January 2004, C821/121/02, Transcript p. 65).

[27] Catherine Lee interviewed by Polly Russell, 14 May 2004, C821/129/05, Transcript p. 97.

[28] Ibid., p. 99.

[29] Ibid., p. 97.

[30] This point was echoed by Andrew McKenzie, a protein buyer at M&S since 1988. Reflecting on the company's core values, he suggested: "We're never going to be cheap, we'll always be, hopefully, value for money" (interviewed by Polly Russell, 4 February 2004, C821/135/03, Transcript p. 51).

[31] Safeway supermarkets no longer exist as they were taken over by Morrison's in 2003.

Yeah, so it was, it was more about cost and volume and managing supplier base and you, you only really got into the agricultural side of things when a supplier let you down. So you know, they'd start talking about bell curves and growing cycles of chickens when a supplier let you down, or when you wanted to change something and it was really difficult for them to do. That was the only time really, when you really got into the agricultural side of it. I mean, it's not that they don't care, that's not it at all, but they, it's the proposition that they're offering is a totally different proposition to what we're trying to offer. We're, they're dealing with a wider spectrum of customers, from the people that just want the cheapest price and they don't care where you got it from – I don't want to think about it, thanks very much – to people that do care. And you know, we'd have quality standards and you know, we'd have an understanding, like we used to have QC, quality checks, and eat the product every week. And you know, we had grown an awareness of, they used to say, you know, happy chickens make happy customers because the meat's tenderer and all of this kind of thing. But I didn't really understand how you could artificially manipulate that so you can tenderise a product, you can mature it in the same way, or you can chicken, in the same way. And I didn't, I understood about killing methods, slightly. Like I, it's not that they didn't show you, it's just, as a buyer I didn't need to know as much. A lot more was left outside of that whereas... and really it was, you thought much more of how am I gonna drive the sales of chicken, but how am I gonna drive it from a – I don't know – from a pure numbers, in the same way you would think how am I gonna drive the sales of pizza. It was that, it was much more an even jump from pizza and sandwiches to chicken. It was thought of in the same way, it's product at the end of the day. It's a finished product, it's a processed product, you know. Whereas here, it's not, it's not so much about that because of the way that M&S puts themselves to the customer in terms of their overall offering.

In explaining the difference between working practices and objectives at Safeway and Marks & Spencer poultry departments, Catherine Lee made no value judgements about which were either ethically or commercially better. Employees in both contexts, she insists, 'care' about product quality and the agricultural system, but different customer bases demanded different emphases and skills from the respective retailers.

Continuing to reflect on the differences between her two experiences of working as a poultry buyer at Marks & Spencer and Safeway, Catherine Lee then describes the difficulties associated with selling 'quality' via notions of livestock farming. In seeking to produce a differentiated product in an otherwise uniform and volume-driven market, Marks & Spencer decided to implement a number of agricultural practices more usually associated with 'alternative' food supply chains. A much celebrated characteristic of alternative supply chains, in particular those associated with farmers' markets and Fairtrade products, has been the transparency associated with them. Speaking about the Soil Association, spokesman Hugh Raven stated that "transparency has always been the organic movement's trump card."[32]

---

[32] Soil Association Press Release 3 March 2006: http://www.soilassociation.org/web/sa/saweb.nsf/ 848d689047cb466780256a6b00298980/e837c6214529ce4a8025711b00562830!OpenDocument.

According to Catherine Lee, however, transparency as it pertains to livestock production does not necessarily appeal to customers and therefore presents a problem for retailers:

> And it comes down to your knowledge of the agricultural system, but customers don't want to know that. Customers don't want to know that at all. And. . . .
>
> *What do customers want?*
>
> Customers want, customers are interested from when they see it in the pack, to what it delivers once they've eaten it. And they're interested in the bit in between that. They don't want to know that that's a dead body sitting in front of them, they don't. You know, and particularly our culture, the British culture doesn't want to know that. And my own theory for it is that we've moved so away, so away from a rural environment and that the majority of the population live in a town, you know, they don't really see a live chicken on a day to day basis any more and therefore they've become squeamish about dealing with the consequences of that and they've become disassociated with it and you know, they all say Britain's a nation of animal lovers and things like that. Well maybe we are, but we, but people really don't want to know and when they think about it, they're put off eating it and so they don't, you know, they don't really want to know about that. And it's very difficult from a chicken perspective to impact anything that happens from the shelf to eating it that will, the customer can really sign on to. So basically, flavour, texture, things like that, it's difficult to change.

According to Catherine Lee's explanation, transparency is a double-edged sword. On the one hand, it provides an opportunity to highlight differences in agricultural production. On the other hand, it risks alienating customers who would rather not be reminded that they are eating an animal. Lee's conviction that customers are primarily concerned with what the product delivers in terms of visual appeal and eating quality suggests that there are limitations to the commercial possibilities offered by reconnecting consumers with modes of agricultural production via the deployment of alternative narratives and practices.

Several other people within the company acknowledged that there were problems articulating the 'points of difference' with Oakham White. Issues like feed and stocking density are not easy points to convey to customers. Mark Ranson felt that "a chicken is much more of a commodity than other protein species".[33] This was particularly true, he felt, from a customer's perspective, once the animal had been slaughtered and processed, where its animal form was no longer clearly recognisable: "From a customer's perspective, once an animal has been cut up, it's lost, in the customer's mind, it's lost what it was . . . in their mind as long as it's a whole bird or a portion, but once it's in a recipe dish the same concerns don't exist that are for the whole piece of meat".[34] He returned to this theme later in the interview when he commented: "when something's being cut up and is not visibly a piece of raw meat, people then tend to lose their [interest in wanting] to understand where it's come from".[35] Asked to reflect upon this, Mark

---

[33] Mark Ranson interviewed by Polly Russell, 15 January 2004, C821/121/02, Transcript p. 56.
[34] Ibid., p. 61.
[35] Ibid., C821/121/06, p. 163.

Ranson equated consumer lack of knowledge with an unwillingness to spend money on food products and a tendency towards wastefulness:

> Ethically, people... people should know where it's come from and how it's being produced, because I think the greater the drive for more, for cheaper food production... if there was an understanding of the, the lengths that had gone into producing that animal, then [pause]. I think there would be a greater appreciation for it in terms of overall rather and again not wasting, wasting product and wasting food and, you know... chickens today are very different to how they were twenty, thirty... around my grandparents' age, so you would have what was called a, you know potentially a dual purpose bird one which was, would lay eggs and you'd kill it and you'd eat, eat chicken, eat the flesh as a, as a bird and chicken post-war was a real treat and often it would be down the local butcher's shop you know... and wasn't available retail wise and if it was it was frozen. But, I mean chicken was a real treat and comparing chicken today to how it was, in terms of flavour, texture and everything else was very, was very, very different....[36]

It is significant that Mark Ranson refers to the past in his attempt to articulate his sense of discomfort about consumer distance from intensive food production. Improving the quality of Marks & Spencer fresh chickens involves a return to some of the practices Mark Ranson associates with his 'grandparents' age' and which are synonymous with 'alternative' poultry production. While appropriating the discursive practices of alternative poultry production presents Marks & Spencer with an opportunity to differentiate their products and promote and market 'value', however, many customers do not want to engage with information about livestock production or may have a limited interest in understanding the origins of the food they eat.

As we have already established, Oakham White is not a distinctive breed of chicken; it is a standard breed (the Ross 508), raised in a different manner and to a different specification (in terms of animal welfare, feed regime etc.). Catherine Lee describes the difference between Oakham White and rival brands as "essentially a marketing message" for what Marks & Spencer do agriculturally and in terms of processing.[37]

As these extracts and marketing material suggest, there is a tension between needing to explain to customers how Oakham White is agriculturally differentiated from other broiler chickens and a concern about customer sensibility pertaining to livestock. The Oakham White brand involves both the implicit and the explicit adoption of practices more closely associated with 'alternative' forms of poultry production. The association of the brand with notions of the rural through reference to an imagined place 'Oakham' and imagined breed 'Oakham White' implicitly connects Marks & Spencer poultry to past practices and places. The adoption of practices which increase poultry growing times and allow for 'natural behaviours' and improve bird welfare explicitly connects Marks & Spencer poultry to alternative poultry supply systems. This raises a question about the limitations and implications of alternative practices and discourses as a means for

---

[36] Ibid., C821/121/02, p. 63.
[37] Catherine Lee interviewed by Polly Russell, 14 May 2004, C821/129/06, Transcript p. 111.

providing transparency as well as a broader question about the ways that the mainstream and alternative food sectors are mutually implicated.

'Alternative' Versus 'Mainstream' Production

The Oakham White brand is ultimately an intensively produced product, albeit with a number of key practices such as lower stocking density and slower growth rate that are associated with free-range and organic poultry production systems. The degree of difference between conventional poultry products and the Oakham White allows the brand to make claims to difference in terms of welfare and also taste. Oakham White thus adopts some of the language and practices associated with 'alternative' poultry supply chains, though it does so through a conventional supply chain. We do not wish to suggest that Marks & Spencer are seeking to mislead their customers through the appropriation of 'alternative' food discourse in the development and marketing of the Oakham White brand but rather that the boundaries between 'alternative' and 'mainstream' are blurred.

As we have demonstrated, conventional systems are adopting the discourses and adapting the practices more closely associated with alternative poultry supply chains because, in part, these systems offer ways to differentiate in the market through claims to quality rather than price. Key to the recent popularity and growth of 'alternative' food systems is an expectation that they are more transparent and trustworthy than conventional production systems.

As others have argued, assuming uncritically that 'alternative' food systems are inherently better is problematic. The difficulties of determining and regulating 'alternative' food systems were highlighted in a recording with a senior FARMA (National Farmer Retail and Markets Association) employee:

> For example, classically, there are some markets where there are a number of farmers taking place, as if in a farmers' market, but they will also allow a local stallholder to come in who might be buying their fruit and vegetables from the wholesale market. In which case, you know, we don't regard that as a proper farmers' market at all. You've got some local authorities where they want to inject a new bit of life into their old traditional market, and so they'll introduce a farmers' market, either in the middle of or associated with their traditional market. And we find that, to some degree, it's quite difficult as well to police. But we're successfully doing that in one or two cases.[38]

Recent media reports have drawn attention to the dubious claims made by some traders at farmers' markets regarding the organic nature of their produce, suggesting that consumers are being 'duped' into buying non-organic meat.[39] Our interview material

---

[38] Anonymous FARMA employee, interviewed by Polly Russell 5 June 2004, C821/1**/02, Transcript p. 18.
[39] A recent article in *The Times* newspaper reported on investigations by the Food Standards Agency into the sale of bogus organic meat at farmers' markets and butchers' shops, claiming that shoppers were being 'duped': http://www.timesonline.co.uk/printFriendly/0,,1-2-2182206-2,00.html (accessed 15 May 2006).

contains some evidence of questionable practices by some traders at farmers' markets, challenging the simplistic view that 'alternative' sources are inherently more trustworthy and transparent than their 'mainstream' competitors.

In the case of one farmers' market supplier we recorded a case could even be made that the production practices of some 'alternative' foods remain as hidden and elusive as products produced by their 'mainstream' counterparts. To explore this issue further, we consider one final life story – that of a small-scale 'alternative' producer who sells free-range chicken through a local farmers' market. The extract begins with the farmer describing changing from intensive to free-range poultry farming:

> Intensively farmed, yes. Yes, that's what we initially started doing. So, we did that. So for the first ten years we were here, we were producing broiler chicken... And it was about five years ago that we couldn't make any more money out of that, and decided it's not what we really wanted to do anyway. So that side of the operation was shut down, and we went free range, and started doing on-farm processing.[40]

Asked to describe his feelings towards intensive poultry farming, the farmer answered in the following manner:

> It just seemed like a good idea at the time. I mean, you go with the flow don't you...
>
> *So the decision to go farmer market or free range, would you say it was primarily economic?*
>
> Yes. Economic probably. Yes, everything, any business is driven by economics isn't it? I mean whatever we do at the end of the day has got to turn a profit, and the bigger the profit the better obviously....[41]

Later in the interview, it became clear that 'free-range' covered a variety of circumstances. The farmer said that he wanted his chickens 'to come outside to get some taste, eat some hayseeds, a few worms, peck about, a varied diet rather than a diet that's completely structured'.[42] But his chicks are reared intensively until they are 5–6 weeks old, when he 'offers them the opportunity to go outside and become free range'.[43] Since some of his chickens are slaughtered within a fortnight of going 'free-range', they will have spent the majority of their lives in intensive conditions and are 'free-range' in little more than name. This does not comply with DEFRA regulations which specify that to be defined as 'free-range' 'the birds have had during at least half their lifetime continuous daytime access to open-air runs'.[44] When asked if his customers know that his birds spend the majority of their lives on a broiler farm, he replies:

---

[40] Anonymous farmer, interviewed by Pally Russell, 12 August 2003, C821/1**/02, Transcript, p. 59. We have anonymised the interview in order to protect the interviewee's privacy.
[41] Ibid., C821/1/02, p. 86.
[42] Ibid., C821/1**/04, p. 156.
[43] Ibid., p. 151.
[44] http://www.defra.gov.uk/foodrin/poultry/epfaq.htm#freerang/.

I don't know. I mean, I always tell people much as I tell you, if they, if they ask me the right questions, I give them, I tell them what, what they want to know. Maybe they assume that they're little fluffy chicks that hatch out and wander outside on the first day, but, I don't think they'd last very long, would they? Certainly the chicks wouldn't anyway [because they need artificial heat to survive].[45]

For producers like this interviewee, farmers' markets provide 'a nice little customer base'.[46] While he promotes the market in terms of traceability and reduced 'food miles' (since all produce sold at the market must come from within a 40-mile radius), there is at least a question about whether customers are getting what they have bargained for:

Well it's more traceable isn't it? If there's a problem with one of my chickens, and I'm only ten miles from market and it's got my telephone number and my name on it, which is something else that has to be on everything you sell, your name and address and telephone number has to be on it, and they've got a chicken that they think might be a bit duff or something, they can ring me up, and we can sort the problem out, and normally say, 'Oh I'll replace it next market.' But if they really [inaudible] I'll say, 'Fine, I'll bring you another chicken then.' You know. And if there's only ten miles down the road, you can do that, but if it's 150 miles, it's a bit more difficult isn't it.

Why is locally best?

Well I think the local people, they prefer to buy local produce really. Well we assume they do. They say they do, so, rather than buy something that's come from 250 miles down the road, if you can get [inaudible] miles down to less than sort of twenty, which we can to most markets, it's got to be good.

For. . . ?

Well good for us, good for the product, good for the bird, you know, good for the public.[47]

But there is at a least a question mark over where the balance of power lies in this particular relationship between the producer and the consumer and it is by no means obvious that this particular version of 'free-range' production serves the consumer interest better than well-regulated intensively reared chicken such as that provided by 'mainstream' retailers. We would suggest that the consumer interest is not automatically better served by one sector or the other (cf. Marsden *et al.* 2000) and that the assumed benefits of an 'alternative' farmers' market poultry product can provide a cover for practices that are at best inconsistent and at worst dishonest. This is not to imply that alternative food systems are inherently less trustworthy than conventional systems, but that they may be subject to individual abuse in ways that raise broader questions about transparency and trustworthiness. Both case studies reveal a lack of transparency

---

[45] Ibid., p. 164.
[46] Ibid., C821/1**/03, p. 117.
[47] Ibid., pp. 141–2.

regarding production practices as neither producer makes explicit the close connection between their brand of chicken and those produced via conventional intensive broiler production. We have focused here on chicken and in particular on the Oakham White brand, but we suggest that our arguments could be applied more broadly to an evaluation of the blurred boundaries between 'alternative' and 'mainstream' food practices.

Conclusion

In this chapter, we have contrasted the example of high-street retailers such as Marks & Spencer who are appropriating the discourse of 'alternative' food producers to market 'mainstream' products with the example of some farmers' market traders whose claims to organic and/or free-range production are dubious and, in some cases reported recently in the media, are currently under investigation by the Food Standards Agency. There are some subtle distinctions at play here, between 'mainstream' producers selling 'alternative' produce and 'alternative' outlets selling goods that have originated from putatively 'alternative' production systems. A similar contrast can be drawn from Jane Dixon's (2002) work on the Australian chicken industry. While the language deployed by 'alternative' and 'mainstream' producers may be similar, her work suggests that another distinction is pertinent to the discussion – between a discourse of food *ethics* (centring on production methods) and a discourse of *aesthetics* (centring on retailing and consumption). Dixon focuses on the re-branding of Kentucky Fried Chicken as 'KFC' in 1993 to demonstrate the conjunction of both discourses: distancing the company from the negative dietary associations of fried food (a social and ethical issue) with a new emphasis on lifestyle and aesthetic issues ("I like it like that"). We make a similar argument in this chapter, demonstrating that there are important ethical issues at stake (in terms of traceability and transparency, for example) in the appropriation of 'alternative' discourses by 'mainstream' food retailers.

The development and marketing of Oakham White chicken was a great commercial success for Marks & Spencer leading to increased sales of fresh poultry across their stores. Poultry buyer Catherine Lee describes herself as immensely proud of the success of Oakham White, talking about her passion for the brand and acknowledging that she is "very protective over it".[48] Yet, we have argued, the success of the Oakham White brand has a wider significance in terms of the way that high-street food retailers, like Marks & Spencer, are responding to the claims made by producers of 'alternative' (free-range, organic and GM-free) food. The launch of the Oakham White brand in March 2003 illustrates the way that 'mainstream' retailers are appropriating the discourse of 'alternative' producers, responding to changing consumer demand and, in particular, to their concerns about food safety and quality. Taking a narrative approach to the development and marketing of Oakham White chicken, we have argued that the process was as much about *manufacturing meaning* as it was about the agricultural, technical and economic development of a new product. We have not argued that Marks & Spencer sought to mislead their customers even when referring (in their press release and elsewhere) to a particular 'breed of bird'. Rather, we have sought to demonstrate the complex cultural

---

[48]Catherine Lee interviewed by Polly Russell, 14 May 2004, C821/129/06. Transcript p. 122.

as well as economic issues that went into the production and marketing of their new brand of chicken. For, in essence, Oakham White is precisely that a brand rather than a breed.

We have argued that the Oakham White story is emblematic of a wider process occurring within the food industry whereby 'mainstream' retailers such as Marks & Spencer are appropriating the language that was formerly associated with 'alternative' producers. Through the development of the Oakham White brand, we argue, companies like Marks & Spencer are trying to get the best of both worlds, combining the reliability and quality assurance that is associated with intensive 'scientific' modes of production with the traceability and concern for tradition and taste that consumers are increasingly demanding. As our example of the questionable claims made by one farmers' market trader suggests, a blurring of boundaries is taking place: between what has previously been defined as 'good' (alternative, organic, free range) and 'bad' (commercial, intensive, industrial) food; and between local, small-scale producers and national (or multinational) corporations, all of whom (to varying degrees and in varying ways) now use arguments about taste, quality, provenance and sustainability. To understand this process, we argue, requires us to go beyond polarised debates about authenticity, 'unveiling' the commodity fetish and revealing the exploitative social relations that are concealed beneath the commodity form. Instead, we have sought to pursue a more complex argument about the discursive process of appropriation. If we are to make normative judgements about commodification, as Castree (2004) and others have argued that we should, we need to understand the cultural processes through which meanings are manufactured as much as the political-economic processes through which alienation and exploitation occur.

References

Bevan, J. (2001). *The Rise and Fall of Marks and Spencer*. London: Profile Books.
Burt, S.L., Mellahi, K., Jackson, T.P. and Sparks, L. (2002). Retail internationalization and retail failure: issues from the case of Marks and Spencer. *International Review of Retail, Distribution and Consumer Research* 12, 191–219.
Castree, N. (2004). The geographical lives of commodities. *Social and Cultural Geography* 5, 21–35.
Compassion in World Farming (2003). *The Welfare of Broiler Chickens in the European Union*. Petersfield: Compassion in World Farming Trust.
Dixon, J. (2002). *The Changing Chicken: Chooks, Cooks and Culinary Culture*. Sydney: University of New South Wales.
Freidberg, S. (2004). The ethical complex of corporate food power. *Environment and Planning D: Society and Space* 22, 513–31.
Gereffi, G. (1994). The organisation of buyer-driven global commodity chains: how U.S. retailers shape overseas production network. In G. Gereffi, M. Korzeniewicz and R. Korzeniewicz (Eds), *Global Commodity Chains and Global Capitalism* (pp. 95–122). London: Greenwood Press.
Gilg, A. and Battershill, M. (1998). Quality farm food in Europe: a possible alternative to the industrialised food market and to current agri-environmental policies: lessons from France. *Food Policy* 23, 25–40.
Goodman, D. (2003). Editorial: the quality 'turn' and alternative food practices: reflections and agenda. *Journal of Rural Studies* 19, 1–7.

Holloway, L., Kneafsey, M., Venn, L., Cox, R., Dowler, E. and Tuomainen, H. (2005). Possible food economies: food production–consumption arrangements and the meaning of 'alternative'. Cultures of Consumption working paper 25 (available at http://www.consume.bbk.ac.uk/publications.html).

Jackson, P. and Russell, P. Life history interviewing. In D. DeLyser, M. Crang, L. McDowell, S. Aitken and S. Herbert (Eds), *Handbook of Qualitative Research in Human Geography*. London: Sage, in press.

Lury, C. (2004). *Brands: The Logos of the Global Economy*. London: Routledge.

Marsden, T., Flynn, A. and Harrison, M. (2000). *Consuming Interests: The Social Provision of Foods*. London: UCL Press.

Mellahi, K., Jackson, P. and Sparks, L. (2002). An exploratory study into failure in successful organizations: the case of Marks & Spencer. *British Journal of Management* 13, 15–29.

Murdoch, J., Marsden, T. and Banks, J. (2000). Quality, nature and embeddedness: some theoretical considerations in the context of the food sector. *Economic Geography* 76, 107–26.

Perks, R. and Thomson, A. (Eds) (2006). *The Oral History Reader*. London: Routledge (second edition).

RSPCA (2001). *Behind Closed Doors: Chickens Bred for Meat*. Horsham: Royal Society for the Prevention of Cruelty to Animals.

RSPCA (2005). *Paying the Price: The Facts About Chickens Reared for their Meat*. Horsham: Royal Society for the Prevention of Cruelty to Animals.

Russell, P. (2003). *Narrative constructions of British culinary culture*. Unpublished doctoral thesis, University of Sheffield.

Seth, A. and Randall, G. (1999). *The Grocers: The Rise and Rise of the Supermarket Chains*. London: Kogan Page.

Somers, M.R. (1994). The narrative constitution of identity: a relational and network approach. *Theory and Society* 23, 605–49.

Sustain (1999). *Fowl Deeds: The Impact of Chicken Production and Consumption on People and the Environment*. Bristol: Sustain – the alliance for better food and farming.

Whatmore, S., Stassart, P. and Renting, H. (2003). What's alternative about alternative food networks? *Environment and Planning A* 35, 389–91.

Chapter 19

Sidestepping the Mainstream: Fairtrade *Rooibos* Tea Production in Wupperthal, South Africa

Tony Binns[1], David Bek[1], Etienne Nel[2] and Brett Ellison[1]
[1]Department of Geography, University of Otago, New Zealand
[2]Department of Geography, Rhodes University, South Africa

Introduction

This chapter examines how the rural community of Wupperthal, in South Africa's West Coast region, a relatively under-developed region in the South African economy (WCII 1997), is responding to external challenges by engaging in a form of production that may be labelled as an 'alternative food initiative' (AFI). With support from a non-government organisation (NGO), a community-based cooperative in the isolated mission community of Wupperthal, is generating employment and income through the production of *rooibos tee* (red-bush tea), which only grows successfully in the specific microclimate and physical environment of the West Coast mountain region (see Figure 19.1). *Rooibos* enjoys a rapidly growing international market due to its health-giving properties, and the community is now able to sell its product internationally by engaging with Fair Trade networks. This particular project has been successful as an economic entity and has also produced various socio-economic dividends for the host community. The project's success is noteworthy in the South African post-apartheid context, since attempts to stimulate community-based development have generally received mixed responses (see Nel and Rogerson 2005).

Building on a theoretical base, which has its roots in developed world economy literature and experience, this chapter explores the applied relevance of the concept of alternative food systems, in terms of community economic survival and development, in the developing world, and in South Africa specifically. At the outset, it seems pertinent to ask two inter-related questions: first, has this project's 'alternativeness', as prescribed by alternative food literature typologies, contributed to its success?, and secondly, how useful is the analytical framework derived from the alternative food literature in evaluating projects such as these in the Global South?

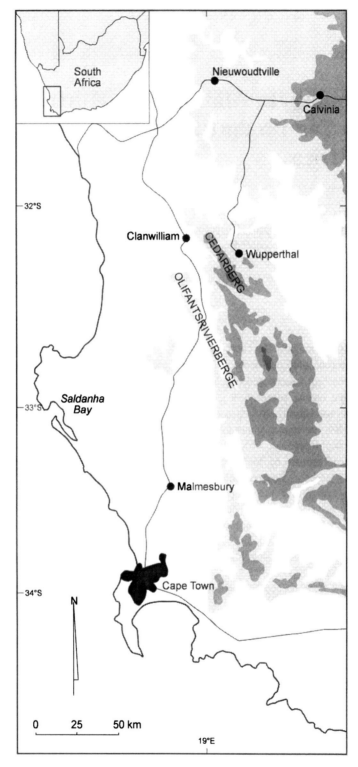

Figure 19.1: South Africa's West Coast Region.

The chapter begins by outlining the theoretical context through which the project is analysed. Our analysis is clearly situated within the terrain of the burgeoning Fair Trade literature, which we supplement with an explicit attempt to draw upon concepts emanating from other parts of the alternative food literature. Specifically, we utilise the notion of 'short food supply chains' (SFSCs) in order to explain the dynamics of our case study. In the second section, the Wupperthal project itself is outlined. An overview of the regional and local spatial context delineates the challenges confronting the inhabitants of Wupperthal in their quest to attain socio-economic upliftment. The ways in which the project can be considered 'alternative' are then outlined. The nature of the supply chain, in particular its shortness, is an important feature, as are the ways in which the project 'sidesteps' mainstream *rooibos* production systems in the region. The project is then evaluated and consideration is given to the reasons why this initiative has enjoyed a measure of success. The chapter concludes with a discussion of the theoretical and practical implications of the findings and a series of avenues for further research are identified.

Theoretical context

In recent years, increasing attention has been given to food production systems that might be labelled as 'alternative' (Goodman 2003; Renting *et al.* 2003; Watts *et al.* 2005). A review of the relevant literature reveals increasing reference to such terms as 'alternative food initiatives' (AFIs) (Allen *et al.* 2003), 'alternative agro-food networks' (AAFNs) (Goodman 2003) and 'alternative food networks' (AFNs) (Watts *et al.* 2005). Individual activities and initiatives that have attracted the label 'alternative' include: farmers' markets, food boxes, urban agriculture, 'slow food' and aspects of organic and Fair Trade production. There is much debate surrounding the question of what is actually 'alternative' in relation to the activities concerned (see Goodman 2004, for e.g. on European AFNs). In essence, their alternative status is derived from the fact that they operate in some way which is counter to the prevailing logic of the broader agro-food industry, which has become increasingly dominated by multinational companies (MNCs). The majority of the most powerful MNCs are based in the developed world and are able to exert considerable power over other actors within their supply chains; the dominance of these firms is such that their ways of conducting business tends to be understood as representing the 'mainstream'.

Ilbery and Maye (2005: 825) observe that "resistance to large producers and retailers" is certainly thought to be a key characteristic of, and motivation for, the initiation of so-called alternatives. Whilst this form of opposition certainly applies to the case study outlined below, we also observe how resistance to *locally* powerful actors and forces is a facet of the project. The spatiality of this resistance is particularly interesting as in the majority of cases outlined in the alternative food literature acts of resistance occur within developed world regions. In this instance, complex networks of power exist linked in with South Africa's colonial and apartheid history and its contemporary status as an emerging economy increasingly exposed to the forces of neoliberal globalisation. It is within such a complex and contested context that alternative approaches such as the one outlined below are now emerging.

Whilst much of the literature on alternative foods is Eurocentric in focus,[1] there has been increasing attention directed towards the role of Fair Trade in improving the socio-economic status of producers and their host communities in less developed regions of the world (Raynolds 2002; Renard 2003, 2005; Jaffee et al. 2004; Smith and Barrientos 2005; Young and Utting 2005). Furthermore, Fair Trade certification and the development of alternative markets are perceived as a challenge to the hegemony of conventional agri-food supply chains as producers benefit from shorter supply chains and hand-in-hand partnerships (see Raynolds 2002). Essentially, Fair Trade certification should ensure certain criteria such as fair prices to producers, fair wages to labourers, environmentally sustainable production practices, democratically run producer cooperatives, and non-exploitative assistance (Jaffee et al. 2004). Worldwide annual Fair Trade sales were valued at US$400 million in 2002, and are growing at an estimated annual rate of 30% (Fair Trade Federation 2005). The Fair Trade phenomenon is attracting an increasing amount of attention from academic research as efforts are made to theorise the dynamics of the Fair Trade movement and to evaluate its potential to contribute to meaningful socio-economic transformation. Here we will focus upon issues related to Fair Trade supply chains and their implications for producers.

The International Fair Trade movement is a somewhat heterogeneous entity, comprising a diversity of organisations and certifications systems. Bezencon and Blili (2006) have identified two main forms of Fair Trade supply chain – the 'Fair Trade Mainstream Type of Distribution Chain' (MTDC), coordinated by the Fairtrade Labelling Organisation (FLO), and the 'Fair Trade Alternative Type of Distribution Chain' (ATDC), largely coordinated by the International Federation for Alternative Trade (IFAT). In the case of the former, commodities are distributed by mainstream distributors such as supermarkets, whereas Alternative Trade Organisations (ATOs) form the key conduit for distribution in ATDCs, with sales usually occurring through dedicated shops. In addition, there are ATOs operating beyond the direct influence of either the FLO or IFAT.

The existence of these different forms of supply chain is important for both ideological reasons, in that some systems more obviously reinforce Fair Trade's inherently oppositional character than others, and also for socio-economic reasons, in that the various dividends and guarantees received by producers are more clear-cut in some systems than others. Bezencon and Blili's (2006) analysis resonates with much recent Fair Trade research, which expresses concern over the extent of mainstreaming occurring within the Fair Trade genre as major companies such as Starbucks, McDonald's and Tesco have developed significant Fair Trade portfolios (Goldie and Peattie 2005; Renard 2005; Smith 2006).

In this instance, we seek to develop Bezencon and Blili's (2006) analysis of Fair Trade supply chains by drawing upon a key concept developed within another part of the fast evolving alternative food literatures – namely SFSCs. A defining feature of many AFIs is that they are deemed to possess SFSCs, which can be more beneficial to rural communities, compared with the longer and more complex conventional supply chains which are typical of the modern agro-food industry (Marsden et al. 2000; Ilbery et al. 2004). This shortness is perceived to operate in different ways (see Marsden et al.

---

[1] See Abrahams, this volume, for further discussion of the limitations of Western derived theory in explicating contemporary processes occurring in the developing world.

2000; Watts et al. 2005). In this chapter, we focus on 'spatially extended' SFSCs. In this case, the final product is sold beyond the boundaries of the region of production, with information about the site and methods of production forming an important component of the consumers' decision-making process. In such cases, environmental, localised or ethical credentials may allow the consumer to make some form of connection with information about the production milieu. This information is important in influencing the consumer's purchasing behaviour, and is often associated with fair labour and production practices (Goodman 2003). As Marsden et al. (2000: 425) suggest:

> It is not the number of times a product is handled, or the distance over which it is ultimately transported, which is necessarily critical, but the fact that the product reaches the consumer embedded with information. It is this which enables the consumer to confidently make connections and associations with the place/space of production and, potentially, the values of the people involved and the production methods employed.

Recent research has further refined the SFSC concept, suggesting that in many cases such supply chains are not 'alternative' in terms of operating purely beyond the mainstream. Instead, many are hybridised, as to a greater or lesser extent there will be dependence upon mainstream actors within some nodes along the supply chain, for example, in terms of upstream and downstream linkages (Ilbery and Maye 2005). The extent and character of hybridisation may be important in determining the socio-economic impacts that will be felt at the local level. Application of the concept of SFSCs and hybridity has the potential to add a further level of sophistication to Fair Trade supply chain analysis, particularly in relation to the dividends experienced by producers. Analysis of such localised impacts is a primary focus of this research. In terms of both spatial and social development, an important question to consider then is the extent to which AFIs can assist historically disadvantaged communities in achieving some degree of socio-economic upliftment (see also Ilbery et al. 2004).

Ideas gleaned from the Fair Trade and alternative food literatures offer novel perspectives from which to evaluate projects that previously might have been viewed solely through the lens of theories emanating from the local economic development literature. The reasonably well-developed African and South African literature on local economic development identifies a clear role for community-based economic development both as a survival strategy in times of crisis and as a mechanism to participate in mainstream activity, particularly where state support is lacking (Gooneratne and Mbilinyi 1992; Binns and Nel 1999; Helmsing 2003; Egziabher and Helmsing 2005). Concerted government and local state endeavours in South Africa to formalise the local economic development process have met with mixed success (Tomlinson 2003; Nel and Rogerson 2005), with the net result that the development mandate must often be assumed directly by affected communities and/or by supportive NGOs (where they are active). The Wupperthal case study helps articulate these perspectives. Overall, it depicts a situation in which, largely through economic desperation and marginalisation, the community with the active support of an NGO has initiated what appears to be a successful community-based Local Economic Development process based on an AFI and the niche market offered by Fair Trade. As such this study contributes and extends the local debates on ways in which to empower and support communities in South Africa.

Rooibos Production in Wupperthal

The empirical material considered in this paper is largely the product of a three-year field-based research investigation in South Africa's Western Cape province. The rural community of Wupperthal, which is located within the province's West Coast region, has been visited on a number of occasions in March and June 2004, March 2005 and March 2006. Interviews were conducted with a range of informants, including Wupperthal community representatives (farmers and their farmers' association, church leaders and the tea court manager), NGO project workers and local government representatives. Semi-formal discussions were also conducted with various representatives from the local commercial *rooibos* industry in the region. This field-based research is forming a springboard for further in-depth investigations into this and other South African-based initiatives, which may be usefully analysed through the lens of the 'alternative food' literatures.

The Regional and Local Spatial Context

Having lagged during the 1960s and 1970s, South Africa's Western Cape province is now a leading province in terms of economic growth rates, having benefited from domestic and international opportunities since the demise of apartheid in 1994 (Wesgro 1999: 2). Currently, Western Cape contributes 14–15% of national Gross Domestic Product, although it contains only 10% of the population. The province is a leading contributor to national agricultural production and exports, with this sector's share of employment and turnover being double the national average per capita. However, Western Cape's economy is dominated by the Cape Metro and the Winelands, which cover only 2% of the land area, but generate 80% of the economic activity (WCII 1997: 2). The West Coast region is a predominantly rural area situated to the north of Cape Town, comprising eight magisterial districts with a total population of 300 000. Agriculture and fishing production and processing, manufacturing and mineral processing dominate the area's economy, which contributes 7.2% of provincial GRP. The region is recognised as being politically conservative, dominated by Afrikaans-speaking White and Coloured communities. Black Africans represent only 5% of the population. The area's economy has long been dominated by small groups of (usually) white business people, who have exercised a stranglehold on business development and have been in a position to influence local policy-making. Innovation and competition in the region have tended to be constrained (Bek *et al.* 2005). The West Coast region was identified by the South African government as an area with 'underutilised economic potential', and as such was designated for developmental support in 1998 through the national Spatial Development Initiatives programme (Jourdan *et al.* 1997; Bek *et al.* 2004).

The Wupperthal community in the high mountain valleys of the Cedarberg was founded in 1827 as a Rhenish mission station. The focal point of the community, the church, was built in 1834. In 1965, the Moravian church took over the mission. Geographically, the area is marginal, with Wupperthal located over 80 km from the nearest town, Clanwilliam, along a single-track gravel road which traverses four mountain passes. The settlement consists of a central village where the church is situated, and 11 'out-stations' or religious meeting points. Some 400 families (c. 2000 people) currently live in the area, including 170 *rooibos* farmers. The rugged topography and arid climate, receiving only 125–300 mm of rainfall annually, place serious constraints on farming, such that out of a total land area of 36 000 ha, only 1000 ha is actually workable. The resident population, almost all of whom are either descended from the original Khoisan inhabitants, or are of mixed race, were discriminated against and marginalised under South Africa's apartheid system. Prior to the start of the *rooibos* project in 1998, the community had very limited income and over 80% unemployment. There was a steady stream of young people leaving the community, as the only available work was in vegetable farming, the local shoe factory (eight jobs), a glove factory (25 jobs), and seasonal work farm labouring.

The Rooibos Industry

Rooibos (*Aspalathus linearis, Fabaceae*) is a legume and part of the genus *Aspalathus* (Wilson 2005). It is unique to the Cape floral kingdom, known locally as the *fynbos*, which has the distinction of being both the smallest of the world's six floral kingdoms, and the only one occurring entirely within the borders of a single country (Editors Inc. 2002). *Rooibos* production only occurs in a region some 200–300 km north of Cape Town, within, and adjacent to, the Cedarberg and Olifantsrivier mountain ranges, and on the Bokkeveld plateau (Oettle *et al.* 2002) where microclimates and soil conditions are particularly favourable (see Figure 19.2). The tea was first identified and used by the Khoi-Khoi tribes of the area, and in 1904, a Russian immigrant, Benjamin Ginsberg, began purchasing the tea from mountain dwellers to sell within South Africa. *Rooibos* has certain health-giving properties, notably it is caffeine free and contains compounds which act as anti-oxidants. As a result, *rooibos* enjoys a rapidly growing international market. Annual production of *rooibos* is currently some 6000 tonnes, of which approximately 30% is exported (Wilson 2005). Exports have increased by a third since 2001.

Until the early 1990s, under the apartheid regime, South Africa's agricultural production was managed by agricultural boards that generally favoured white producers whilst

Figure 19.2: *Rooibos* tea growing in the wild.

marginalising black farmers.[2] All production and processing of *rooibos* was under direct control of the state-appointed *Rooibos* Tea Board, based in Clanwilliam, the primary service centre for the production area. *Rooibos* production was controlled by a strict quota system, and small-scale producers were only minor suppliers. However, in 1991, the industry was deregulated, allowing a number of new commercial producers and marketing companies to enter the market. Unfortunately, this move did not allow small producers with limited capital and skills to find a market niche, even though some new firms sought to provide market access to community producers and those producing the tea organically (Wilson 2005). The dominant player in the *rooibos* industry is a firm named *Rooibos* Ltd, who have an overall market share of 70% and a domestic market share of 90%. The real market break for community producers came when they were targeted for support by an empowerment and development focused NGO. As a direct result of this association, funding support for community economic development became more significant for *rooibos* production, and international Fair Trade organisations offered to distribute the crop internationally.

The Wupperthal Community Rooibos Project

Limited sales of Wupperthal grown *rooibos* tea to commercial producers in Clanwilliam started in the early 1900s, and a 'tea barn' for storage purposes operated in the village between 1952 and 1962. The foundation of the current initiative was laid in 1997 following a 'needs assessment'. The community and church, as custodians of the land, formally decided to start the *rooibos* tea project, which was identified as having considerable commercial potential that could generate much-needed local employment and income. A key player in the success of the Wupperthal project has been ASNAPP (Agribusiness in Sustainable Natural African Plant Products), a USAID-funded non-profit agricultural support organisation. ASNAPP operates in nine African countries and focuses on the development of high-value natural plant products that enable African agribusinesses to compete in local, regional and international markets. ASNAPP (ASNAPP 2005) works through partnerships with research and higher education institutions, and in South Africa is based at the University of Stellenbosch near Cape Town. ASNAPP assisted in the participatory-based community needs assessment in Wupperthal, and key outcomes were the identification of the potential of *rooibos* as a driver for development and the significance of local indigenous knowledge and skills. ASNAPP subsequently assisted with the development of a business plan, and training and research into the sustainable harvesting of the product (ASNAPP 2005). ASNAPP has also provided invaluable technical support in plant agronomy, improving production and quality, encouraging sustainable growing and harvesting practices, strengthening the organic aspects of production, and facilitating access to overseas markets and higher prices.

In establishing the *rooibos* initiative, a distinctive feature of Wupperthal was the strong pre-existing levels of social capital. Well-developed local networks and a good

---

[2]We use the term 'black' as inclusive of people of all colour (incorporating those referred to as 'coloured', 'Indian' and 'African'), whilst being mindful of the political sensitivities associated with such terminology. Such terminology represents the system most commonly used in government literature and government surveys, although subtle variations do exist (see Lester *et al.* 2000: 26/7 for a fuller discussion of racial terminology in South Africa).

level of interaction exist among the community, factors which have been crucial in establishing a collaborative business venture. The entire community shares a common language (*Afrikaans*) and attends the local church. Families have lived in the valley for generations, and have developed remarkable levels of self-sufficiency and resilience which have fostered the building of localised types of social capital based around trust. The church owns the land, whilst the church council, comprising the community elders and leading farmers, oversees access to and use of the land. Farmers have a 99-year leasehold agreement for use of the land. In the absence of a local government office, the church and church council play a dominant role in managing community life.

To provide the necessary infrastructure and equipment to initiate the project, ASNAPP helped the community-based Farmers Association apply to the provincial Reconstruction and Development Programme (RDP) for R290 000 in 1997.[3] The grant was subsequently awarded and administered by LANOK (now re-named Cape Agency for Sustainable Integrated Development in Rural Areas – CASIDRA), the provincial government's rural development agency, which seeks to support community-driven projects. The funds were used in 1998 to resuscitate the old 'tea court' in the village, to buy a tractor, to lay a new sheet of concrete in the tea-drying area and to build a new store. A second grant of R300 000 was successfully secured in 1999 from the provincial Department of Social Services, and was used to enlarge the drying floor to $1000\,m^2$, and to extend the storage shed (LANOK 2005; CASIDRA 2005). These moves are important, since they enabled the community to process their tea locally rather than having to transport the raw material to Clanwilliam. The Association (recently re-structured as a cooperative) now functions independently of external funding and is increasingly independent in terms of its general operations.

Production of *rooibos* in Wupperthal follows a regular annual pattern. Seedlings are obtained from Clanwilliam, or a local supplier in the village, and planting takes place in winter, between June and August. The growing plant needs little attention, but it can only be harvested when mature, usually after 3 years. Organic production is assured through the addition of animal manure, and use of natural pesticides derived from a local plant known as *khakibos*. Harvesting takes place in mid-summer (January). At the present time, 10 tonnes of *rooibos* are harvested from wild plants and over 70 tonnes from plantations. An individual farmer's output ranges from 200 kg to 2 tonnes, with each farmer employing up to eight labourers, ensuring significant employment benefits in the local community.

After processing at the tea court (Figure 19.3), the bulk of the tea is sent to Clanwilliam for sterilisation and packaging, though the latter aspect will now be modified as detailed below. Most farmers sell directly to the cooperative, although some sell privately to 'Rooibos Ltd.' in Clanwilliam. The first major customer of the Association, however, was a Dutch-based Fair Trade firm, which approached Wupperthal in 1997. Most sales were initially made through a wholesaler to this firm in Europe, which certified the product as being organically produced. Prices were favourable, since the high altitude of Wupperthal ensures good quality tea. The Association is optimistic about the future, and hopes to double its income by initiating sales directly to European retailers. To facilitate this expansion, the Association has recently attained FLO accreditation, which

---

[3]Exchange rate: 29 March 2006, £1 = SA Rand 11.00.

Figure 19.3: *Rooibos* tea harvested and drying in a tea court.

will facilitate direct sales from the tea court to a wider range of European retailers including UK-based firm Equal Exchange who already market FLO-accredited *rooibos* from the nearby Heiveld Cooperative. There is also potential for upgrading production levels in response to the increasing market demand for organic produce.

Alternative Characteristics

The alternative characteristics of the Wupperthal Cooperative can be defined in two main ways. First, the community-based production system contrasts markedly with local and national mainstream agri-production systems. Secondly, the cooperative exhibits some of the characteristics of a SFSC, albeit one that is a fluid hybrid of the alternative and mainstream. We would argue that through a combination of these alternative characteristics, which are outlined in more detail below, the cooperative's activities are in effect endeavouring to resist structural constraints that operate locally and globally.

It is important to recognise the formidable structural barriers faced by individuals and black communities in seeking to formalise and expand productive operations in regions such as the northern West Coast. On the one hand, the legacies of the apartheid era have left local communities with acute deficits in terms of skills, resources and external network connections. On the other hand, the influence of white farmers, with their well-established economic and political networks, remains strong. In common with the rest of the South African agricultural sector, the *rooibos* industry is dominated by white commercial farmers utilising plantation-style production systems. Income for local populations is largely derived from wages earned as labourers. Working conditions and wages in South African land-based industries are notoriously poor (Brown *et al*. 2003; du Toit 2003). However, improvements have been witnessed since the end of apartheid, but pressures imposed by global trade regimes have exacerbated some aspects of inequality. Externalisation and casualisation of the labour force have occurred, whereby permanent workers are retrenched and their positions replaced by casual and seasonal workers often recruited via labour brokers (Barrientos and Kritzinger 2004; Ewert and du Toit 2005). Such employment practices act to drive down pay and working conditions. Thus, downward pressures exerted by global supply chain management are generating a transition which has increasingly negative socio-economic impacts on an agricultural labour force which numbers some 216 510 employees, or 14% of the Western Cape labour force (Department of Agriculture 2005).

It is within this context that one facet of the alternativeness of the Wupperthal project can be observed. The institutional arrangements under which *rooibos* tea is produced contrast markedly with mainstream production systems in the region. Opportunities for non-white people to gain ownership of the means of production are rare, both within the *rooibos* industry and the regional economy. This particular initiative has provided an opportunity for a community to circumnavigate such constraints. The starting point for such circumnavigation was the establishment of the specific project as an association (the initiative is now run as a cooperative). By establishing the venture the community was able to formalise an alternative structure of local cooperation (WRTA 2005). In addition, the community has actively networked with another small-farmer cooperative in the Northern Cape, as well as making use of traditional nodes within the distribution chain, albeit nodes which are seeking to engage in ethical trading practices. For example, the Wupperthal initiative's processed *rooibos* requires sterilisation prior to packaging, a process contracted to Cape Natural Tea Products in Cape Town. However, even this company pursues a practice of 'social responsibility' as it has in recent years actively engaged in sponsoring other small-farmer cooperatives within the Western Cape (CNTP 2006). Therefore, the community has developed networks which have bypassed traditional distribution chains. The services of an internationally operating NGO have been sought in order to overcome some of the barriers faced by the community in formalising and expanding their operations.

Furthermore, the initiative exhibits many characteristics of a 'SFSC'. Distinct efforts have been made to connect with niche consumer markets through organic certification and various links with Fair Trade ATOs. In some cases the tea has been explicitly marketed as being sourced from Wupperthal, with the unique characteristics of the locale and its impacts upon taste and quality being prominent in the packaging. The majority of sales have occurred via alternative outlets rather than mainstream retail multiples. These outlets have included European and North American Fair Trade distributors such as TopQualiTea, EqualExchange, SERRV, and CTM. In contrast, commercially produced Cedarberg *rooibos* is widely sold in UK supermarkets, supplied by the Redbush Tea Company and Twining's. These brands do not as yet offer Fair Trade versions. However, even in these mainstream examples there is an 'alternative twist', as the packaging emphasises the purported health giving properties and the unique geographical character of the Cedarberg.

Despite the links with ATOs, the supply chain has inevitably exhibited signs of hybridity, with a dependence upon mainstream agencies within various nodes of the supply chain. However, an interesting facet of this case study is the way that the community has gradually sought to take control over an increasing number of nodes. In the first place, the community has established its rights to the means of production in terms of land access. Then, by constructing a new and larger tea court it was able to take greater control over the processing of the raw material. Full control of this process was established with government support and increasing revenue reinvested into machinery for cutting harvested *rooibos*, and essential equipment which means that the tea can be processed, dried, sifted, and stored at the Wupperthal tea court. As mentioned earlier, Cape Natural Tea Products handles the sterilisation of the tea after leaving Wupperthal, further illustrating the hybrid nature of this project. Most recently, they have extended their control over the value chain by participating in the establishment of a *rooibos* packaging venture in Cape Town. This participation is very much an investment along alternative and ethical lines, as the community has equal shares in this company with another small-farmer cooperative from the Northern cape, and a private individual who has been actively engaged in the Fair Trade market for several years. In this way, Wupperthal-sourced *rooibos* operates through an increasingly separate supply network to that of the vast majority of Cedarberg *rooibos*. Comparisons may be drawn here with the Day Chocolate company, in which cooperatively organised producers have taken ownership of the processing, marketing and retail units (see Doherty and Tranchell 2005; Smith 2006). On this last point, Smith (2006: 1) concludes that

> Though relatively few in number, these initiatives illustrate a model of fair trade that seeks to compete with conventional companies rather than change them. Its strength is that it offers a more empowering and long-term solution to producers in developing countries.

Evaluating the Project

The Wupperthal initiative has achieved a significant level of success, such that since 1998, the number of farmers has increased from 25 to 170, while *rooibos* production has risen from 60 tonnes to up to approximately 100 tonnes in good years. Cooperative members are now able to add value to the product that they grow by utilising their own tea court for processing (shown in Figure 19.3) and, as mentioned above, the recent establishment of a packaging venture is now cementing the community's control over downstream elements of the value chain. Most farmers employ up to four labourers, and an additional 14 people work at the tea court. Importantly, the outflow of young migrants has been stemmed to a degree, as some are prepared to experiment with *rooibos* farming due to the improved prospects offered by Fair Trade linkages. In addition, labourers are now able to gain employment locally without having to circulate around other farming regions in search of seasonal work. Anecdotal evidence suggests that local seasonal work is better remunerated than on neighbouring white-owned farms.

A significant source of income has been derived from the venture, which has clearly impacted on the community, local employment and incomes. Funding from *rooibos* sales have also been invested in local community facilities, including farming equipment and the local school, enhancing the overall impact of supplying a niche market crop to consumers internationally. Furthermore, the community's sense of self-determination has increased substantially through the ongoing successes of the project. Mutually beneficial links have been forged with another *rooibos* producing community in Heiveld, situated on the South Bokkeveld plateau, east of the Cedarberg. Exchange visits for the purposes of sharing knowledge and expertise are a feature of these inter-community linkages. Even more significant, the two communities have joined forces to take a two-thirds ownership of the above-mentioned *rooibos* packaging plant, Fair Packers, based in Cape

Town, which opened in March 2006. This plant has already generated employment for Wupperthal and Heiveld migrants, as well as providing producers with an increased share of the value of their product. The initiation of this venture is indicative of the extent to which communities are progressively becoming empowered. However, it should be noted that the community's efforts to extend its operations further into the value chain have faced various forms of direct and indirect resistance from other actors within the value chain and the broader economy. Such resistance has its roots within the dynamics of the region's own variant of the apartheid legacy, as well as the generic asymmetries found within global supply chains.[4]

Field-based research has revealed that the success of the Wupperthal *rooibos* initiative is due to a number of critical factors, most notably, the strength of localised forms of social capital which exists among this small, relatively isolated, mission community, and the ability to produce a commodity which is spatially and ecologically limited, but has a steadily increasing international market. The community recognises the rare combination of these factors, with one leader suggesting that: "Wupperthal is unique" (Cooperative representative 30/3/05). The community-driven initiative has undoubtedly led to tangible benefits for most residents, and, as a representative of the Cooperative asserts: "a community needs to seize the potential that is offered by local resources and take full ownership of the development process" (personal communication 30/3/05). Wupperthal also benefits from a market-oriented approach, and has tapped into an alternative trade network which is supplying a product for which an international demand exists. Support for the project has come from a well-funded and effective NGO (ASNAPP), with considerable experience of supporting projects that engage with global markets. At the local level, and supported by the strong social capital and unity provided by the church, which also controls the land, a cohesive and collaborative institutional framework has supported and facilitated development. Of particular interest in this initiative is how pre-existing forms of social capital have become extended and broadened via access to international markets facilitated by the support of ASNAPP.

Concluding Observations

Viewed through the analytical lens of local economic development literatures, the Wupperthal Cooperative can be heralded as a success story owing to the benefits attained in raising community living standards and also the broader economic sustainability of the project itself. Sadly, such success stories are all too infrequent in post-apartheid South Africa. In the context of this study, the alternative foods literature clearly provides a useful conceptual framework for evaluating ongoing processes in food production in developing countries. It is somewhat surprising therefore, that there have been so few attempts to apply ideas derived directly from the alternative food literature to experiences in countries characterised by widespread poverty, and where such initiatives could possibly provide a valuable opportunity for uplifting livelihoods and local economies. Indeed, as Goodman (2003: 4) observes more generally: "... the literature on networks, quality

---

[4]In the light of the sensitivity of the issue, and in order to protect informant confidentiality, it is not possible to expand further upon the precise nature of the resistance being faced.

conventions and artisanal production and marketing systems is narrowly ethnocentric in its focus on North America and Western Europe".

The Wupperthal project's alternativeness has certainly helped the community to reap benefits from tapping into international markets. The project also demonstrates the importance of enabling factors such as appropriate institutional support mechanisms and community cohesion. Overall, the Wupperthal case clearly demonstrates that such initiatives can play an important role in challenging poverty and inequality in rural communities. In the remainder of this chapter, we return to the questions posed in the introduction relating to the utility of concepts derived from the alternative foods literature. Discussion then ensues concerning some of the issues raised by the Fair Trade dimensions of the project. As Watts *et al.* (2005) observe, further research is needed to understand the broader outcomes of alternative forms of agri-production. Throughout this final section, issues meriting further research are identified, pertaining to Wupperthal, Fair Trade and the alternative foods genre more broadly.

Without doubt the alternative elements of the project have contributed to its successes, in terms of economic sustainability and improved living standards. By largely sidestepping mainstream value chains, the cooperative is able to attain far greater economic dividends and forms of empowerment than would be possible via involvement in mainstream *rooibos* production. Indeed, within the mainstream system farmers would probably only be able to act as suppliers to middlemen and would be subject to various price and demand uncertainties. Under the alternative model that has emerged, farmers are able to benefit from favourable prices, guaranteed markets and other social dividends. Furthermore, the very alternativeness of the project, and in particular the fact that it is enhancing the transformation process in South Africa, is a key selling point in itself since it broadens the ethical appeal of the product within European markets. In this sense, the concept of a SFSC is reinforced by these direct links between producers and consumers.

Modes of analysis drawn from the alternative food literature do, we believe, offer much potential for shedding light upon the precise dynamics of projects such as the Wupperthal Cooperative and its potential for challenging poverty and exclusion. Notions of opposition and resistance draw attention to the broader political economy within which these types of project evolve. Local and national issues inherited from the apartheid era are particularly strong in this case, and by casting this initiative as oppositional, important questions can be asked about the ways in which the initiative circumnavigates local constraints that are strictly controlled by long-established actors.

By analysing these production systems using the SFSC concept, it is possible to focus attention upon elements that are critical drivers of success. As outlined above, connections between producers and consumers are important in generating a market, whilst the ability of producers to gain control over several elements of the value chain is important in increasing financial dividends. The concept of hybridity is equally valuable in drawing attention to the characteristics of the whole supply chain and the relationships between mainstream and alternative elements. Mapping out such chains would appear to be a useful tool for evaluating their stability and sustainability from the producer's perspective. In this way it may be possible to evaluate the extent to which 'short chains' actually facilitate increased wealth production at the producer end and to consider whether concepts of hybridity can be utilised to further differentiate between different types of supply chain, such as those identified by Bezencon and Blili (2006).

There is clearly a need for more investigation into the different experiences of AFIs in developing countries, in order to assess both their desirability and their levels of success. Whilst, understandably, each case study will have its own unique characteristics, further research may reveal certain features and processes which are common to a number of initiatives, and which could possibly lead to both a better understanding of the reasons for varying degrees of success, and might also assist in devising and implementing future initiatives. Our research thus far indicates that AFIs, including Fair Trade models, have barely registered on the radar of South African government agencies. We would strongly argue that this situation requires rapid rectification. There is much potential for AFIs of one form or another to contribute to socio-economic transformation. The process at present is ad hoc, and critical engagement by policy makers, recognising the perils, as well as opportunities, of these types of initiatives, could lead to the identification of mechanisms that assist communities in making innovations within agri-production systems. There is an important role here for praxis that is well informed by theory. In this sense, it is vital that concepts such as 'alternative' and 'mainstream' are problematised and their utility beyond the Western world critically interrogated. Ultimately, the considerable power and influence of dominant economic actors can only be challenged successfully if the processes at work are fully explicated from the perspective of different actors within the relevant value chains (Cook and Harrison 2003).

The Wupperthal Cooperative's experience of Fair Trade raises a number of interesting issues, three of which are outlined as follows. First, concerns have been raised about the activities of some self-styled ATOs, many operating outside of internationally recognised certification systems, which were keen to access South African-sourced goods in the mid-to-late 1990s. The extent to which the terms of engagement were consistent with the ethos of the broader Fair Trade movement is certainly open to question (Kruger 2004). Since the inception of the project in the late 1990s, the Wupperthal Cooperative has engaged with various ATOs claiming Fair Trade credentials before attaining FLO accreditation in 2005. The latter offers producers greater transparency and consistency in terms of contracts and supply arrangements, and certainly opens up more market avenues and sets minimum standards in terms of economic dividends. Wupperthal's experiences with diverse ATOs seem to have been largely positive and it will be interesting to monitor the cooperative's observations concerning the ongoing impacts of FLO accreditation.

It should be noted that concerns have been articulated concerning the FLO paradigm, and the fact that its principles are overly driven by western concerns, and as such are insufficiently flexible in accounting for different producer contexts around the world (Kruger and Hamman 2004; Low and Davenport 2005; Bezencon and Blili 2006). Interestingly, there is much ongoing work within South Africa trying to ensure that the 'southern voice' is heard in the evolution of Fair Trade networks, such that mainstreaming and the imposition of western values upon local production systems are being resisted (Kruger 2004). The success of such endeavours could have a very significant bearing upon the future sustainability of initiatives such as Wupperthal.

A second concern relates to the expansion of Fair Trade accredited sites. At present, cooperative leaders in Wupperthal are delighted with their ability to gain international Fair Trade recognition and observe that they have 'sewn up the Fairtrade *rooibos* market'. However, we would urge caution, as Fair Trade has fast become something of a 'must-have designer accessory' within South African agri-industries and applications to FLO have been rising exponentially in recent years. Other communities are in the process of considering approaching FLO with regard to their *rooibos* production, and it is plausible that mainstream white-owned businesses could meet FLO requirements if they so desired. Thus, while Wupperthal and their collaborators in Heiveld are presently sitting pretty, it is possible that they could face increased competition in the future. Oversupply is not out of the question in the medium term, and there are issues here that merit consideration by proponents and facilitators of Fair Trade concerning the extent to which new small producers and cooperatives should be encouraged to meet FLO criteria when there is a risk of oversupply and thus no guaranteed market. This very dilemma faces ASNAPP, who are presently developing links with another *rooibos* producing community in the Western Cape. Such ethical concerns may not register in the promotion of mainstream production, but it can be argued that in the case of alternative ventures, predicated upon an ethical dimension, every effort should be made to ensure that existing and potential producers are fully aware of the state of the market and its future potential.

Thirdly, Fair Trade has been a fast evolving entity in recent years, and it is not inconceivable that *rooibos* will join the list of products being dragged back into the mainstream by suppliers and retailers. If this happens to Wupperthal-sourced *rooibos*, what would be the likely implications for the Cooperative, its cohesion and the management of the *rooibos* initiative? What importance will be given to these issues by decision makers, as AFIs face the probability of being progressively mainstreamed by large-scale European supermarket chains seeking closer involvement? Furthermore, recent research indicates that even within supply chains there is considerable differentiation in socio-economic outcomes and empowerment according to the business strategy of the big players within the supply chain (Smith 2006). This is a complex terrain to navigate, most acutely for communities whose knowledge and understanding of entities such as Fair Trade and niche overseas markets may understandably be limited. Further research into these different types of supply chain and their impacts is an absolute necessity.

Whilst the Wupperthal *rooibos* initiative has undoubtedly brought considerable benefits to the local community, similarly successful projects will need to flourish in many more of South Africa's impoverished rural communities if there is any possibility of ameliorating the legacies of apartheid-era policies. Realistically, this is unlikely to happen, and it will take generations to alleviate poverty and reduce the stark inequalities that are a feature of South African society and space. In this chapter, whilst celebrating the success of the Wupperthal *rooibos* initiative, we have also endeavoured to indicate some possible threats to project sustainability and have considered a range of issues associated with community engagement with international trade networks. Perhaps most important, there is an urgent need for similar investigations to be undertaken in other developing countries, where the quest for initiatives that might lead to meaningful poverty alleviation and livelihood upliftment should be placed high on the agenda.

Acknowledgements

This research was made possible through the generous support of the British Academy (award no. SG-36776). Brett Ellison would like to acknowledge the New Zealand Agency for International Development and the University of Otago for generously providing financial support. The authors would like to thank the Geography Department cartography office at Otago University for producing the map. Assistance from our respondents is gratefully acknowledged, as is the advice of Damian Maye, Lewis Holloway, Moya Kneafsey and the anonymous reviewers.

References

Allen, P., FitzSimmons, M., Goodman, M. and Warner, K. (2003). Shifting plates in the agrifood landscape: the tectonics of alternative agrifood initiatives in California. *Journal of Rural Studies* 19, 61–75.
Agribusiness in Sustainable Natural African Plant Products (2005). Country Programme: South Africa. Available at: www.asnapp.org/country-progs/sa.html (accessed 3/4/2005).
Barrientos, S. and Kritzinger, A. (2004). Squaring the circle – global production and the informalisation of work in South African fruit exports. *Journal of International Development* 16, 81–92.
Bek, D., Binns, T. and Nel, E. (2004). 'Catching the development train': perspectives on 'top-down' and 'bottom-up' development in post-apartheid South Africa. *Progress in Development Studies* 4 (1), 22–46.
Bek, D., Binns, T. and Nel, E. (2005). Regional development in South Africa's West Coast: dividends on the process side? *Tijdschrift voor Economische en Sociale Geografie* 96 (2), 168–183.
Bezencon, V. and Blili, S. (2006). Fair trade channels: are we killing the romantics? *International Journal of Environmental, Cultural, Economic and Social Sustainability* 2 (1), 187–196.
Binns, J.A. and Nel, E.L. (1999). Beyond the development impasse: local economic development and community self reliance in South Africa. *Journal of Modern African Studies* 37 (3), 389–408.
Brown, M., du Toit, A. and Jacobs, L. (2003). *Behind the Label: A Workers' Audit of the Working and Living Conditions on Selected Wine Farms in the Western Cape*. The Western Cape: Labour Research Service, Women on Farms Project and the Programme for Land and Agrarian Studies, University of the Western Cape.
Cape Agency for Sustainable Integrated Development in Rural Areas (2005). *Casidra: Activities: Agriculture and Agro Industries*. Available at: www.casidra.co.za/activities_agriculture.shtml (accessed 12/4/2005).
Cape Natural Tea Products (2006). *Social Responsibility, Honeybush Plantation Community Projects – Haarlem and Ericaville*. Available at: http://www.rooibostea.co.za/honeybush-tea-projects.html (accessed 1/3/2006).
Cook, I. and Harrison, M. (2003). Cross over food: postcolonial economic geographies. *Transactions of the Institute of British Geographers* 28 (3), 296–317.
Department of Agriculture, Western Cape Province (2005). *A Profile of the Western Cape Province: Demographics, Poverty, Inequality, and Unemployment*. The Provincial Decision-Making Enabling Project, Background Paper Series, 2005: 1(1).
Doherty, B. and Tranchell, S. (2005). New thinking in international trade? A case study of the Day Chocolate Company. *Sustainable Development* 13, 166–176.
du Toit, A. (2003). Hunger in the valley of fruitfulness. Globalization, 'social exlcusion' and chronic poverty in Ceres, South Africa. Paper presented at conference on, 'Staying Poor: Chronic Poverty and Development Policy', University of Manchester, 7–9 April 2003.
Editors Inc. (2002). *SA 2002–3: South Africa at a Glance*. Johannesburg: Editors Inc.
Egziabher, T.G. and Helmsing, A.H.J. (2005). *Local Economic Development in Africa*. Maastricht: Shaker.

Ewert, J. and du Toit, A. (2005). A deepening divide in the countryside: restructuring and rural livelihoods in the South African wine industry. *Journal of Southern African Studies* 31 (2), 315–332.
Fair Trade Federation (2005). Fair trade facts. Available at: http://www.fairtradefederation.com/ab_facts.htm (accessed 8/11/2005).
Goldie, K. and Peattie, K. (2005). In search of a golden blend: perspectives on the marketing of Fair Trade coffee. *Sustainable Development* 13, 154–165.
Goodman, M. (2003). The quality 'turn' and alternative food practices: reflections and agenda. *Journal of Rural Studies* 19, 1–7.
Goodman, D. (2004). Rural Europe redux? Reflections on alternative agro-food networks and paradigm change. *Sociologia Ruralis* 41 (1), 3–16.
Gooneratne, W. and Mbilinyi, M. (1992). *Reviving Local Self-Reliance*. Nagoya: UNCRD.
Helmsing, A.H.J. (2003). Local economic development: new generations of actors, policies and instruments for Africa. *Public Administration and Development* 23, 67–76.
Ilbery, B. and Maye, D. (2005). Alternative (shorter) food supply chains and specialist livestock products in the Scottish-English borders. *Environment and Planning A* 37, 823–844.
Ilbery, B., Maye, D., Kneafsey, M., Jenkins, T. and Walkley, C. (2004). Forecasting food supply chain developments in lagging rural regions: evidence from the UK. *Journal of Rural Studies* 20, 331–344.
Jaffee, D., Kloppenburg, J.R. Jr and Monroy, M.B. (2004). Bringing the 'moral change' home: fair trade within the North and within the South. *Rural Sociology* 69 (2), 169–196.
Jourdan, P., Gordhan, K., Arkwright, D. and de Beer, G. (1997). *Spatial Development Initiatives (Development Corridors): Their Potential Contribution to Investment and Employment Creation*. DTI Pretoria, unpublished document.
Kruger, S. (2004). *Fairtrade in South Africa: Growth, Empowerment and Fairness in the South African context*. Unpublished consultation paper, Programme for Land and Agrarian Studies, University of the Western Cape, South Africa.
Kruger, S. and Hamman, J. (2004). *Guidelines for FLO's Empowerment Strategy in South Africa*. Unpublished consultation paper, Programme for Land and Agrarian Studies, University of the Western Cape, South Africa.
LANOK (2005). LANOK: we open doors: Projects: Agriculture. Available at: www.lanok.co.za/html/rooibos.html (accessed 11/4/2005).
Lester, A., Binns, T. and Nel, E.L. (2000). *South Africa, Past, Present and Future: Gold at the End of the Rainbow?* Harlow: Pearson.
Low, W. and Davenport, E. (2005). Postcards from the edge: maintaining the 'alternative' character of Fair Trade. *Sustainable Development* 13, 143–153.
Marsden, T., Banks, J. and Bristow, G., (2000). Food supply chain approaches: exploring their role in rural development. *Sociologia Ruralis* 40, 424–438.
Nel, E. and Rogerson, C.M. (2005). *Local Economic Development in the Developing World: The Experience of South Africa*. New Brunswick: Transactions Press.
Oettle, N. Koelle, B. Arense, A. and Mohr, D. (2002). *Juweel van die Berge*. Cape Town: Environmental Monitoring Group.
Raynolds, L.T. (2002). Consumer/producer links in fair trade coffee networks. *Sociologia Ruralis* 42 (4), 404–424.
Renard, M.C. (2003). Fair trade: quality, market and conventions. *Journal of Rural Studies* 19, 87–96.
Renard, M.C. (2005). Quality certification, regulation and power in fair trade. *Journal of Rural Studies* 21, 419–431.
Renting, H., Marsden, T. and Banks, G. (2003). Understanding alternative food networks: exploring the role of short food supply chains in rural development. *Environment and Planning A* 35, 393–411.

Smith, S. (2006). Fair trade in the mainstream. Available at: www.ids.ac.uk/news/SmithStevens FairTrade.html (accessed 23/3/2006).
Smith, S. and Barrientos, S. (2005). Fair and ethical trade: are there moves towards convergence? *Sustainable Development* 13, 190–198.
Tomlinson, R. (2003). The local economic development mirage in South Africa. *Geoforum* 34, 113–122.
Watts, D.C.H., Ilbery, B. and Maye, D. (2005). Making reconnections in agro-food geography: alternative systems of food provision. *Progress in Human Geography* 29 (1), 1–19.
Wesgro (1999). Business prospects 2000: Accelerated growth in South Africa's Western Cape. *Western Cape Economic Monitor*, November 1999.
West Coast Investment Initiative (1997). *West Coast Investment Initiative Appraisal Document.* Document produced for Investors Conference 26–27 February 1998.
Wilson, N.L.W. (2005). Cape Natural tea products and the U.S. market: *Rooibos* rebels ready to raid. *Review of Agricultural Economics* 27 (1), 139–148.
Wupperthal Rooibos Tea Association (2005). *Wupperthal News*. May 2005, 1 (1).
Young, W. and Utting, K. (2005). Editorial: fair trade, business and sustainable development. *Sustainable Development* 13, 139–142.

Index

Abruzzo, Italy, 83
Access to food, 108–9
Actor Network Theory, 12, 189–90
'Adopt-a-sheep', 87–9
Aesthetics:
  in Buy Local Food campaigns, 260–3, 264, 268
Aesthetics *see* Food, and aesthetics
African American:
  selective patronage, 266
Agri-environmental governance, 223–5, 231–4
Agri-environmental schemes, 140, 222, 225, 227, 229
Agri-food partnership, Wales, 178
Agribusiness in Sustainable Natural African Plant Products, 338–9, 343, 346
Agricultural restructuring, 4–5
Allen, Patricia (alternative food practices), 31
Alternative agri-food movement:
  activists in, 266, 270
  food system localisation as a uniting factor, 255–6
  sustainability goals of, 263
Alternative agri-food organisations:
  in California, 255
  food system localisation platforms in, 255
  and territorial markers and place-of-origin labels, 270
  transferring resources and commitments to, 257
  in US, 256
Alternative capitalist organization, 265
Alternative economies, 4–5

Alternative Energy Resources Organization, 259
Alternative food, 56–7, 60, 69–70
Alternative food economy, definition of, 4–9, 309–10
Alternative food geographies, 1–9, 116–17, 121, 128–9, 150, 268
  see also New geographies of food
Alternative food networks, 2, 23, 78, 98–9, 115–18, 120, 185, 204, 256, 273, 278, 289, 305, 333–4
  in the South, 106, 106–7
Alternative food systems, 1, 2, 241
Alternativeness, 79–80, 268, 278, 289, 309–10
Analytical fields (for food projects), 10, 78, 80–1, 85, 87, 89–90
Animal rights, 136
Animal welfare, 314–15, 317
Anorexia nervosa, 249
Anti-hunger perspective, 260
Apartheid (legacy of), 331, 333, 336, 337, 340, 343–4, 346
Appalachian Sustainable Agriculture Project, 259
Apprentice farmers, 45–7
Argentina (yerba mate), 122
Auditing, 225, 232–4
Australia (agricultural policy), 224–6, 230

Backcasting, 241–2
Barriers to adoption, 231–2
Barriers to entry, 32
Baton Rouge Economic and Agricultural Development Alliance, 259
Be a Local Hero/Buy Locally Grown, 258–9

Best value, 186–8
Bioregionalism, 263
Body image/shape, 248–51
Boltanski, Luc (convention theory), 118
Boston Tea Party, 257
Boundary politics, 28
Bovine Spongiform Encephalopathy (BSE), 280
 crisis, 203, 206
Bovine tuberculosis (Mycobacterium bovis), 210
Boycotts, 256–7, 266–7
Brands and branding, 14, 16, 176, 178, 205, 312–27
 see also Marketing
Brazil (yerba mate), 10–11, 121–9
 government, 128–9
Bringing the Food Economy Home, Norberg-Hodge, et al., 263
BSE see Bovine Spongiform Encephalopathy (BSE)
Burgerville, 268
Business Alliance for Local Living Economies, 259
Buttel, Fred (Hightowerism), 26–7
Buy American, 266–8
'Buy Local Food' campaign, the US, 13, 256–68
Buy Local Food campaigns:
 aesthetics in, 260–2, 264, 268
 in California, 262, 270
 community in, 260–2, 264, 267–8
 economics in, 265–6, 268
 environment in, 263–5, 268
 health in, 260–2, 264–5, 268
 in Pennsylvania, 270
 programmatic objectives of, 256, 261–2, 266, 269
 programmatic promotions of, 256–8, 260–2, 267–70
 projects, 259
 social equity in, 260–3, 264, 266–8
 in the US, 255–70
 websites of, 258, 269
Buycotts, 257, 266–7

California:
 alternative agri-food organisations in, 255
 alternative food initiatives, 27–8, 29
 Buy Local Food campaigns in, 262, 270
 farm-to-school programmes in, 266
 organic foods, 241–52
 Santa Cruz, 262
Canada, 273
Capitalocentrism, 4–5
Center for Energy and Environmental Education, 259
Certification, 232, 235
Chatham County Center, 259
Chicken, 310–28
CISA see Community Involved in Sustaining Agriculture (CISA)
Co-operative, 331, 334, 339–43
Co-regulation, 176
Codification, 25, 273, 275, 281–2
Common Agricultural Policy (CAP), 6
Community, 7–8, 98, 102, 112
 in Buy Local Food campaigns, 260–2, 263, 267–8
Community Alliance with Family Farmers, 259
Community food security, 251, 255
Community Involved in Sustaining Agriculture (CISA), 258–9
 Harvesting Support for Locally Grown Food: Lessons from the 'Be a Local Hero, Buy Locally Grown' Campaign, 258
Community Supported Agriculture (CSA):
 as a form of food system localisation, 27, 31, 34–5, 67, 85, 255, 263
Consumer:
 anxieties (about food), 3, 13–14, 79, 82, 203–4
 connecting farmers and the, 258, 261, 267
 preferences (yerba mate), 127–8
 profile, 110

squeamishness, 323
yerba mate, 127–8
Consumerism:
　economic, 265
　green, 266
　political, 256, 260, 265–7
Consumption, 98
Content analysis, 260
Convention theory, 11, 116–21, 125, 128–30
Conventional agriculture, 223–6, 230–1, 235
Conventional food networks, 115–18, 122, 128
Conventionalisation, 24–5, 242, 246
Cooperation, 43–4, 51–2, 297
Cornucopia Project, 257
Cornwall Food Programme, 188, 191–7
CSA *see* Community Supported Agriculture (CSA)
Cultural food networks, 103–4, 106–9
Curry Report, 152, 160–1, 165, 187

Decentralisation, 150–3, 158, 162–3, 255
Defensive localism, 267
DeLind, Laura, 34
Detroit, MI, 267
Developing world, 96–8, 109
Diet, 11, 17, 133–42, 248–50
Direct food networks, 103, 108
Distributive justice, 43
Diverse economy, 265
Don't Buy Where You Can't Work, 266

Earthbound farms, the US, 245, 247
Earthshare CSA, 83–5
Eco-localism, 263
Eco Trust, 259
Economic consumerism, 265
Economic rent, 31–2
Economics:
　in Buy Local Food campaigns, 260–6, 268
Elitist, 97, 110
Endogenous development, 187
England Rural Development Plan, 295
Entrepreneurialism, 27, 34

Enviromeat, 227–8, 232–3
Environment:
　in Buy Local Food campaigns, 260–5, 268
Environmental Management Systems, 12–13, 223–4, 226
Equity *see* Social equity
Ethical consumerism *see* Political consumerism
Ethical consumption, 82, 256
Ethical trading practices, 341–2, 346
Ethics, 133, 136–9, 310, 324
Europe:
　local food initiatives in, 255

Fairtrade, 8, 14, 17, 331, 333–5, 338–9, 341–2, 344
　limits to, 345–6
Fairtrade Labelling Organisation, 334, 339–40, 340, 345, 346
Family farmers, 23–4, 264–5, 268
Farm-to-school, 266
　the US, 241, 252
Farmers:
　connecting consumers and, 258, 261, 267
　family, 265–6, 269
　small-scale, 255, 259, 263, 265–6
Farmers' markets, 29, 34, 265, 269, 327–30
　as a form of food system localisation, 255, 263, 265, 267
　and public food assistance programmes, 267
Farming Connect, Wales, 179
Fast food chains:
　and sourcing local foods, 268
Fear, 273–5, 277–83
Following, 17
Food:
　and aesthetics, 242, 261–2, 268
　and health, 85–6
　and science, 12, 276
　deserts, 8
　first, 268
　insecurity, 108
　labelling, 6, 32

Food miles *see* Transportation, and local foods
Food quality, 14, 97, 206–7, 225, 244, 309
  see also Quality turn
Food relocalisation, 6–8, 152–4
  see also Localisation
Food safety, 12, 14, 203–5, 215–16, 273–4, 275–6
Food Safety Authority of Ireland (FSAI), 204, 215–16
Food security:
  as a benefit of local food, 263
Food Standards Agency, the UK, 207, 215
Food supply chains, 6–8, 12, 17, 152–3, 185–6
Food system localisation *see* Local food systems
FoodRoutes, 258–9, 261–2
  Buy Fresh Buy Local "toolbox", 265
  Harvesting Support for Locally Grown Food: Lessons from the 'Be a Local Hero, Buy Locally Grown' Campaign, 258
  projects of, 259
Foodshed, 275, 284
Free range, 319, 326
Free trade, 122
  see also Neoliberalism

Gibson-Graham, J.K. (diverse economies), 5, 80, 265
Gippsland Environmental Management System, 224, 226–7, 229, 231–2, 240
Global climate change:
  and food system localisation, 268
Global value chains, 120
Globalisation:
  of the food system, 255, 260, 270
Globalised food system, 5, 15, 134, 140
Good food box, Toronto, 30
Goodman, David (alternative food networks), 4, 9, 78–9, 96
Green consumerism, 266
Growing Small Farms, 259

Halaal, 103
Harvesting Support for Locally Grown Food: Lessons from the 'Be a Local Hero, Buy Locally Grown' Campaign, Food Routes and CISA, 258
Harvey, David (militant politics), 28
Haskins Report, 152, 165, 179
Health, 137–9
  in Buy Local Food Campaigns, 260–2, 263–5, 268
Hightowerism, 26–7, 31
Hope, 282–5
Hybrid consumption, 141

Identity politics, 49
Indigenism, 27
Industrialisation of food, 275, 277–9, 283–4
Informal economy, 107, 109
Inland Northwest Community Food Systems Association, 259
Institutional maps, 155–6
Institutional thickness, 12, 162–3, 172–3, 177–8
Institutions, 172–3
Integrity, 14, 295, 298–9, 302–3
Internal regulation, 177
Ireland (raw milk cheese), 207–9

Johannesburg, 98–9
Justification (regimes of), 121, 122, 124–5, 127–9

Labels, 257, 267, 270
Labour, California, 242, 246, 251
Labour unions *see* Unions
Lagging rural regions, 154
Land Stewardship Project, 259
Land values, California, 243, 247, 251
LEADER, 88
Leyshon and Lee (alternative economies), 79–80
Life histories, 14, 310–12
Livelihood, 30–4
  approach, 9, 23, 32–4

Local businesses:
  support of, 262, 264, 269
Local Exchange Trading Systems (LETS), 80, 82, 84
Local food networks, 24, 28–35, 102–4, 106
Local food systems, 255–70
  and buycotts, 257
  eco-localism and bioregionalism in, 263
  economic benefits of, 260–1, 265–70
  environmental benefits of, 263, 265, 266, 268, 270
  in Europe, 255, 265
  and global climate change, 268
  and large corporations, 267–8
  social equity benefits of, 262, 266–8
Local food, 7, 11–12, 149–52, 164, 187–8, 191, 255, 292–3
  and institutional support, 158–64
Local knowledge, 41
Localisation, 187
  see also Food relocalisation
Localism, 28–9
Locality foods, 6, 12, 153, 164, 292
Low-income:
  benefits of local food systems to, 267

McDonaldisation, 268
Maine Organic Farmers and Gardeners Association, 259
Manufacturer (of food), 282
MAP (Mycobacterium avium subsp. paratuberculosis), 214
Market Oriented Initiatives (MOIs), 223–6, 225, 234–5
Marketing, 317–28
  see also Brands and branding
Marks & Spencer, 310–29
Marsden, Terry, 170, 207
Massachusetts, US, 258
Massey, Doreen (power geometries of place), 33–4
MERCOSUL, 121–9
Mesclun, 13, 244–5
  see also Organic salad mix
Michigan Department of Agriculture, 259

Michigan Integrated Food and Farming Systems, 259
Middlemen, 265
Midwest US, 266
Miele, Mara, 140–1
Migrant labour, 45–7
Minorities:
  benefits of local food systems to, 267
Mission Market of Ronan, 259
Monbiot, George, 136
Multi-level governance, 170, 178
Multifunctionality, 225, 230, 234
Multinational corporations:
  and sourcing local foods, 266–8

Narrative approach, 310–12
National Health Service, 188, 190, 191–6, 198
National organic standards, 269
Nature, 278, 280, 282–5
Neoliberalism, 5
  see also Free trade
Networking, 14, 295, 298
Networks, 41, 331, 333, 338, 340–2
New geographies of food, 56–9, 66, 70
  see also Alternative food geographies
New realism, 30–2
New regionalism, 150–2, 154
  see also Regionalisation
New rural development dynamic, 6–7
Non-capitalist organization, 265
Non Governmental Organisations (NGOs), 310, 317, 331, 335–6, 338, 341–3
Non-importation movement, 257
Normative localism, 29
North America, 273–80, 283, 285
North Carolina Cooperative Extension, 259
North East England, 151, 154
Northumberland, England, 154–5
Nouvelle cuisine, 244, 248–50
Nutrition ecology, 135

Oakham White (chicken), 310–29
Objective One, 191–6
OPEC oil embargo, 257

Oppositional, 96
Order of worth, 118–19, 121, 129
Ordinary consumption, 140
Organic:
  foods, 257, 261, 269
Organic farming, 42
  the United States, 24–8
Organic food, 8, 13, 242
Organic salad mix:
  consumption, California, 248–51
  production, California, 244–8
  see also Mesclun
Othering:
  in relation to local food systems, 267

Paraguay (yerba mate), 122
Participatory methods, 17
Partnerships, between institutions, 178
Pasteurisation, 208–9
Pathogens, 204, 209
Pennsylvania:
  Buy Local Food campaigns in, 259, 270
Pennsylvania Association for Sustainable Agriculture, 259
Political consumerism, 256, 260, 265–6
Post 9–11 era, 267
Poor (urban poor), 104–5, 107
Post organic, 24–5
Poverty, 343–4, 346
Power geometries of place, 33–4
Power relations, 79–80, 89, 294–5
Practical Farmers of Iowa, 255, 259
Precautionary principle, 218
Price premiums, 234
Princeton Future, 259
Process and place alternatives, 3–4, 7–9
Processor, 274–5, 278–83
Product and place alternatives, 3–4, 5–7
Productivist agriculture see Conventional agriculture
Programmatic objectives:
  of Buy Local Food campaigns, 258, 261–2, 266, 267
Programmatic promotions:
  of Buy Local Food campaigns, 256–8, 260–3, 267

Protected Designation of Origin (PDO), 6
Protected Geographical Indication (PGI), 6
Public goods, 17
Public policy (food), 11–13, 16
Public sector food procurement, 8, 12, 185–88, 193

Quality:
  of local food, 261, 263
Quality turn, 6
  see also Food quality

Racism:
  as a part of Buy American campaign, 267
Raw milk
  cheese, 204–5, 207–9
  consumption of, 208–9, 213
  microbiology of, 209, 213
Reflexive consumption, 140
Reflexive localism, 29
Reflexivity, 140
Regard, 4, 82, 291
Regional development, 6, 172, 176, 179–82
Regional Development Agencies (RDAs), 151–2
  One NorthEast, 158–9, 160–1
Regional governance, 150–2
Regional spaces, 154
Regionalisation, 6, 11–12
  see also New regionalism
Regulation, 205–7, 212, 218, 225, 227, 229, 248, 252
Research, Education, Action and Policy on Food, 259
Retailing, 309–29, 327–30
Risk, 205–207, 209, 218
Rodale Institute, 257
Rooibos (Red Bush) Tea, 14, 331, 336, 337–42
  Rooibos industry, 337–8, 340–2
  Rooibos Project, Wupperthal Community, 331, 338–43
Roots of Change, 240–1

Rural development:
  in the North, 79, 88, 175, 185–7, 205
  in the South, 331, 336–7, 338–40, 343, 345
Rural policy, the UK, 152–4
Rural Roots, 259

Salop Drive market garden, 85–7
Santa Cruz, CA, 262
Scales of food governance, 156
Scaling up, 16, 30
Science, role of in food safety, 205–7, 218
Secession, 29, 257
Sodexho, 268
  boycotts, 256–7, 266–7
  85 Broads, 266
  Buy American, 267–8
  Buy Local Food campaigns, 255–70
  Don't Buy Where You Can't Work, 266
  non-importation movement, 257
Senior nutrition programmes, 267
Short Food Supply Chain, 6–7, 334–5, 340–1, 344
Slow food, 255
Small-scale farmers, 251, 255, 266–7
Social capital, 338–9, 343
Social construction, 275, 283
Social equity:
  in Buy Local Food campaigns, 260–3, 266–8
  in local food systems, 260
  in selective patronage campaigns, 266
Social justice, 9–10, 39, 42–4, 48, 51, 79, 266, 267
  see also Social equity
South Africa, 331–3, 335–37, 340, 343–6
Southeastern Massachusetts Agricultural Partnership, 259
Spaces of regionalism, 154
Speciality foods, 104, 108, 161, 163–4, 169–71
Standards, 124, 127, 129, 207, 225, 229, 242, 245
Storper, Michael (institutions), 173
Supermarkets (in the South), 99, 105–8
Survival, 98, 103, 106

Sustainability, 39, 41, 43, 58, 59, 69, 186, 188, 224, 260, 263, 270
Sustainability transition, 56, 69
Sustainable, 10, 133–4, 137, 139
Sustainable agriculture, 41, 43
  organisations, 255
Sustainable Agriculture Movements (SAMs), 23, 25–7, 30–1
Sustainable Connections, 259
Sustainable food, 58–60, 61, 64–5, 66–71
Sweden, 263

Tacit knowledge, 282
Tariffs, 121–2, 126
Taste, 245, 248–9
Technologically-led vision (of organics), 26–7
Thévenot, Laurent (convention theory), 118
Thirsk, Joan, 15
Transformative potential, 25–8
Transparency (of supply chains), 322
Transportation:
  and local food, 263–5, 268
Trust, 14, 293–5, 297, 305
Tuberculin test, 210–11, 215
Tuscarora Organic Growers, Pennsylvania, 44–7

UK. *see* United Kingdom (UK)
Unions:
  labour, 267
United Farm Workers (UFW), 266
United Kingdom (UK), 265
United States (US):
  "Buy American" campaigns in, 267
  Buy Local Food campaigns in, 256–63
  "Don't Buy Where You Can't Work" movement, 266
  "good jobs" in, 267–8
  Midwest, 266
  national organic standards, 266, 269
  organic certification legislation in, 266
  Pacific Northwest, 268
  unions in, 267
Unterman, Patricia, 250
Urban agriculture, 77, 104, 106, 108

Urban food systems, 10, 34–5, 98, 274
  Johannesburg, 95–111
US *see* United States (US)
USDA Organic Standards, 24

Vegan, 136, 138, 142
Vegetarianism, 11, 133–42
Vermont Agency of Agriculture, Food and Markets, 259

W.K. Kellogg Foundation, 258
Waste (food), 324
Websites:
  of Buy Local Food campaigns, 260–1, 269
Welfare *see* Animal welfare
West Midlands region, the UK, 86, 289, 295

Whole Foods Market, 268
Women:
  benefits of local food systems to, 268, 267
  buying power of, 266
  farmers, 47, 49
  Infants and Children (WIC), 267
Women's Agricultural Network, 47–51
World Trade Organisation (WTO), 5
Wupperthal, South Africa, 331, 333, 335, 338–6

Yerba mate, 115
  industry, 127–8
  processors, 123–8
  producers, 126–7
Yuppie chow, 13, 241, 250

Lightning Source UK Ltd.
Milton Keynes UK
UKOW031522140313

207648UK00001B/82/P

9 780080 450186